8th International Multi-Conference on Engineering and Technology Innovation (IMETI 2019)

Recent Advances in Intelligent Technology Innovation

IOP Conference Series: Materials Science and Engineering
Volume 1113

Kaohsiung, Taiwan
15 - 19 November 2019

ISBN: 978-1-7138-3402-1
ISSN: 1757-8981

Printed from e-media with permission by:

Curran Associates, Inc.
57 Morehouse Lane
Red Hook, NY 12571

Some format issues inherent in the e-media version may also appear in this print version.

Additional copies of this publication are available from:

Curran Associates, Inc.
57 Morehouse Lane
Red Hook, NY 12571 USA
Phone: 845-758-0400
Fax: 845-758-2633
Email: curran@proceedings.com
Web: www.proceedings.com

TABLE OF CONTENTS

IOP Conf. Series: Materials Science and Engineering 1113 (2021) 011001 doi:10.1088/1757-899X/1113/1/011001

Preface

The 8th International Multi-Conference on Engineering and Technology Innovation 2019(IMETI2019) scheduled from November 15-19, 2019in Kaohsiung, Taiwan is an expansion of the International Conferences on Engineering and Technology Innovation (ICETI2011, ICETI2012, ICETI2014), and International Multi-Conference on Engineering and Technology Innovation (IMETI2015, IMETI2016, IMETI2017, IMETI2018).

The conference is mainly organized by Taiwan Association of Engineering and Technology Innovation. The theme of the conference is set as "Recent Advances on Intelligent Technology Innovation". IMETI2019 consists of two sub-conferences (ICBEI2019and ICECEI2019), 15 regular sessions, and 29 special sessions. This is a great opportunity for the engineering and technology community to present their new research results, exchange ideas and become familiar with new trends and directions in the fields of Science and Engineering Innovation. In addition, it is also a unique opportunity for us to form an international collaboration framework for promoting engineering and technology innovation.

We have with us today participants from research funding organizations, universities, as well as other research agencies. I hope that this five-day conference -a platform for thought-leaders, academics and researchers to share their ideas and views on common engineering and technology innovation issues -will challenge all participants, in particular the heads of institutions, to think more about the technology challenges and responses in engineering innovation which may in turn inspire new and practicable standards in the field.

The papers submitted from 8 countries on various topics are presented in various parallel oral and poster sessions in the conference. To all members of the Technical and Organizing Committees, we would like to take this opportunity to thank all of them for their tremendous efforts to organize this conference successfully.

The organizing committee of the conference thanks all the participants for their fruitful work and personal contribution to the development of these conference proceedings.

IOP Conf. Series: Materials Science and Engineering 1113 (2021) 011001 doi:10.1088/1757-899X/1113/1/011001

Committees

Program Chairman

Wen-Hsiang Hsieh, President, Taiwan Association of Engineering and Technology Innovation
Professor, National Formosa University, Taiwan

Program Co-Chairman

- Chun-Pin Lin, Asia University, Taiwan
- Ing-Jr Ding, National Formosa University, Taiwan
- Jia-Shing Sheu, National Taipei University of Education, Taiwan
- Yu-Fen Chen, National Formosa University, Taiwan
- Jian-Ming Lu, National Center for High-Performance Computing, National Applied Research Laboratories, Taiwan
- Van-Tsai Liu, National Formosa University, Taiwan

Technical Committee Chairman

- Chih-Ta Yen, National Formosa University, Taiwan
- Tao-Hsing Chen, National Kaohsiung University of Science and Technology, Taiwan
- Yi-Chang Wu, National Yunlin University of Science & Technology, Taiwan

Program Committee

- An Cheng, National Ilan University, Taiwan
- Chan Young Park, Hallym University, South Korea
- Chang-Chun Lee, Chung Yuan Christian University, Taiwan
- Chang-Huang Chen, Tungnan University, Taiwan
- Chen Lu, Beihang University, China
- Chun-Hsiung Tseng, Nanhua University, Taiwan
- Chung-Chi Huang, Far East University, Taiwan
- David Pei-Cheng Lin, Chung Shan Medical University, Taiwan
- Dae-Jin Kim, Kyung Hee University, South Korea
- Hsiang-Chen Hsu, I-Shou Univerity, Taiwan
- Hsiao-Han Liu, I-Shou University, Taiwan
- Hsin-Hung Lee, Center for Measurement Standards, Industrial Technology Research Institute, Taiwan
- Kuan-Chen Lin, Far East University, Taiwan
- Jeong-Dong Kim, Korea University, South Korea
- Jie-Tong Zou, National Formosa University, Taiwan
- Jih-Gau Juang, National Taiwan Ocean University, Taiwan
- Jium-Ming Lin, Chung Hwa University, Taiwan
- Minvydas Ragulskis, Kaunas University of Technology, Lithuania
- Shih-Sung Lin, National Defense University, Taiwan
- Tai-Ping Chang, National Kaohsiung first University of Science and Technology, Taiwan
- Te-Kwei Wang, Ming Chi University of Technology, Taiwan
- Wei-Ting Lin, National Ilan University, Taiwan
- Yi-Cheng Huang, National Changhua University of Education, Taiwan
- Yih-Chuan Lin, National Formosa University, Taiwan

- Yit-Jin Chen, Chung Yuan Christian University, Taiwan
- Yu Guo, Kunming University of Science and Technology, China
- Yung-Cheng Lee, WuFeng University, Taiwan
- Yung-Hui Chen, Lunghwa University, Taiwan
- Zongyu Chang, Ocean University of China, China

International Scientific Committee

- A. Gottscheber, University of Heidelberg, Germany
- Amir Abidov, Tashkent State Technical University, Uzbekistan
- Aylie Han, University of Diponegoro, Indonesia
- A. Walton, University of Edinburgh, UK
- Buntara S. Gan, Nihon University, Japan
- R. S. Beniwal, National Institute of Science Communication and Information Resources, India
- C. Warwick, University College London, UK
- E. Manske, TU Ilmenau, Germany
- G. Jäger, TU Ilmenau, Germany
- Giovanni Pau, Kore University of Enna, Italy
- Hyeungsik Choi, Korea Maritime and Ocean University, Korea
- Igor Meglinski, University of Oulu, Finland
- J. Wood, Goldsmiths College, University of London,UK
- J. Carlos, Metrolho Instituto Politecnico de Castelo Branco Castelo Branco, Portugal
- Jong-Han Lee, Daegu University, South Korea
- J.S. Dai, Kings of college London, UK
- Kihong Shin, Andong National University, South Korea
- M. Woolley, University of Coventry, UK
- P.F. Savanna, College of Art and Design, USA
- R.X. Du, The Chinese University of Hong Kong, HK
- S.G. Hanson, Technical University of Denmark, Denmark
- S.D. Prior, University of Middle Sex, UK
- T. Froelich, TU Ilmenau, Germany
- Uldis Strautins, University of Latvia, Latvia
- WooSeok Kim, Chungnam National University, South Korea
- Zhengchao Xie, University of Macau, Macau

Advisory Committee

- B.W. Huang, Cheng Shiu University, Taiwan
- C.C. Wen, Metal Industries Research & Development Centre, Taiwan
- C.H. Huang, Wufeng University, Taiwan
- Chun-Huy Wang, Nan-Jeon University of Technology, Taiwan
- C.M. Lin, National Taitung University, Taiwan
- C.T. Lin, Metal Industries Research & Development Centre, Taiwan
- F. Long, Xiamen University, China
- H.S. Lin, National Kaohsiung University of Science and Technology, Taiwan
- I.T. Tang , National University of Tainan, Taiwan
- J.C. Zhong, Changchun University of Science and Technology, China
- J.F. Yao, Xiamen University, China
- J.X. Xie University of Science and Technology, Beijing, China
- K.H. Wen, Fuzhou University, China
- K.S. Shu, Chia Nan University of Pharmacy and Science, Taiwan

- L.W. Ji, National Formosa University, Taiwan
- M. Chang, Chung Yuan Christian University, Taiwan
- M.H. Wang, National Chin-Yi University of Technology, Taiwan
- Q.K. Xue, Tsinghua University, Beijing, China
- S.J. Chang, National Cheng Kung University, Taiwan
- S.J. Lin, National Kaohsiung University of Science and Technology, Taiwan
- S.T. Shen, National Formosa University, Taiwan
- Sungjin Kim, Kumoh National Institute of Technology, South Korea
- T.H. Meen, National Formosa University, Taiwan
- W. Water, National Formosa University, Taiwan
- W.Z. Guo, Shanghai Chia-Tung University, China
- Wonho Suh, Hanyang University, South Korea
- X.L. Du, Beijing National Laboratory for Condensed Matter Physics, China
- Y.C. Wang, National Yunlin University of Science & Technology, Taiwan
- Y.L. Hwang, National Formosa University, Taiwan
- Y.L. Yeh, Far East University, Taiwan
- Y.Q. Zhou, China University of Petroleum, China
- Y.X. Sun, China University of Petroleum, China

Award Committee and Technical Committee

- Chung-Chi Huang, Far East University, Taiwan
- Dae-Jin Kim, Kyung Hee University, Korea
- Dong-Ha Lee, Kangwon National University, Korea
- Dyi-Cheng Chen, National Changhua University of Education, Taiwan
- Hyeung-Sik Choi, Korea Maritime and Ocean University, Korea
- Jia-Shing Sheu, National Taipei University of Education, Taiwan
- Kimio Oguchi, Seikei University, Japan
- Shyh-Chour Huang, National Kaohsiung University of Science and Technology, Taiwan
- Wei-Ting Lin, National Ilan University, Taiwan
- Wen-Chung Kuo, National Yunlin University of Science & Technology, Taiwan
- Wonho Suh, Hanyang University ERICA Campus, Korea
- Yenming J. Chen, National Kaohsiung University of Sciences and Technology, Taiwan
- Yih-Chuan Lin, National Formosa University, Taiwan
- Zhang-Yan Zhao, Wuhan University of Technology, China

IOP Conf. Series: Materials Science and Engineering 1113 (2021) 011002 doi:10.1088/1757-899X/1113/1/011002

Peer review declaration

All papers published in this volume of *IOP Conference Series: Materials Science and Engineering* have been peer reviewed through processes administered by the Editors. Reviews were conducted by expert referees to the professional and scientific standards expected of a proceedings journal published by IOP Publishing.

- Type of peer review: Single-blind

The reviewers (at least two for each paper) received the manuscript with all the details of the authors and a form, prepared for the conference, with certain questions and a field for comments. The authors, at the next step, received anonymous reviews with the following possible recommendations: accept as is/minor revisions/major revisions/reject. The paper was revised by the authors accordingly and then final decision was taken by the editors.

- Conference submission management system:

The papers were submitted to an e-mail, handled by the editor. The handling editor managed the correspondence between reviewers and authors.

- Number of submissions received: 39

- Number of submissions sent for review: 39

- Number of submissions accepted: 29

- Acceptance Rate (Number of Submissions Accepted / Number of Submissions Received X 100): 74.4%

- Average number of reviews per paper: 2

- Total number of reviewers involved: 67

IOP Conf. Series: Materials Science and Engineering 1113 (2021) 011002 doi:10.1088/1757-899X/1113/1/011002

•Any additional info on review process (ie plagiarism check system):

There was a pre-screening process of the abstracts by the editors, prior to the submission of the final papers.

•Contact person for queries:

imeti2019@imeti.org and imeti2019@gmail.com

IOP Conf. Series: Materials Science and Engineering 1113 (2021) 012001 doi:10.1088/1757-899X/1113/1/012001

A study on technologies that enable high-speed communication regardless of the interface standard between the host devices and peripheral devices

So Ki Jung

Head of R&D, Infra Innovation Team, SKbroadband, Korea

E-mail: mkgg0107@naver.com

Abstract. Recently, I/O interfaces for connecting a computer to peripheral devices have emerged to provide 10Giga services for communication companies. Therefore, Intel has developed Thunderbolt cables, which enable electrical signals and high-speed data transfer to facilitate 10Giga services. Thunderbolt is a type of input–output interface that is used to connect computers and peripherals. It is capable of high-speed input/output(I/O), incurs lower overheads than USB, and supports data, images, and voice simultaneously. As these are very recent standards, they have not been applied to many computers. Moreover, no device exists to use interfaces having a different protocols while differentiating between Thunderbolt and USB devices. Therefore, if both interfaces are mutually incompatible, the user must utilize Thunderbolt or USB devices separately. For this reason, a device that automatically identifies Thunderbolt or USB and automatically adapts to the user's environment is required. In this study, Thunderbolt MAC and a MAC are alternately connected to each other via an electrical Multiplexer to auto detect link information and connect to Host. Then, based on the determined result, the Thunderbolt to PCIe bridge or the USB to PCIe bridge is electrically connected to the PCIe BUS to which the Multiplexer is connected.

1. Introduction

Recently, as a variety of services such as 5G, 8K UHD, and OTT have emerged, the IP and Internet traffic increases per person have surpassed the level of existing gigabit Internet service in the last decade. Consequently, subscriber traffic has increased rapidly Globally, the monthly IP traffic will increase from 16 GB per person in 2017 to 50 GB person in 2022, and the Internet traffic will increase from 13 GB per person in 2017 to 44 GB per person in 2022 [1]. Also, By 2020, forecasts suggest that there will be around 6.58 network connected devices per person around the globe. With a total world population of over 7.5 billion people, this means that there could be nearly 50 billion network connected devices by 2020[2]. As network connection devices have greatly increased to provide ultra-high definition, ultra-low delay and 8K video technology, increase interconnection bandwidth requirements between end devices and the interconnect technology is determined by the device model used by customers [3]. Therefore, devices to devices, 5Gigabit, 10Gigabit per second (Gbps) Internet connection technology is required to provide stable, large capacity, and high quality service such as hyper-realistic UHD media, virtual reality (VR), and augmented reality (AR) services and 802.11ax. New interface standards have been continuously researched and developed to improve inter-device communication speed and performance. Therefore, Intel Corp. and Apple Inc. have jointly developed Thunderbolt, which is an ultra-high speed PC interface standard between devices, to facilitate high speed data transmission between PC and LAN card in order to utilize 10Gbps-class Internet [4]. The technologies of inter-device interface standards facilitate interconnection between independent devices having mutually different input and output technologies to ensure high speed communication speed. This paper suggested a novel technique that detects link information to connect to a host such as s computer by alternately-connecting the data signals of a USB connector to Thunderbolt MAC and USB MAC through an electrical multiplexer to facilitate high-speed communication between a computer and a peripheral regardless of interface standards.

Furthermore, interface enhancement between a computer's CPU and LAN card is crucial for providing broadband Internet service. However, as computing has shifted to emphasize on becoming mobile-oriented, light, and small, there is insufficient space for I/O connector arrays for various applications [4].

In addition, as notebook PCs become thinner, the physical space to accommodate interface connectors is restricted. In order to maintain the current interface functions and secure expandability at the same time, more than one existing interface component must be integrated into one. Thunderbolt has been proposed to solve this problem [5]. Thunderbolt combines PCI Express (PCIe) and DisplayPort (DP) into two serial signals, and additionally provides DC power, all in one cable. Up to six peripherals may be supported by one connector through various Topologies [6-7]. Thunderbolt is an I/O interface cable that is used to connect a computer with peripherals. Because it facilitates high speed input/output, it has lower overheads than USB while supporting data, video, and voice simultaneously. Few computers support Thunderbolt because it is a latecomer, and the majority of computers still support USB. However, because the two interfaces of Thunderbolt and USB are incompatible with each other, the devices should be used separately when connecting a computer to external peripherals. Moreover, there is currently no device that distinguishes computer interfaces having mutually different protocols, such as Thunderbolt, USB, and power supply, in the device aspect. Therefore, because device interfaces are incompatible with each other, devices for Thunderbolt and devices for USB must be differentiated for the user. To solve the compatibility problem, a technology is required to automatically connect interfaces between devices according to the user environment.

This study designed a technique that detects link information to connect to a host such as a computer by alternately-connecting the data signals of a USB connector to Thunderbolt MAC and USB MAC through an electrical multiplexer to facilitate communication between the computer and a peripheral regardless of interface standards. Furthermore, a docking station device was developed to electrically connect a Thunderbolt-to-PCIe bridge or a USB-to-PCIe Bridge to PCIe BUS to a multiplexer based on the determined result. This study describes the patent application and device development conducted jointly with Dayou Plus Co., Ltd. for a study on the automatic compatibility of interfaces between devices. This paper is structured as follows. Section 2 describes the components and operating principle characteristics of 10Gbps-class Internet networks and multiplexers. Section 3 Research and Experiment, and lastly, Section 4 presents the conclusion.

2. Operating principle of compatibility between devices using thunderbolt

2.1. Interface between devices using thunderbolt

Recently, computing has become mobile-oriented and consequently, a considerable proportion of notebook computers to be released in future are expected to be in the ultrabook form. Thus, notebook computers will become thinner and have limited physical space for accommodating interface connectors [4]. To maintain the current interface functions while simultaneously possessing scalability, two or more existing interface components should be integrated into one. To address this issue, Intel Corp. and Apple Inc. jointly developed and proposed Thunderbolt, an ultra-high speed PC interface standard between devices [4]. Thunderbolt is an I/O interface cable used to connect a computer and peripherals, and it was developed by Intel. Because Thunderbolt facilitates high speed input/output, it has lower overheads than USB and supports data, video, and voice simultaneously. However, because the standard was a latecomer, few computers currently support it.

Fig. 1 shows that communication is facilitated by connecting a Thunderbolt cable through a dockin g station between an optical modem and interface hardware of various hosts or computers. Because co mputers are becoming thinner, lighter, and smaller, there is almost no space available for an I/O conne ctor array [3]. Thunderbolt cables are classified into two types: copper and optical. However, this stud y was conducted based on copper cables to supply electrical power as well between a host and end dev ices. Based on the USB standard, copper is used by limiting the cable length to 5 m for 480 Mb/s USB 2, 3 m for 5 Gb/s USB2, and 1 m for 10Gb/s USB 3.1 Gen2 [3]. Thunderbolt is a dual protocol I/O tec hnology, developed for performance, simplicity, and flexibility, that can transmit data using a longer a nd more expensive cable by combining PCI Express and display port with a new serial data interface [8]. Peripheral component interconnect (PCI) is a high speed bus developed by Intel in the early 1990s t o replace low speed bus technologies such as EISA, ISA, MCA, and VESA [9]. Thunderbolt is based on the Mini Display Port connector developed by Apple and is implemented in various customer devic es such as laptops by using small connectors [8]. As the Thunderbolt3 cable has been released recently , USB3.1 Gen1/Gen2, DisplayPort1.2, and PCIe Gen3 tunneling are all included in a PC, thereby supp orting high speed I/O protocols [3]. Fig. 1 shows a docking station that interconnects all Thunderbolt a nd USB type interfaces.

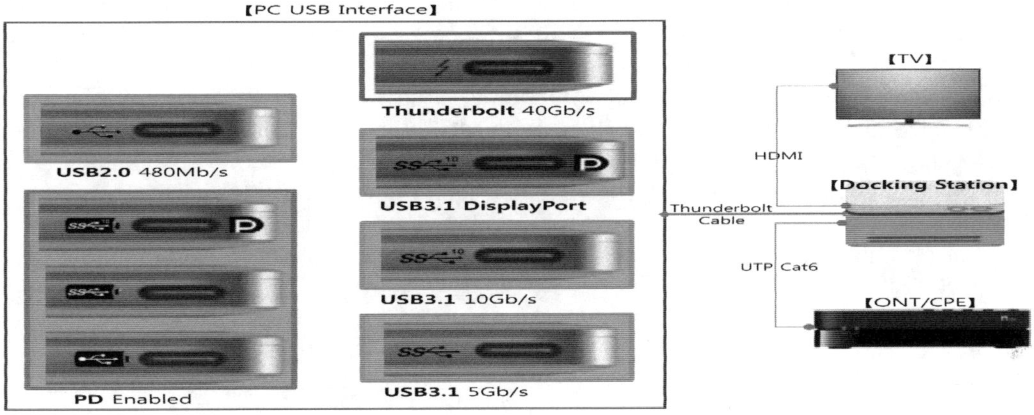

Figure 1. USB connector interface standards

2.2. Operating method of thunderbolt and docking station

In Fig. 2, the Thunderbolt cable, unlike a bus-based I/O architecture, provides the entire bandwidth of t he link in both directions without sharing the bandwidth between ports or the upstream and downstrea m directions when communicating with a computer. A Thunderbolt connector can provide two full du plex channels. Each channel provides bidirectional high speed 10Gbps bandwidth. Physical speed prov ide down load 20Gbps x 2, Upload 20Gbps x2. And Logical speed provide download 40Gbps, Upload 40Gbps. A computer's Thunderbolt connector is connected to a Thunderbolt product or a DisplayPort device via a cable [6]. In Fig. 2, the Thunderbolt cable has a built-in dedicated transceiver IC in the co nnectors at both ends and compensates for time delays by amplifying signals to transmit high speed 10 Gbps signals. Furthermore, for internal high speed communication, lanes are composed with a dedicat ed upload pair and a download pair to allow upload and download transmissions to be performed inde pendently and simultaneously [2]. Because Thunderbolt's Interconnect technology can support 40 Gb/ s throughput per connector interface, it is leading the adoption of interconnect technologies for various devices with respect to 8K videos and Internet speed [7]. Thunderbolt cable provides four times the d ata and twice the video bandwidth of any other cable, while also supplying power, 8 lanes of DisplayP ort 1.2, supports two 4k displays(4094x2160 30bpp @60Hz)[10].

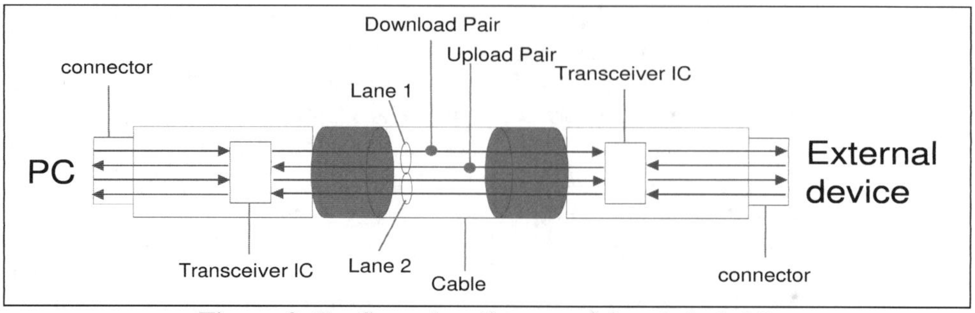

Figure 2. Configuration diagram of thunderbolt [4]

USB, as detailed in Table 1, is a typical interface standard used for connecting a host device, comp uter, and peripheral devices because it is economical and exhibits excellent performance. A main contr oller is present at the uppermost part, and two USB ports are provided through the root hub. Therefore, if there is an insufficient number of USB ports, more USB ports can be prepared by connecting a hub . As shown in Table 1, USB-C type connectors have been used since USB 3.1 Gen1. As shown in Tabl e 1 and Fig. 2, Thunderbolt provides a 10Gb/s bidirectional transmission channel on two lanes, thereby facilitating much faster high speed data transmission than those of other standards [4]. A Thunderbolt connector can provide two full duplex channels. Furthermore, it uses a protocol that transmits data an d video through only one cable by providing PCI Express for data and a display port for video data [1 1]. The maximum delay time is set to 8 ns to process video data smoothly [4]. Because Thunderbolt is a standard that came out late, the majority of host computers still support USB. Moreover, because the USB and Thunderbolt interfaces are incompatible with each other, a user must purchase a relevant dev ice after checking whether USB and Thunderbolt are supported when connecting the peripheral device to a host device or computer, which is inconvenient. Therefore, to resolve this inconvenience and allo w simultaneous use of both devices, this study has developed a docking station, as depicted in Fig. 3, t o facilitate inter-communication regardless of whether an interface standard is conformed.

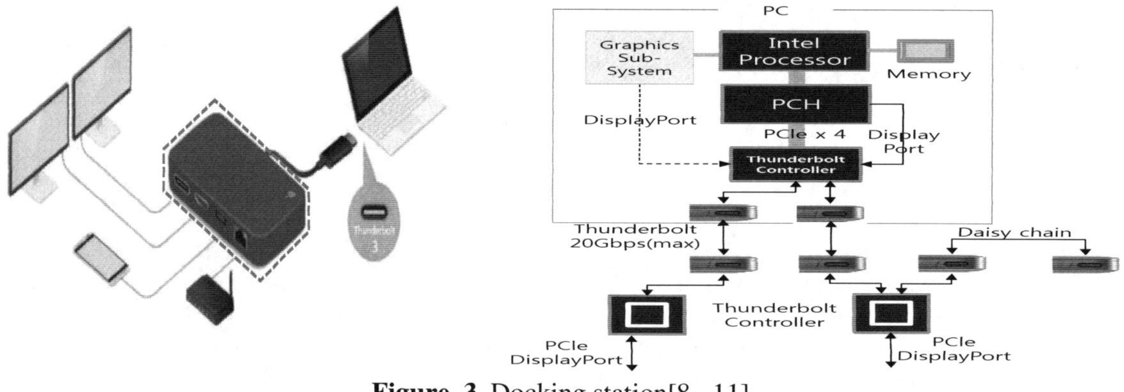

Figure 3. Docking station[8, 11]

Table 2 Compares the USB3.1 and Thunderbolt™3 features. Thunderbolt™ 3 brings Thunderbolt to USB-C at speeds up to 40 Gbps, creating one compact port that does it all – delivering the fastest, most versatile connection to any dock, display, or data device. For the first time, one computer port connects to Thunderbolt devices, every display, and billions of USB devices. A single cable now provides four times the data and twice the video bandwidth of any other cable, while also supplying power. It's unrivaled for new uses, such as 4K video, single-cable docks with charging, external graphics, and built-in 10GbE networking [5].

For the docking station represented by the red dotted lines in Fig. 3, a data-providing port was deve loped to facilitate 10Gb/s and 5 Gb/s transfer simultaneously. Furthermore, it was designed to detect li nk information and connect to a host such as a computer by alternately connecting the data signals of a

USB connector to both Thunderbolt MAC and USB MAC through an electrical multiplexer. Accordi ng to the final result determined, the docking station device was developed to electrically connect the Thunderbolt-to-PCIe bridge or USB-to-PCIe bridge with the PCIe BUS connected to the multiplexer.

Table 1. Device interface standards [4]

Standard	Thunderbolt (USB 3.1 Gen2)	USB2.0	USB3.0(Gen1)
Transmission speed	Bidirectional 10Gb/s 2 lanes	Bidirectional 480 Mb/s 1 lanes	Bidirectional 5 Gb/s 1 lanes
Cable length	Max. 3 m(electrical) Max. 30 m(optical)	Max 5 m	Max 3 m
Feature	PCIe and DisplayPort Daisy-chained Power 10 W	Data transmission Tree connection Power 2.5 W	Data transmission Tree connection Power 4.5 W
Established	2011	2000	2008

Table 2. Analysis of Thunderbolt™3 and USB3.1 [5]

Capability	USB 3.1	Thunderbolt™ 3
Small and reversible	●	●
Connects to USB Devices	●	●
Speed	Up to 10Gb/s	Up to 40Gb/s
Fastest data, video, and power	○	●
Two 4K displays	○	●
External graphics	○	●
Single-cable docks with two 4k displays	○	●

Each channel provides bidirectional 10Gbps bandwidth. A Thunderbolt connector on a computer is capable of connecting to a Thunderbolt product or a DisplayPort device through a cable [11].

3. Research and experiment

Fig. 4 depicts a 10Gbps network by classifying an apartment building and a detached house. XGS-PON installed in the Internet network is a device that provides 10Gbps speed upstream/downstream to a customer's Home ONT device through an optical cable and a splitter. The XGS-PON technology is a technical standard that was presented by the Full Service Access Network (FSAN) group in 2010 [12]. The tree architecture is very fiber efficient as only a single feeder fiber is used to connect a central office to the local service area, where a passive optical splitter provides onward connectivity to each subscriber [13]. In a home network, broadband Internet service is provided through an integrated docking station on a PC that provides 10Gbps speed by detecting Thunderbolt, power delivery, and USB3.x interfaces and connecting to a USB-C connector via an electrical multiplexer. Internet, IPTV, and WiFi services are provided to B2C customers of wired communication by targeting apartment buildings and detached houses; according to the needs of customers, speeds of 1, 2.5, 5, and 10Gbps are provided. To receive such a variety of speeds, an interface between devices should be used by connecting to different devices. For example, when a customer wants high-speed 10Gbps service, the network should be designed with a high-speed serial-bus and memory interface to overcome the problems of various interfaces and standards such as USB3.1, Thunderbolt3, DisplayPort, HDMI, and PCIe on a PC [14]. Furthermore, to ensure high-speed communication, the I/O interface technology between devices should facilitate inter-device communication by connecting different devices with each other.

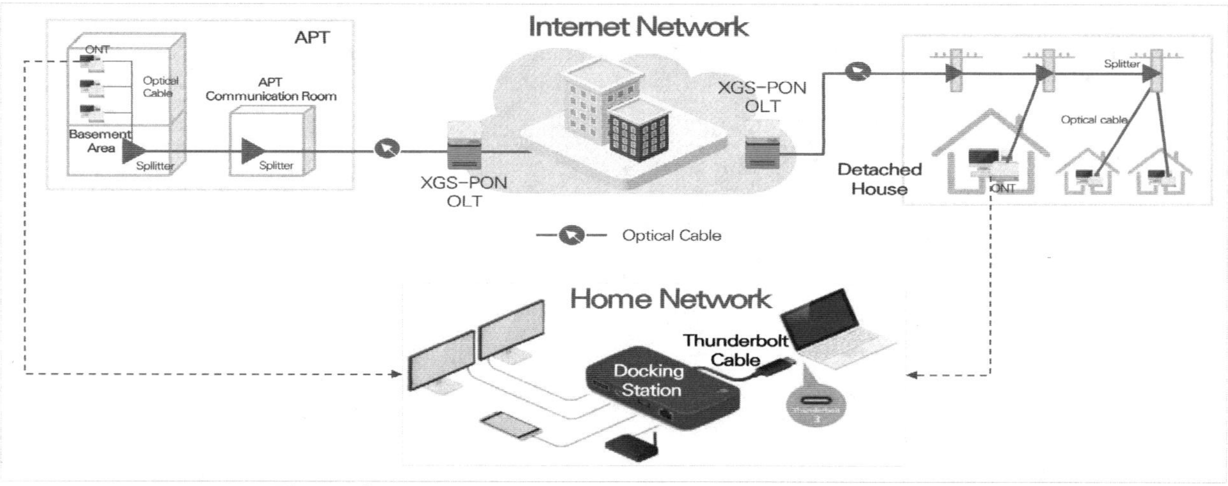

Figure 4. High-speed home network using docking station

Even if the cable connectors in a USB-C interface connected to a computer in a home network are identical, the interface standards may be different. In such a situation, a docking station is required to use high-speed 10Gb/s or 5Gb/s Internet by freely connecting the computer and peripheral devices without differentiating Thunderbolt, USB, and power delivery devices. The docking station connects a host device such as a computer to peripheral devices, and when a new peripheral USB device is connected to a connection port of the device, it determines whether the interface standard of the connected computer has been identified. If the computer's interface standard has not been confirmed, a certain signal is processed as signals of USB and Thunderbolt standards and transmitted alternately to the computer.

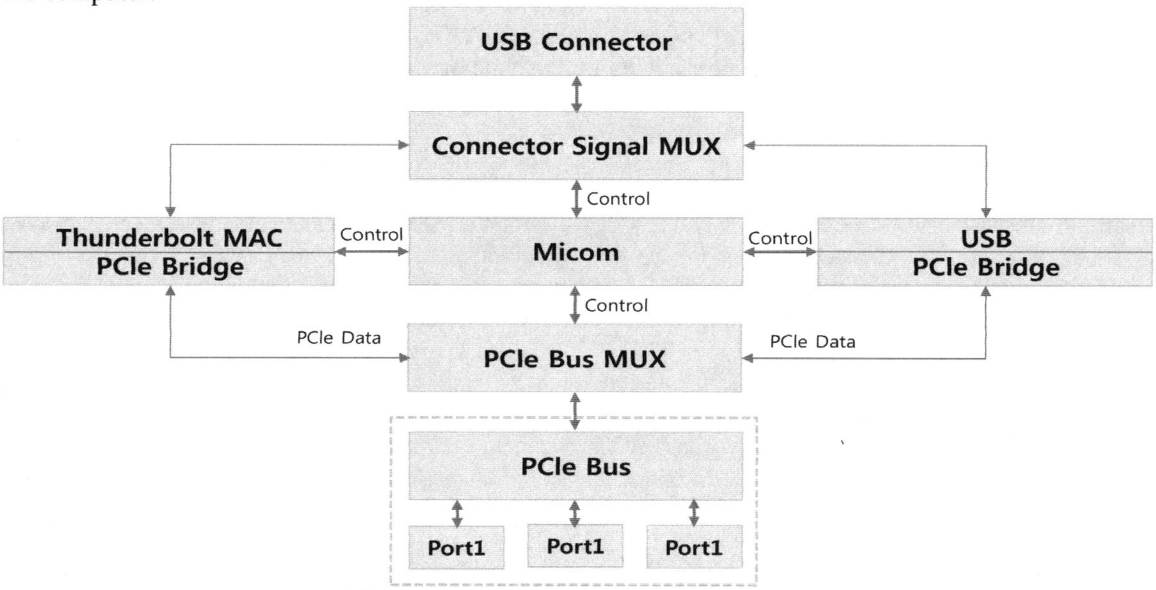

Figure 5. Diagram of the docking station

Fig. 5 depicts the solution that facilitated the simultaneous use of Thunderbolt, USB, and power delivery, as tested in this study. As shown in the docking station diagram in Fig. 5, the connector signal Multiplexer enables a connection between a USB connector's connection port and a USB PCIe bridge; subsequently, it is controlled to make a connection between a USB connector's connection port and a Thunderbolt MAC PCIe bridge. Furthermore, after a certain signal generated in the docking station is processed as a USB standard signal at the USB PCIe bridge, it is transmitted to Thunderbolt;

IOP Conf. Series: Materials Science and Engineering 1113 (2021) 012001 doi:10.1088/1757-899X/1113/1/012001

then, after processing the certain signal as a Thunderbolt standard signal at the Thunderbolt Mac PCIe bridge, it is transmitted to Thunderbolt. The docking station allows the connector signal Multiplexer to establish a connection between the USB connector's connection port and the USB PCIe bridge; afterwards, it is controlled to make a connection between the USB connector's connection port and the Thunderbolt Mac PCIe bridge. Furthermore, it allows the PCIe bus Multiplexer to connect the PCIe bus to the USB PCIe bridge while the connector signal Multiplexer makes a connection between the USB connector's connection port and the USB PCIe bridge. In addition, while the connector signal Multiplexer connects the USB connector's connection port to the Thunderbolt Mac PCIe bridge, the connection between the PCIe bus and the processing part is controlled such that, after processing a certain signal, i.e., a signal of a USB device at the USB PCIe bridge, it is sent to the computer; then, after converting the signal into a Thunderbolt standard signal at the Thunderbolt Mac PCIe bridge, it is sent to the computer. Furthermore, according to the determined result, the docking station device was developed to electrically connect the Thunderbolt-to-PCIe bridge or USB-to-PCIe bridge to the PCIe bus, which is connected to the multiplexer. It was developed to facilitate automatic compatibility of interfaces between devices.

After developing the diagram in Fig. 5, the hardware and software of both the host computer and Thunderbolt should be prepared prior to testing to facilitate 10Gbps-class high-speed communication between the computer and peripheral devices. First, the notebook computer should be capable of using Thunderbolt3 because it must use a mainboard with an 8th generation or higher Intel CPU and a chipset supporting 40 PCIe lanes at most. In other words, in terms of satisfying the PCIe version and Thunderbolt lane requirements in the experiment, PCIe x1 implies that the bandwidth per lane based on the PCIe ver. 1.0 is 250 MiB/s (half-duplex) because, in the case of x16 sockets, which are typically used for graphic cards, 16 PCIe x1 lanes are allocated with a bandwidth of 250*16 = 4 GiB/s (half-duplex). The latest version of the PCIe standard released is ver. 3.0, and although a device using a higher version of PCIe can be mounted on a slot of a lower version, the bandwidth is restrained by the lower one among the two. Therefore, depending on the mainboard, there may be 16 lanes slot in shapes while filling only four or eight lines for the actual contact points. An opposing case also exists; a slot can be inserted regardless of device-side lane because the end parts of lane slots such as 1x, 4x, and 8x are opened. In both cases, a slot and a device operate with an electrically connected lane alone. The host computer is recommended to utilize a built-in graphic function on the CPU if possible, and although main boards may all look similar, there may be a limitation in the number of PCIe lanes. Particularly, in the case of AMD, the Ryzen2 Mainboard Chipset supports only 24 PCIe lanes at most. Furthermore, it cannot use a 10Gbps LAN card because the lanes consist of x16 for graphic cards, x4 for SSD, and x4 for CPU. For a case of minimal use for AMD, the built-in graphic function on the CPU is used, and 10Gbps LAN card is mounted on the x16 slot. Sometimes, a main board having two x16 slots should be used, and in this case also, there is usually a constraint that the primary slot is for x16 and the second slot is for x4. Furthermore, if there is no forced allocation function of the x8/x8 type in the ROM BIOS, the graphic card must be inserted into the second x4 slot and the 10Gbps LAN card in the primary x16 slot even if the graphic card's performance drops. If the graphic card is inserted into the primary x16 slot, the graphic card automatically gains control over the x16, thereby creating collisions with the second slot. In terms of CPU performance requirements, two cores or more is sufficient, but if running heavy programs such as a Web browser, it will contribute to speed fluctuation. Although PCIe 2.0 can also be used, it is based only on a theoretical maximum speed and its performance deteriorates. Therefore, it is advised to use a main board with PCIe version 3.0 if possible. In the case of Windows, background programs such as firewalls and Windows Defender should be stopped. When testing the 10Gbps full speed, the CPU cannot read the Rx buffer value of the 10Gbps LAN card and packet buffer losses occur while the abovementioned functions are being operated. Therefore, in the case of a slow CPU, the 10Gbps card buffer should be increased, the interrupt should be minimized, and the FlowControl function should be avoided. Based on these test environment conditions, a block diagram of the docking station was designed, as shown in Fig. 5.

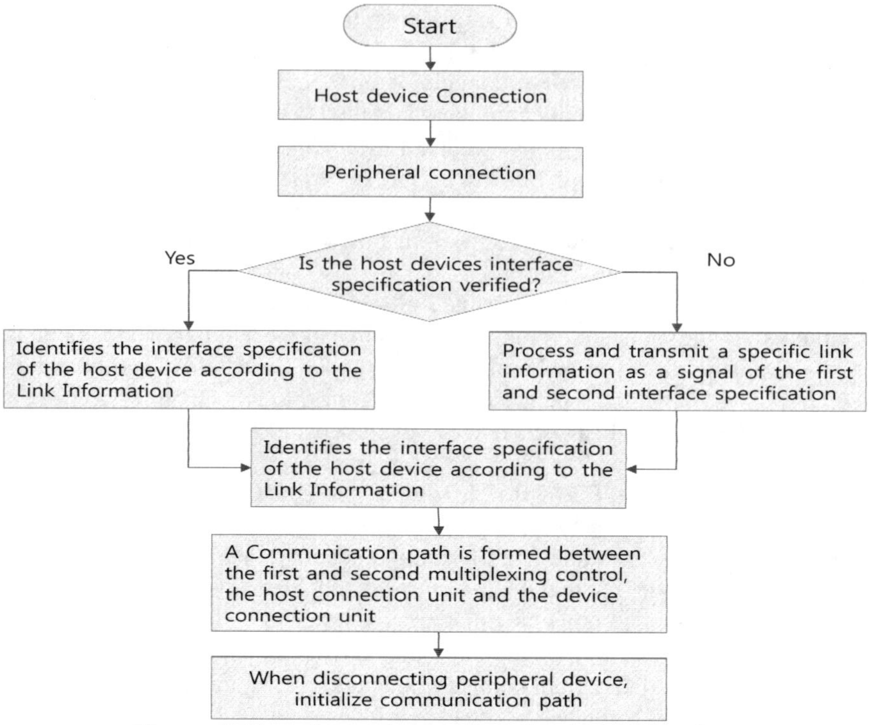

Figure 6. Experimental procedure for docking station

Fig. 6 shows a flowchart depicting the experimental procedure that processes a signal at the Thunderbolt MAC and USB MAC by identifying the interface standards when a host computer and a peripheral device are connected with a variety of communication devices. In other words, for a variety of communication devices, a communication path is formed at a port between the host device's connector and the peripheral device's connector alternately through the peripheral device's connecting PCIe bus, which connects the USB connector of the connector signal Multiplexer and the Thunderbolt signals. In the first step, the interface standard of the host device is checked and a signal is processed as the signals of Thunderbolt MAC and USB MAC interface standards or transmitted alternately. In the second step, the time information is included in the link information so that the interface standard of the host device can be identified during the process of connecting the host computer with the MAC layer processing functions of Thunderbolt and USB, respectively. In the third step, when the interface standards of various communication devices are identified, the connector signal Multiplexer and the PCIe bus Multiplexer are controlled to establish communication paths between the host device and the peripheral connectors.

4. Conclusion

This study performed research and development for the automatic compatibility of computer and peripheral devices according to new interface standards while improving the speed and performance of wired communication. When Thunderbolt capable of high-speed I/O is used as an interface, the overheads are lesser than those of USB, and data, video, and voice are supported simultaneously. However, there are few host devices supporting Thunderbolt, and presently, although most hardware supports USB connections, interconnection between USB and Thunderbolt is incompatible with each other. Therefore, when connecting peripheral devices to a host computer, the user should check the computer and the peripheral devices and utilize them by distinguishing suitable Thunderbolt- and USB-supporting devices. This study facilitated intercommunication among Thunderbolt, USB, and power delivery devices regardless of interface standards between the computer and peripheral devices. To accomplish this, the link information was detected and a connection was made with the host computer by alternately connecting data signals of a USB connector with the Thunderbolt MAC and the USB MAC through an electrical multiplexer. Furthermore, a docking station device was developed to implement an electrical connection of Thunderbolt-to-PCIe bridge or USB-to-PCIe bridge and PCIe BUS connected to the multiplexer based on the determined result. Consequently, the users no longer faced the inconvenience of identifying the interface standards of a computer individually and distinguishing Thunderbolt, USB, and power delivery devices during use. Moreover, users can expect an increasing degree of freedom of interface between devices and improved convenience. Therefore, this docking station overcomes the limitations of conventional technology and has sufficient potential for the commercial release or sale of devices using the relevant technologies. Moreover, it can be utilized efficiently in the high-speed communication industry.

Acknowledgments

This paper was prepared by the SK Broadband Consortium as part of the "2019 10Giga Internet Promotion business Project," which is a joint project of the Ministry of Science and ICT and the National Information Society Agency (NIA).

References

[1] Barnett T Jr and Jain S and Andra, U and Khurana T 2018 Cisco Visual Networking index(VNI) Complete Forecast Update 2017-2022 americas/EMEAR Cisco Knowledge Network(CKN) Presentation.

[2] Statista, Number of network-connected devices per person around the world from 2003 to 2020 https://www.statista.com/statistics/678739/forecast-on-connected-devices-per-person/

[3] Cheng H, Gao J, Wu H C, Liu G, Lau E, Yuan L and Krause C *Optics vs. copper—From the perspective of Thunderbolt 3 interconnect technology* (2016 China Semiconductor Technology International Conference. IEEE, pp 1-3)

[4] Sakurai W, Hayashista T, Watanabe Y, Takahashi T, Tsukui K, Umetsu H, Takahashi H, Tamura M and Chigusa Y 2012 *SEI Technical Review* **75**(121)

[5] Intel 2019 THUNDERBOLT $_{TM,}$ https://thunderbolttechnology.net//, Intel Coporation

[6] Frakes D 2019 What you need to know about Thunderbolt. MacWorld. Retrieved 17

[7] Cunningham Andrew 2015 USB 3.1 and Type-C: The only stuff at CES that everyone is going to use

[8] Meghana C R, Ramya M and Deepak S S 2011 *THUNDERBOLT-LIGHT PEAK* First National conference – CCSO

[9] Halvorsen O H and Clarke D 2011 *OS X and iOS Kernel Programming*. Apress pp 173-204

[10] Gao J, Cheng H, Wu H C, Liu G, Lau E, Yuan L and Krause C *2X25G low power optical IC for Thunderbolt optical cable technology* (2016 IEEE International Symposium on Electromagnetic Compatibility (EMC), pp 896-900)

[11] Singh V and Singh S 2014 *International Journal of Enhanced Research in Science Technology & Engineering* ISSN: 2319-7463 **3**(3) pp 259-264

[12] 10-gigabit-capable passive optical network (XG-PON) systems ITU-T G.987.x Series of Recommendations

[13] Nesset D 2017 *IEEE/OSA Journal of Optical Communications and Networking* **9**(1) pp. A71-A76

[14] Nelson R 2017 *EE-Evaluation Engineering* **56**(1) pp 8-11

Application of sequential quadratic programming for obtaining optimal obstacle avoidance posture and end position of a 7-axis robotic manipulator

T Y Huang and W J Chen

Deparment of Industrial Education and Technology, National Changhua University of Education, Changhua, Taiwan

Email: wjong@cc.ncue.edu.tw

Abstract. This paper used sequential quadratic programming (SQP) method to obtain the optimal obstacle avoidance posture and end position of a 7-axis robotic manipulator. By applying restraining conditions, obstacle information, and finishing point coordinates, a homogeneous transformation matrix was firstly used to construct a space for limited movement for the first and sixth axes of the robotic manipulator. Subsequently, we determined whether the terminal point was within reach of the robotic manipulator and calculated the angle of each axis. Finally, the MATLAB software was used to conduct a simulation and verification. In this study, the SQP method was used to introduce inequality constraints into the computation and conducted the iterative method. The simulation results revealed that the method can obtain avoidance of obstacle postures and select optimal values from the iteration results of multiple initial values to avoid the local optimal problem. Overall, the research findings are that the functions and applicability of the 7-axis robotic manipulator can be extended in industry when applied in future intelligent manufacturing.

1. Introduction

With the advent of smart manufacturing (Industry 4.0) in recent years, robotic manipulators have become indispensable. Redundant manipulators with many degrees of freedom (DOF) can not only position and orientate themselves but also avoid obstacles and singularities. Redundant manipulators with six DOF are widely used in various industries, manufacturing, agriculture, and electronics. However, the use of a manipulator with six or fewer DOF on a production line will not offer any advantage in terms of space-saving. Moreover, workspaces can be limited and size and contain obstacles. Therefore, several studies on optimizing the location of an end-effector have been conducted. Novitarini et al [1] used Denavit–Hartenberg (D–H) parameters to kinematically analyze a four-DOF manipulator. Their research highlighted the percentages of error of two different kinematics in the manual, simulation, and reality. Iliukin et al [2] attempted to control a five-DOF manipulator by using a brain–computer interface. They modeled the manipulator, solved the associated forward and inverse kinematics and simulated manipulator position in MATLAB. Filipovich et al [3] applied two kinematic schemes to a SCARA-type manipulator and analyzed the manipulator's reachable space in each scheme as well as the modeling method and its characteristics. Sun et al [4] addressed the inverse kinematics of a six-DOF robot by using D–H parameter modeling, a homogeneous transformation matrix, and forward kinematics. Shehab Mohamed [5] analyzed each rotating axis of a seven-axis serial robotic arm with forward and inverse kinematics; they analyzed the singularity configuration and workspace to control an end-effector to avoid obstacles and maintain joints within constraints while following straight trajectories. Busson et al [6] utilized a seven-DOF manipulator to maximize Cartesian rigidity. To find the stiffest configuration, they projected the gradient of the rigidity cost function onto the null space of the manipulator Jacobian applied for the redundant robot. Kim et al [7] proposed a rapidly exploring random tree path-planning algorithm for a seven-DOF robot arm. The occlusion-free and collision-avoiding path-planning constraints were determined through MATLAB and V-REP simulation, respectively. Cao et al [8] improved the rapidly exploring random tree algorithm to generate a collision-free path. A genetic algorithm was used to smooth out sharp points and reduce the redundant points in the path. These methods were applied to a six-DOF serial robot arm. Chao Mal et al [9] analyzed each angle of a robot's joints with inverse kinematics. A spatial path-planning method based on the RRT algorithm was explored for a 6R serial manipulator. Mahdavian et al [10] proposed a method for optimizing energy consumption and avoiding moving obstacles in trajectory generation for a four-DOF arm. Yi Cao et al [11] proposed a path-planning method for an obstacle-free workspace and a 3R manipulator. Computer graphics were used to depict the intersection between the workspace and an obstacle through the RRT method within the nonobstacle workspace to generate a collision-free path.

Most aforementioned have achieved obstacle avoidance for only the end-effector. When a manipulator moves from a starting point to a terminal point, each link in the manipulator is likely to collide with an obstacle. Therefore, this study proposes a method for generating a collision-free end-posture for each link of a seven-DOF manipulator. First, the kinematic analysis of the manipulator with constraints is imposed on the first and sixth axes. The sequential quadratic programming (SQP) method in MATLAB applies equality and inequality constraints in the iterative computation of the angle of each joint of the robotic manipulator. The modeling, simulation, and verification of the optimized end-posture and obstacle avoidance strategies for the entire robotic manipulator are presented.

2. Modeling seven-DOF manipulator

As illustrated in Fig. 1, the design parameters of the seven-DOF serial manipulator are as follows: θ_1–θ_7 represent the angle of rotation of each joint of the manipulator, α_1–α_7 the difference in the rotation angle of each joint with respect to their former joints, $d_{(i=1.3.4.7)}$ the distances along axis $\overrightarrow{Z_i}$ between axes $\overrightarrow{x_i}$ and axis $\overrightarrow{x_{i+1}}$, and I_i the length of each link. Here, d_1= 184 mm, I_2= 96 mm, d_{34}= 200 mm, I_5= 78 mm, and d_6= 96 mm; d_{34} consists of d_3 and d_4. The limitations of each joint are as follows: $-180° \leq \theta_1 \leq 180°$, $-90° \leq \theta_2 \leq 90°$, $-90° \leq \theta_3 \leq 90°$, $-90° \leq \theta_4 \leq 90°$, $-90° \leq \theta_5 \leq 90°$, $-90° \leq \theta_6 \leq 90°$, and $-90° \leq \theta_7 \leq 90°$. The coordinate equations of the individual joints were derived using homogeneous transformation matrices and D–H parameters. To model the end-effector, the orientation of the manipulator, coordinates, and parameters of each axis listed in Table 1 are essential. The 4 × 4 homogeneous transformation matrix consists of a rotation matrix and the coordinates of the end-effector, $P = [Px\ Py\ Pz]^T$, in the world coordinate system.

Table 1. D–H parameters

	Θ(Joint angle)	d(offset)	a_i(Link length)	α_i(Twist angle)
Link1	0	d_1	0	$\pi/2$
Link2	$\pi/2$	0	I_2	0
Link3	$-\pi/2$	0	0	$-\pi/2$
Link4	0	d_{34}	0	$\pi/2$
Link5	$\pi/2$	0	I_5	$-\pi/2$
Link6	$-\pi/2$	0	0	$-\pi/2$
Link7	0	d_6	0	0

(a) Three-dimensional manipulator (b) Local coordinates of each joint

Figure 1. Model of seven-DOF manipulator

The following steps are used to determine the relationship between two sequential axes. First, rotate the *Z*-axis by an angle θ. Second, translate vertically distance *d* along the *Z*-axis. Third, translate a distance a_i along the *X*-axis. Finally, rotate the *X*-axis by an angle α_i (in degrees). The homogeneous transformation matrix is expressed as follows:

$$\mathrm{Rot}(Z_i,\theta)\cdot\mathrm{Trans}(0,0,d)\cdot\mathrm{Trans}(a_i,0,0)\cdot\mathrm{Rot}(X_i,\alpha_i)$$

$$=\begin{bmatrix} cos\theta & -sin\theta cos\alpha_i & sin\theta sin\alpha_i & a_i\cdot cos\theta \\ sin\theta & cos\theta cos\alpha_i & -cos\theta sin\alpha_i & a_i\cdot sin\theta \\ 0 & sin\alpha_i & cos\alpha_i & d \\ 0 & 0 & 0 & 1 \end{bmatrix} = {}^{i-1}_{\ \ i}T \quad (1)$$

The matrices ${}^{i-1}_{\ \ i}T$ (i = 1,2,3, …7) represent the homogeneous transformation matrices of each joint from the first axis to the seventh axis. Through substitution of all variables from the Table 1 into the matrix, each coordinate of each joint is obtainable.

$${}^{0}_{7}T = \begin{bmatrix} A_X & B_X & C_X & P_{7X} \\ A_Y & B_Y & C_Y & P_{7Y} \\ A_Z & B_Z & C_Z & P_{7Z} \\ 0 & 0 & 0 & 1 \end{bmatrix} \quad (2)$$

In (2), *Ax*, *Ay*, *Az*; *Bx*, *By*, *Bz*; and *Cx*, *Cy*, *Cz* represent the coordinate system of the seventh axis (O_7, X_7, Y_7, Z_7) relative to the world coordinate system (O_0, X_0, Y_0, Z_0), as illustrated in Fig. 1(b). To simplify the relationship, assume that the world coordinate system (O_0, X_0, Y_0, Z_0) and the coordinate system of the first axis are the same. The coordinates of the end-effector of the seven-DOF manipulator are P_{7X}, P_{7Y}, and P_{7Z}.

3. Deriving constraints and limiting method

By applying the coordinates of the obstacle, inequality and equality constraints, restraining conditions, and terminal point coordinates, not only end-effectors but also other links of the manipulator can be made to avoid obstacles by attaining optimal posture.

3.1. First constraint

To avoid the area *OABC* (as shown in Fig. 2), it is essential to define θ_1, which is the permissible rotation angle of the first axis; θ_R denotes the angle from the X_1-axis to the shortest line ($\overline{OB}$) between the origin, coordinate *O*, and the center of the obstacle. θ_o is the angle between $\overline{OB}$ and the border of the obstacle ($\overline{OA}$ or $\overline{OC}$). θ_1 is limited within two ranges of (180° to $(-\theta_R+\theta_o)$) and (−180° to $(-\theta_R-\theta_o)$), as depicted in yellow and green respectively in Fig. 2.

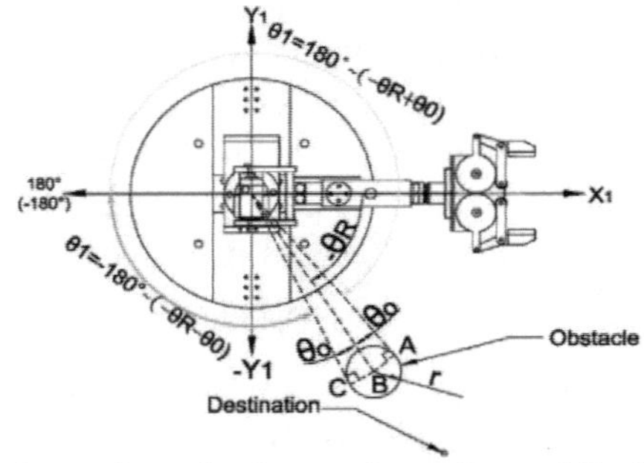

Figure 2. Top view of manipulator, obstacle, and limited ranges of θ_1

3.2. *Second constraint*

Constraining the coordinates of the sixth axis (P_{6X}, P_{6Y}, and P_{6Z}) in the movable area can limit the end-posture of the manipulator. The coordinates of the obstacle are O_X, O_Y, and O_Z. The radius of the obstacle is r. When $(P_{6X} - (O_X))^2 + (P_{6Y} - (O_Y))^2 > r^2$, the coordinates of the sixth axis (green arrow) cannot collide with the obstacle (red area represents the obstacle's area). Because this constraint has no height limitation, the border conditions of the obstacle are only two-dimensional, as shown in Fig. 3. Adding the height constraint ($P_{6Z} \leq O_z$) can limit the height of the sixth axis to be lower than that of the obstacle. The second constraint is expressed in (3).

$$(P_{6X} - (O_X))^2 + (P_{6Y} - (O_Y))^2 - (rod + r)^2 > 0$$

$$P_{6Z} \leq O_z$$

(a) Three dimensional diagram (b) Top view diagram

Figure 3. The schematics of the second constraint

3.3. *Third constraint*

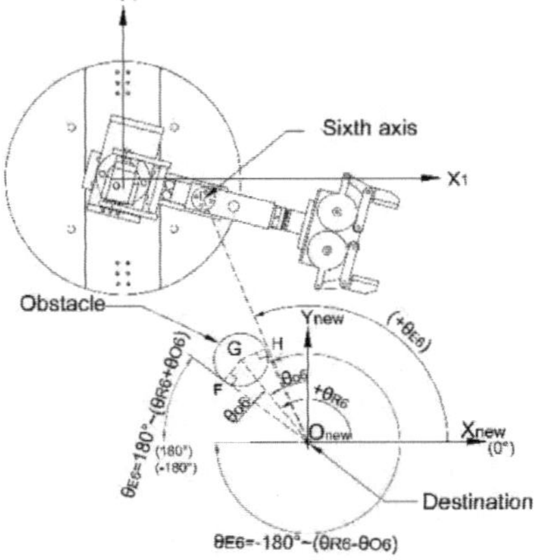

Figure 4. Position of third constraint of reference point

By limiting the coordinates of the sixth axis within a certain region, we can avoid collisions between the sixth axis and area $FGHO_{new}$, as depicted in Fig. 4. To satisfy this new constraint, a new coordinate system $(O_{new}, X_{new}, Y_{new}, Z_{new})$ is formed. The angle (θ_{R6}) is defined by the relationship between the new coordinate system and obstacle; that is, (θ_{R6}) is formed between the shortest line from the center of the obstacle to the destination $(\overline{GO_{new}})$ and the X_{new} axis. The angle θ_{E6} is formed between the shortest line from the sixth axis to the destination $(\overline{IO_{new}})$ and the X_{new} axis. The angle θ_{O6} is formed between the border $(\overline{FO_{new}}$ or $\overline{HO_{new}})$ of the obstacle and $\overline{GO_{new}}$.

By utilizing the angles θ_{R6} and θ_{O6}, we can limit θ_{E6} to within the rotatable angle region (180° to $(\theta_{R6}+\theta_{O6})$) and (−180° to $(\theta_{R6}-\theta_{O6})$). The rotatable angle region of θ_{E6} is $(\theta_{R6}+\theta_{O6}) \leq \theta_{E6} \leq 180°$ (light blue area) and $(\theta_{R6}-\theta_{O6}) \geq \theta_{E6} \geq -180°$ (purple area).

3.4. Fourth constraint

The Euler angles of the seventh coordinate system relative to the world coordinate system can be obtained from the matrix in (2). To realize this constraint, the end-effector and the coordinates of the sixth axis must be presented at the same height, as expressed by (4).

$$P_{6Z} = P_{7Z} \tag{4}$$

Utilizing the equation $\theta_{Yaw} = 2\tan^{-1}(\frac{A_X}{A_Y})$ (from matrix in (2)), the yaw of the seventh coordinate system can be determined. This constraint is expressed as (5).

$$2tan^{-1}\left(\frac{A_X}{A_Y}\right) = \pm 90 \tag{5}$$

4. SQP and karush-kuhn-tucker(KKT)

4.1. SQP

By utilizing the gradient and given inequality and equality equations in SQP, the minimum solution from objective functions can be acquired in this method. In this study, SQP was used to calculate the angle of each joint to optimize obstacle avoidance behaviour and the manipulator's end-posture. $f(x)$ is the objective functions. $g_u(\mathrm{x})$ and $c_n(\mathrm{x})$ are the inequality and equality equations, as shown in equation(6).

$$\begin{gathered} \min f(x) \\ \mathrm{s.t.}\, g_u(\mathrm{x}) = 0 \;\; \mathrm{u} = 1, \ldots\ldots, \mathrm{v} \\ c_n(\mathrm{x}) \leq 0 \;\; \mathrm{n} = 1, \ldots\ldots, \mathrm{k} \end{gathered} \tag{6}$$

4.2. Karush-Kuhn-Tucker condition

The Karush–Kuhn–Tucker (KKT) condition is essential for optimizing nonlinear programming in optimization theory. The KKT condition applies equality constraints with Lagrange multipliers to derive corresponding inequality equations. By applying Lagrange multipliers to the equality equation $g_u(\mathrm{x})$, we can calculate λ_i and x_i.

$$\mathrm{f}(x_i, \lambda_i) = \begin{bmatrix} \nabla \mathrm{f(x)} - \mathrm{G(x)}^T \lambda \\ g_u(x) \end{bmatrix} \tag{7}$$

$f(x)$ can be transformed to $f(x_i, \lambda_i)$ by utilizing the Lagrange equation, as in (7). $G(x) = [\nabla g_1(x), \nabla g_2(x), \ldots \nabla g_u(x)]$; $f(x) = g_u(x) \cdot \lambda_u$. It can be written as (8) by substituting (7) into (6).

$$f(x_i, \lambda_{u,} m_n) = f(\mathrm{x}) + \sum_{u=1}^{v} \lambda_u g_u(x) + \sum_{n}^{k} m_n c_n(k) \tag{8}$$

The Lagrange multiplier of the inequality equations in the KKT condition given in (6) is m_n, as expressed in (8). To satisfy the constraint, m_n should less than or equal to zero. Equation (9) can be obtained by expanding the objective functions with Taylor's second-order expansion and expanding the equality and inequality equations with Taylor's first-order expansion. Let $\mathrm{P} = \mathrm{x} - x^i$, applying the Newton's method , $\mathrm{P} = x_i - \frac{f(x_i)}{f'(x_i)}$.

$$\begin{gathered} \min f(x^i) + \nabla f(x^i)^T(\mathrm{P}) + \frac{1}{2}(\mathrm{P})^T \cdot \nabla^2 f(x_i, \lambda_{u,} m_n)(\mathrm{P}) \\ \mathrm{s.t.}\, g_u(x^i) + \nabla g_u(x^i)^T \mathrm{P} = 0 \\ c_n(x^i) + \nabla c_n(x^i)^T \mathrm{P} \leq 0 \end{gathered} \tag{9}$$

$f(x^i) = [\theta_1, \theta_2, \dots\dots, \theta_7]$. By applying the homogeneous transformation matrix (${}^0_7T$), the coordinates $[P_{7X}, P_{7Y}, P_{7Z}]^T$ can be obtained, as given in (2). The objective functions can be written as $f_{P7X}, f_{P7Y}, f_{P7Z}$. The inequality and equality equations can be written as (10) and (11), respectively.

$$\begin{gathered} c_n(\mathrm{x}) = [+180° \geq \theta_1 \geq (-\theta_R + \theta_o) \text{ and } -180° \leq \theta_1 \leq (-\theta_R - \theta_o), \\ (P_{6X} - (O_X))^2 + (P_{6Y} - (O_Y))^2 - (rod + r)^2 > 0, P_{6Z} \leq O_Z, \\ (\theta_{R6} + \theta_{O6}) \leq \theta_{E6} \leq 180° \text{ and } (\theta_{R6} - \theta_{O6}) \geq \theta_{E6} \geq -180°] \end{gathered} \tag{10}$$

$$g_u(x) = [P_{6Z} = P_{7Z}, 2tan^{-1}\left(\frac{A_X}{A_Y}\right) = \pm 90°] \tag{11}$$

5. Research process

This research considers each rotation angle to be limited. By applying the given coordinates of the terminal point and four constraints, all joint angles corresponding to the optimal posture are calculated through iterative SQP computation. The process followed in this study is illustrated in Fig. 5.

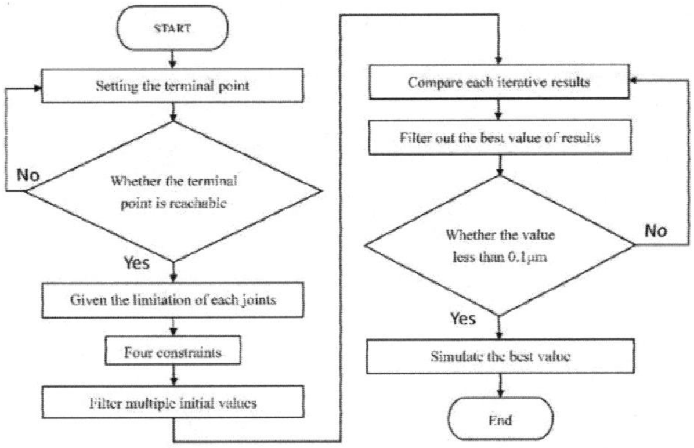

Figure 5. Research flowchart

6. Simulation results

This research optimizes posture for obstacle avoidance through SQP with four constraints. After obtaining the best solution from the iterative computation, a program verifies each angle by simulating the manipulator and obstacle. The first iterative result converges at the 48th iteration, 4.978343×10^{-8} m, as illustrated in Fig. 6.

Table 2 showed the angles of the motors and the position of the end-effector. The result is the position error $\sqrt{X^2+Y^2+Z^2}$. The position errors in each dimension are X_{error}= 6.34796742948751×10^{-7} cm, Y_{error}= −4.665374547796475×10^{-6} cm, and Z_{error}= 1.617160555467123×10^{-6} cm. MATLAB was used to simulate the starting point of the end-effector X= 35.068 cm, Y= 0 cm, Z= 30.899 cm, as shown in Fig. 7(a). The blue cylinder represents an obstacle of height 26.5 cm and radius 2 cm, and the coordinates of the obstacle are X=−10cm and Y=20cm. The widest radius of the manipulator is 5 cm (rod). The entire manipulator avoids the obstacle. The end-effector denoted by the gray lines reaches the terminal point X= −18.9 cm, Y= 35.5 cm, Z= 12.5 cm, as depicted in Fig. 7(b).

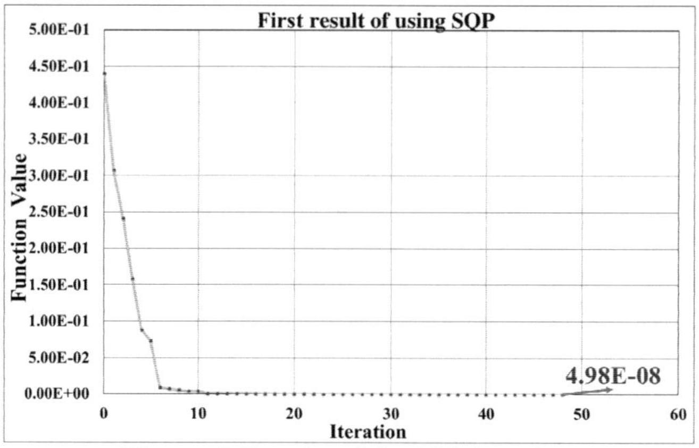

Figure 6. Results of the 48th iteration

Table 2. The results of the 48th iteration for position and rotational angles of end-effector

	Obtained angles	Terminal position (cm)		
θ_1	85.24553689°	X	Y	X
θ_2	-53.05491743°	-18.9	35.5	12.5
θ_3	-82.12917794°			
θ_4	-9.697441977°	Obtained position (cm)		
θ_5	64.024917°	X	Y	Z
θ_6	69.28116133°	-18.8999993652033	35.4999953346255	12.500001617606
θ_7	-70.39163178°			

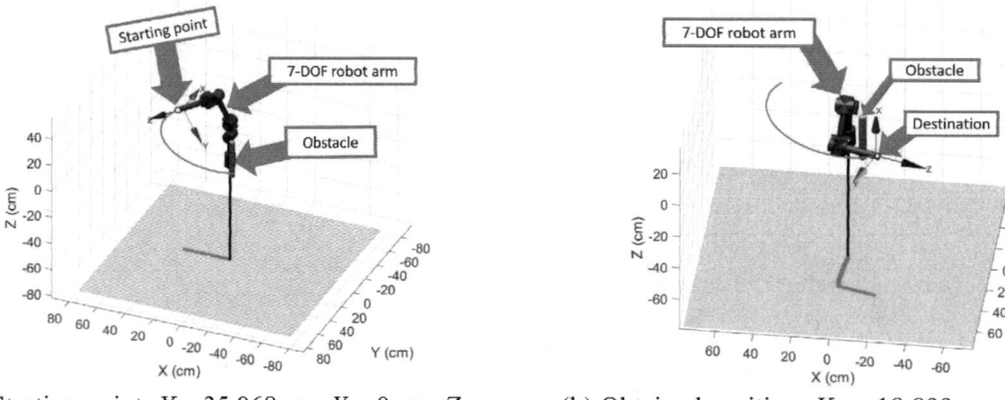

(a) Starting point: X = 35.068 cm, Y = 0 cm, Z = 30.899 cm

(b) Obtained position: X = −18.899 cm, Y = 35.499 cm, Z = 12.500001 cm

Figure 7. The position of the end-effector converging at the 48th iteration

The same terminal point is used in the second iterative computation, which converges at the 52nd iteration, $5.634721444921398 \times 10^{-8}$ m, as illustrated in Fig. 8. Table 3 showed the angles of the motors and the position of the end-effector. The position errors in each dimensions are $X_{error} = 1.168242738658343 \times 10^{-6}$cm, $Y_{error} = 3.668677750479077 \times 10^{-6}$cm, and $Z_{error} = 4.114133958510458 \times 10^{-6}$ cm. The starting point of the end-effector was $X = 35.068$ cm, $Y = 0$ cm, $Z = 30.899$ cm, as indicated in Fig. 9(a). The end-effector reaches the same terminal point $X = -18.9$ cm, $Y = 35.5$ cm, $Z = 12.5$ cm, as depicted in Fig. 9(b), in which the gray line denotes the trajectory of the end-effector.

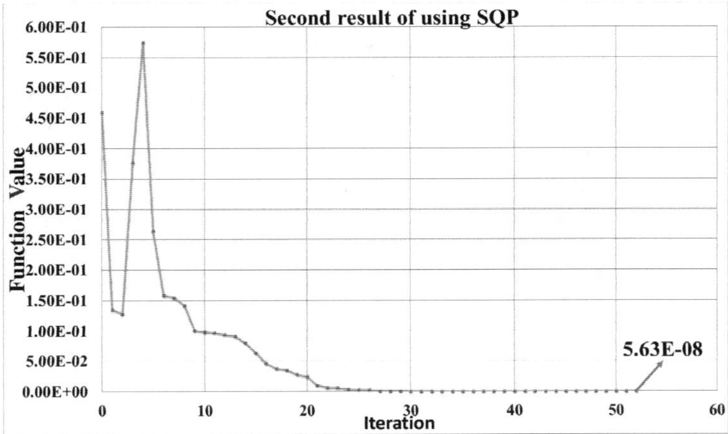

Figure 8. Results of the 52nd iteration

Table 3. The results of the 52nd iteration for position and rotational angles of end-effector

	Obtained angles	Terminal position (cm)		
θ_1	161.95705129699°	X	Y	X
θ_2	-42.31190250384°	-18.9	35.5	12.5
θ_3	-90.00000000028°			
θ_4	75.887357775075°	Obtained position (cm)		
θ_5	79.21578562672°	X	Y	Z
θ_6	-4.080974708412°	-18.8999988317573	35.5000036686777	12.500004114134
θ_7	44.028418816509°			

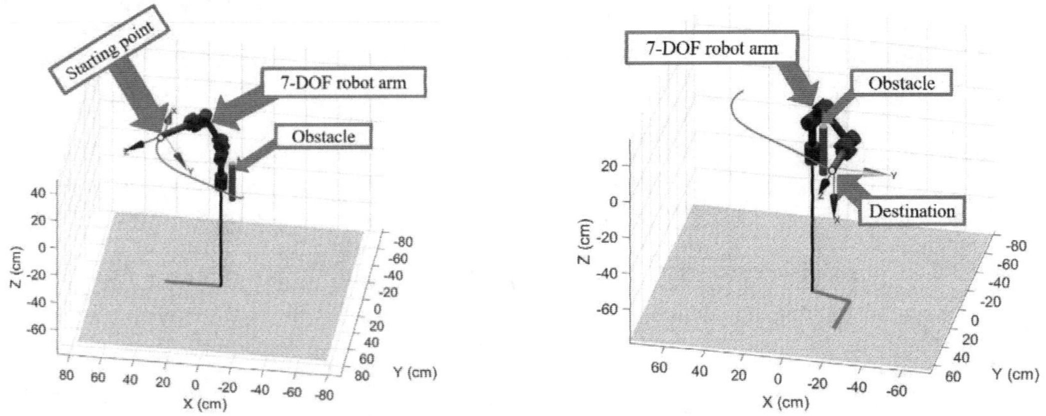

(a) Starting point: $X = 35.068$ cm, $Y = 0$ cm, and $Z = 30.899$ cm

(b) Obtained position: $X = -18.899$ cm, $Y = 35.500003$ cm, and $Z = 12.50004$ cm

Figure 9. The position of the end-effector converging at the 52nd iteration

7. Conclusion

SQP with multiple initial values can improve local solutions. Sampling the optimal value from several local solutions avoids the possibility of the solution being trapped in one local optimum. By using the computational scheme and the constraints outlined in this study, the optimal angles for obstacle avoidance posture of a manipulator can be obtained. This paper discussed the application of the SOP algorithm with constraints on the first and sixth axes to a seven-DOF manipulator. Simulation results indicate that entire manipulator can avoid obstacles. Moreover, the end-effector can reach its destination precisely. Thus, convergence is achieved, with a three-dimensional position error smaller than 0.1 μm.

Acknowledgment

The authors gratefully acknowledge the financial support of the Ministry of Science and Technology of Taiwan under Grant No. MOST 107-2221-E-018-008-.

References

[1] Novitarini A, Aniroh Y, Anshori D Y and Budiprayitno S 2017 *A Closed-Form Solution of Inverse Kinematic for 4 DOF Tetrix Manipulator Robot* (Int. Conf. on Advanced Mechatronics, Intelligent Manufacture, and Industrial Automation, Surabaya, pp 25-29)

[2] Iliukhina V N, Mitkovskiib K B, Bizyanovaa D A and Akopyana A A 2017 *Procedia Engineering* **176** pp 498-505

[3] Filipovich O, Chalenkov N and Balakin A 2018 *The analysis of the kinematic characteristics of the multifunctional module based on the SCARA type manipulator* (Int. Multi-Conf. on Industrial Engineering and Modern Technologies, FarEastCon, pp 1-6)

[4] Sun J D, Cao G Z, Li W B, Liang Y X and Huang S D 2017 *Analytical Inverse Kinematic Solution Using the D-H Method for a 6-DOF Robot* (Int. Conf. on Ubiquitous Robots and Ambient Intelligence (URAI), Korea: Jeju, pp 714-716)

[5] Mohammed S, VT 2016 *Kinematic Motion Planning for a 7-Axis Robotic Arm* Master's thesis (Sewden, Umeå University)

[6] Busson D, Bearee R and Olabi A 2017 *Task-oriented rigidity optimization for 7 DOF redundant manipulators* (Int. on Federation of Automatic Control (IFAC), France: Toulouse, pp 14588-14593)

[7] Kim Y J, Wang J H, Park S Y, Lee J Y, Kim J J and Lee J J 2014 *A RRT-Based Collision-Free and Occlusion-Free Path Planning Method for a 7DOF Manipulator* (Proc. Int. Conf. on IEEE Mechatronics and Automation, China: Tianjin, pp 1017-1021)

[8] Cao X, Zou X, Jia C, Chen M and Zeng Z 2019 *Computers and Electronics in Agriculture* **156** pp 105-118

[9] Ma C, Zhang Y, Zhao Q and Bai K 2016 *6R Serial Manipulator Space Path Planning Based on RRT* (Int. Conf. on Intelligent Human-Machine Systems and Cybernetics(IHMSC), China: Hangzhou, pp 99-102)

[10] Mahdavian M, Shariat-Panahi M, Yousefi-Koma A and Ghasemi-Toudeshki A 2015 *Optimal Trajectory Generation for Energy Consumption Minimization and Moving Obstacle Avoidance of a 4DOF Robot Arm* (3rd RSI Int. Conf. on Robotics and Mechatronics(ICROM), Iran: Tehran, pp 353-358)

[11] Cao Y, Guo M and Li Y 2017 *Obstacle-free Workspace based Path Planning for Serial Manipulator* (Proc. on Chinese Control Conference(CCC), China: Dalian, pp 6897-6900)

IOP Conf. Series: Materials Science and Engineering 1113 (2021) 012003 doi:10.1088/1757-899X/1113/1/012003

Digital tracking control of the boost converter: prediction-based digital redesign approach

Cheng Shion Shieh

Department of Electrical Engineering, Far-East University, Tainan, Taiwan

E-mail: csshieh@mail.feu.edu.tw

Abstract. Several continuous-time tracking controllers of the DC-to-DC boost converters had been addressed, but few literature touched the digital tracking control because that was a challenging theoretical problem which consists of nonlinear, non-minimum phase and saturated control. Given a continuous-time nonlinear boost converter with continuous tracking control, this paper proposes a new sampled-data feedback laws from the available analog counterpart via digital redesign such that the state trajectories of the sampled-data closed-loop systems close to the continuous-time one in the sense of state-matching. An existing boost converter circuit has been demonstrated to visualize the feasibility of the proposed methodology.

1. Introduction

DC-to-DC converter and DC-to-AC inverter are very important topics in power conversion and are usually accomplished by some classical continuous circuits. The power transfer efficiency of power conversion often is influenced by used control methodology. Hence, the control issue of power conversion is still a major concern. The control of an inverter is closely related to the tracking control problem of the basic converter [1]. Recently the converter or inverter device is often nonlinear, non-minimum phase system with saturated control. Therefore, the control problem is much more complicated and is a challenging problem even in the regulator case (DC-to-DC conversion) due to its dynamical characteristics.

Analysis of the dynamical properties of the boost converter with control strategies had been addressed [2-3]. Most of the previous work dealing with the continuous control of converter/inverter. However the sampled-data systems are widespread more and more because most controlled plants encountered in engineering applications are continuous while controls are digitally implemented using computers. In order quickly to transfer continuous system to sampled-data one, a digital redesign is an efficient design tool of digital control [4-6]. That is, a continuous controller is designed first, and discretized afterward to obtain an equivalent digital controller. The advantages are that the existing continuous-time methods are more intuitive and better established than their discrete-time counterparts and, furthermore, the sampling period can be selected based on the known continuous-time closed-loop bandwidth [7]. That is just what this paper wants to construct a digital tracking controller for DC-to-DC converter by digital redesign.

This paper is organized as follows: Section 2 formulates the digital redesign problem of the tracking control of the booster converter. The proposed digital control is developed in Section 3, which control performance the DC-DC conversion state matching is satisfied between continuous system and sampled-data one. An illustrative example is shown in Section 4. Finally, in Section 5, conclusions are presented.

2. Problem formulation

The solved diagram of the boost converter with digital control is shown in Fig. 1. It is well known that when the switching frequency is high, it is possible to model the average behavior of this power converter. The given DC-to-DC converter expression is shown as [1]:

$$\mathrm{L}\dot{x}_{1c} = -u_c x_{2c} + v_{in} \tag{1a}$$

$$\mathrm{C}\dot{x}_{2c} = u_c x_{1c} - \frac{x_{2c}}{R} \tag{1b}$$

$$y_c = x_{2c} \tag{1c}$$

where the v_{in} is supposed to be known constant voltage, the x_{1c} is the current across the inductor, x_{2c} is the voltage in the capacitor, and the control variable $u_c \in [0 \quad 1]$, known as the duty cycle, is the average value within a communication cycle. Throughout this paper, the state symbols represent average values.

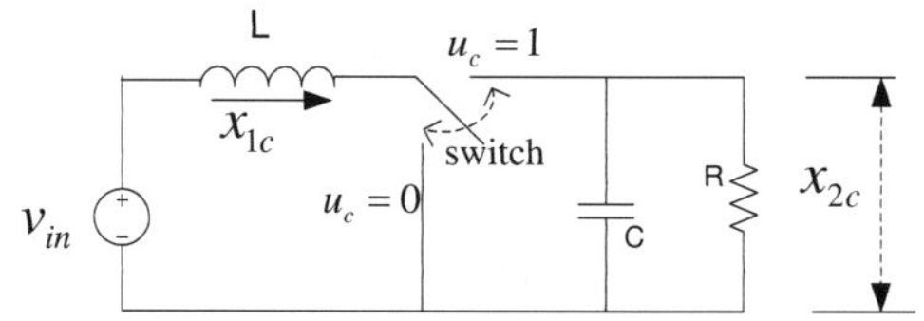

Figure 1. Schematic diagram of the boost converter [1]

We rewrite system (1) in general form

$$\dot{x}_c = f(x_c) + g(x_c)u_c \tag{2a}$$

$$y_c = Cx_c \tag{2b}$$

where

$$x_c = \begin{bmatrix} x_{1c} \\ x_{2c} \end{bmatrix}; f(x_c) = \begin{bmatrix} f_1(x_c) \\ f_2(x_c) \end{bmatrix} = \begin{bmatrix} v_{in}/L \\ -x_{2c}/RC \end{bmatrix}; g(x_c) = \begin{bmatrix} -x_{2c}/L \\ x_{1c}/C \end{bmatrix}; C = [0 \quad 1]$$

The Cortes et al. [1] had proposed an easy-to-implement continuous control law shown as

$$u_c(\mathrm{t}) = \frac{v_{in} + K(x_{1c}(t) - x_{1r})}{x_{2c}(t)} \tag{3}$$

where K is a positive constant and x_{1r} is given by [1]

$$x_{1r} = \frac{x_{2r}^2}{R v_{in}} \tag{4}$$

where x_{1r} and x_{2r} represent the steady-state value of the inductor current and the voltage in the capacitor. The Cortes et al. also proposed the following proposition to define the range of parameters R, L and C in (1).

The proposition [1]: If $x_{2r} = a + b\sin(wt)$ then

$$\frac{v_{in}(a+b-v_{in})}{(a+b)\Delta_{x_{1c}}} T_s < L < \frac{R v_{in}^2}{2bw(a+b/2)} \tag{5}$$

$$\frac{a+b-v_{in}}{R\Delta_{x_{2c}}}T_S < C < \frac{E_\varepsilon}{Rbw(a+b/2)} \tag{6}$$

where $\Delta_{x_{1c}}$, $\Delta_{x_{2c}}$ and T_s denote the maximum allowed ripple in the inductor, the capacitor voltage and the switching period, respectively. E_ε satisfies the inequality $\|x_2^2 - x_{2d}^2\| < E_\varepsilon$. In short, a given continuous DC-to-DC boost converter system with continuous controller (3), we should construct $u_d(t)$ piecewise-constant input to replace the original continuous control $u_c(t)$, which is expressed as

$$u_d(t) = u_d(kT) \text{ for } kT \leq t < (k+1)T \tag{7}$$

where T is the sampling period. A zero-order hold is utilized in (7).

Fig. 2 shows the traditional digital design schema but it is hard to do digital modeling because the input function $g(x)$ is non-constant. This paper offers a different issue of digital tracking control methodology based on the proposed modified digital redesign diagram, described in Section 3, to solve the digital tracking control of the DC-to-DC boost converter.

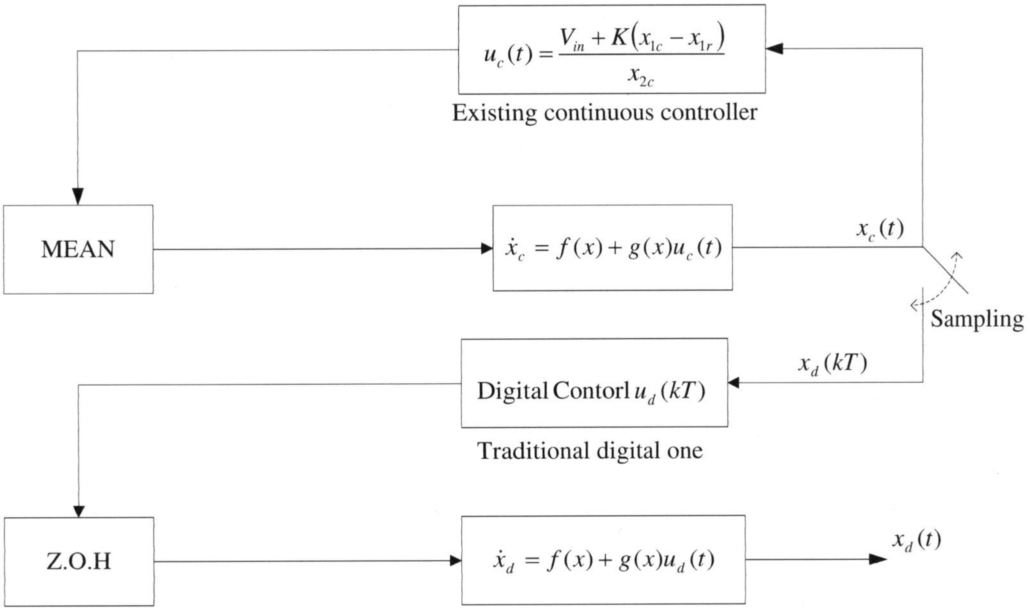

Figure 2. Traditional digital redesign block for the boost converter [1]

3. Digital redesign of boost converter

Because the digital controller needs feedback digital state of the controlled plant, the digital modeling of the boost converter (2) must be solved. However, it is hard to achieve a digital state from the model (2a) for g(x) non-constant. Here we represent the modified digital redesign block diagram Fig. 3 to replace the traditional one in Fig. 2 for solving the digital states. The proposed design diagram is described as bellows.

Rewrite the continuous model (2a) in the following:

$$\begin{bmatrix}\dot{x}_{1c}\\ \dot{x}_{2c}\end{bmatrix} = \begin{bmatrix}0 & -u_c/L\\ u_c/C & -1/RC\end{bmatrix}\begin{bmatrix}x_{1c}\\ x_{2c}\end{bmatrix} + \begin{bmatrix}1/L\\ 0\end{bmatrix}v_{in} \tag{8}$$
$$\triangleq \dot{x}_c = Ax_c + Bv_{in}$$

where $A \triangleq \begin{bmatrix} 0 & -u_c/L \\ u_c/C & -1/RC \end{bmatrix}$; $B = \begin{bmatrix} 1/L \\ 0 \end{bmatrix}$. The system matrix A is floating according to the given continuous control u_c and input matrix B is constant. The solution of the continuous-time controlled system (8), $x_c(t)$, at a time $t = t_v = kT + vT$ for a design parameter $0 \leq v \leq 1$ is found to be

$$x_c(t_v) = e^{A(t_v - kT)} x_c(kT) + \int_{kT}^{kT+vT} e^{A(kT+vT-\tau)} B v_{in} d\tau \tag{9a}$$

which, since B and v_{in} are constant, reduces to

$$x_c(t_v) = e^{A(t_v - kT)} x_c(kT) + \int_{kT}^{kT+vT} e^{A(kT+vT-\tau)} B v_{in} d\tau = G^v x_c(kT) + H^v v_{in} \tag{9b}$$

where $G^v = \left(e^{AT}\right)^v = \begin{bmatrix} G_{11} & G_{12} \\ G_{21} & G_{22} \end{bmatrix}^v$ and $H^v = (G^v - I)A^{-1}B = \begin{bmatrix} h_1 \\ h_2 \end{bmatrix}^v$. While the solution to the sampled-data controlled system, $x_d(t_v)$ at $t = t_v = kT + vT$, is obtained as

$$x_d(t_v) = G^v x_d(kT) + H^v v_{in}$$
$$\triangleq \begin{bmatrix} x_{1d}(t_v) \\ x_{2d}(t_v) \end{bmatrix} = \begin{bmatrix} G_{11} & G_{12} \\ G_{21} & G_{22} \end{bmatrix}^v \begin{bmatrix} x_{1d}(kT) \\ x_{2d}(kT) \end{bmatrix} + \begin{bmatrix} h_1 \\ h_2 \end{bmatrix}^v v_{in} \tag{10}$$

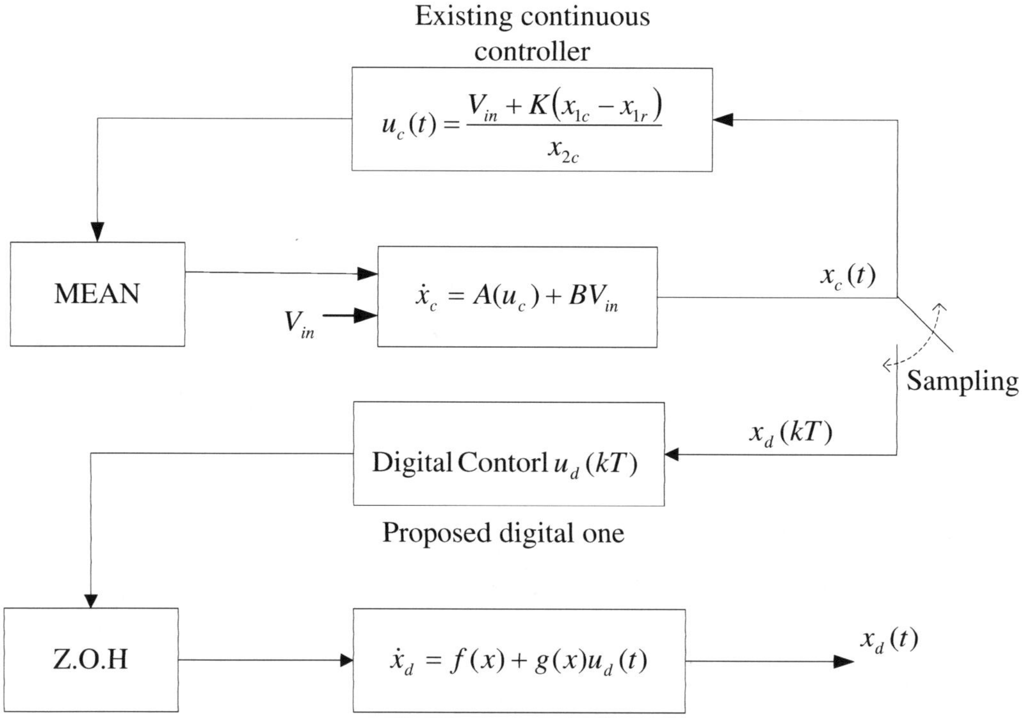

Figure 3. Modified digital redesign block for the boost converter [1]

The continuous control (3) at time t_v is shown as

$$u_c(t_v) = \frac{V_{in} + K(x_{1c}(t_v) - x_{1r})}{x_{2c}(t_v)} \tag{11}$$

IOP Conf. Series: Materials Science and Engineering 1113 (2021) 012003 doi:10.1088/1757-899X/1113/1/012003

From the viewpoint of digital redesign principle, state matching requirement $x_c(t_v) = x_d(t_v)$ under the assumption that $x_c(kT) = x_d(kT)$, it is necessary to have $u_c(t_v) = u_d(kT)$. This makes the following prediction-based digital controller:

$$u_d(t_v) = u_c(t_v) = \frac{V_{in} + K(x_{1c}(t_v) - x_{1r})}{x_{2c}(t_v)} = \frac{V_{in} + K(x_{1d}(t_v) - x_{1r})}{x_{2d}(t_v)} \tag{12}$$

After substitution (10) into (12), we obtain the digital form of (12) based on v=1 shown as

$$u_d(kT) = \frac{V_{in} + K(G_{11}x_{1d}(kT) + G_{12}x_{2d}(kT) + h_1V_{in} - x_{1r})}{G_{21}x_{1d}(kT) + G_{22}x_{2d}(kT) + h_2V_{in}} \tag{13}$$

Fig. 3 shows the proposed digital control algorithm schematically for the nonlinear boost converter (2) and (3). It is important to note that the matrix A in (8) is obtained via the existing continuous control u_c. Then calculate the gains G_{11}, G_{12}, G_{21} and G_{22} in (9) to obtain the predicted discrete-time controller $u_d(kT)$.

4. Illustrative example

We use the proposed digital controller for the boost converter to generate a reference sinusoidal signal $x_{2r}(t) = 100 + 30\sqrt{2}\sin(20\pi t)$ over a load of 220Ω. A DC-power supply of V_{in}=48V and a switch commutation frequency $f_c = \frac{1}{T_c}$ of 30kHz are used. The maximum admissible error is $E_\varepsilon = 0.1\|x_{2r}\|$, and the maximum ripples are $\Delta_{x_1} = 0.05\|x_{2r}\|^2/(Rv_{in})$ and $\Delta_{x_2} = 0.02\|x_{2d}\|$. (5) and (6) are taken the form

$$11.04\text{mH} < L < 16.34\text{mH},\ 5.02\mu\text{F} < C < 28.53\mu\text{F} \tag{14}$$

The continuous control gain K=20, capacitor L=12mH and inductor C=15μF in (8) are set same as reference [1].

Numerical simulations are carried out on Simulink of Matlab using the proposed method with a sampled-hold period 0.005 seconds. The 0.005 second period is based on the sampling theorem that the sampling frequency is greater than 2 times the controlled frequency. The frequency controlled in this paper is f = 10 for 2πf=w=20π. Its corresponding period is 0.1 seconds. Based on the sampling theorem, the sampling frequency of this computer simulation is at least 20 or higher, that is, the best digital simulation can be obtained by a cycle greater than 0.05 seconds. Fig. 4 shows the state trajectories for applying both continuous-time and digitally redesigned controller to the boost converter circuit. Furthermore, the control laws for both the continuous-time and the digitally redesigned controllers are presented in Fig. 5. The simulation result illustrates $x_d(t)$ is very close to $x_c(t)$.

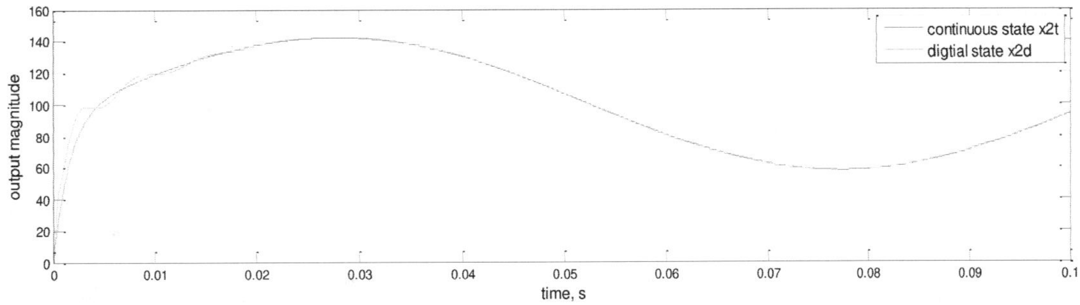

Figure 4. Trajectories of the boost converter under the digitally redesigned controller and original continuous controller for the boost converter circuit [1]

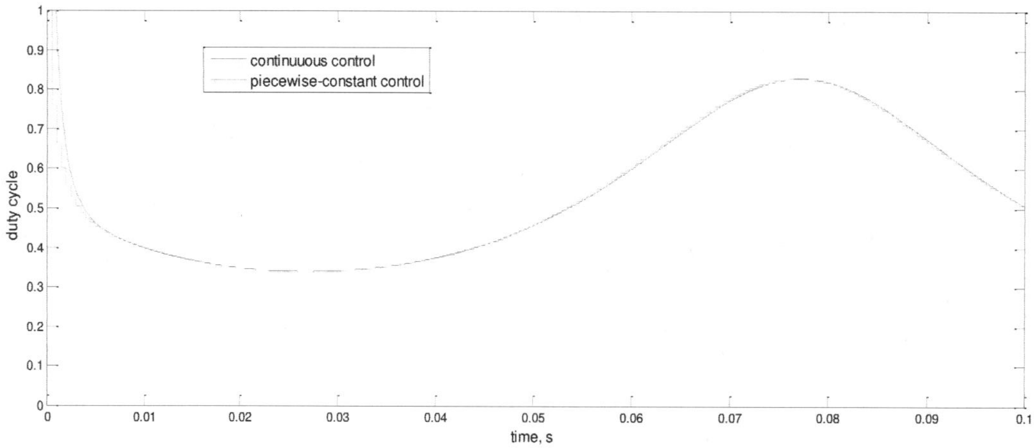

Figure 5. Continuous-time and digitally redesign controller for the boost converter circuit [1]

5. Conclusion

A prediction-based digital redesign technique has been proposed for finding a dynamic digital control law in (13) from the given analog counterpart in (11) for the boost converter system, and it can be implemented using microcomputers, which satisfies a state-matching requirement. The simulation result, state-matching, demonstrated the effectiveness of the developed method.

References

[1] Cortes D, Alvarez Jq, Alvarez J and Fradkov A 2006 *IEE Proc.-Control Theory Appl.* **151** pp 218-224
[2] Fossas-Colet E and Olm-Miras J M 2002 *Discrete Continuous Dyn. Syst.-Ser. B* **2** pp 295-307
[3] Cortes D Alvarez Jq and Fradkov A L *Tracking on the boost converter using standard regulation techniques* (Proc. 2001 American Control Conf., Arlington, VA, pp 2167-2172)
[4] Sung H C, Park J B and Joo Y H 2012 *IEE Proc.-Control Theory Appl.* **6** pp 1842-1850
[5] Tsai J S H, Chen F M, Yu T Y, Guo S M and Shieh L S 2012 *ISA Transactions* **51** pp 81-94
[6] Sung H C, Kim D W, Park J B and Joo Y H 2010 *Fuzzy Sets and Systems* **161** pp 919-933
[7] Ababneh M, Barajas-Ramirez J G, Chen G and Shieh L S 2007 *Chaos, Solitons and Fractals* **31**, pp 1149-1164

Different plastic-bag type photobioreactor for biomass production of *Chlorella* species

Yu Pei Chen[1,2,3,4], Yao Hsin Huang[5] and Hsiang Chin Huang[4,5]

[1]Department of Medical Technology, Xiamen Medical College, Xiamen, China
[2]Application Technique Engineering Center of Natural Cosmeceuticals College of Fujian Province, Xiamen Medical College, Xiamen, China
[3]Xiamen Key Laboratory of Traditional Chinese Medicine Bioengineering, Xiamen Medical College, Xiamen, China
[4]Department of Biological Science and Technology, China University of Science and Technology, Taipei, Taiwan
[5]ELEZA-HYC Company, Limited, Taipei, Taiwan

E-mail: elezasean@gmail.com

Abstract. Microalgae were generally recognized as safe and a promising source for proteins, polyunsaturated fatty acids, carbohydrates, and other bioactive substances in the industrial application such as foods, health supplements, biofuels, and pharmaceuticals. For mass production and cost down, plastic-bag has been utilized for microalgal growth. In this study, three plastic-bag types (plastic-bag, vertical plastic-bag, and horizontal plastic-bag) were designed to shape the liquid circulation for improving the mass transfer. Microalgae were cultivated under different weathers including the sunshine times and average global radiation as 46.7 h and 9.15 MJ/m^2, 13.7 h and 4.8 MJ/m^2, 38.2 h and 8.07 MJ/m^2, and 37.1 h and 11.25 MJ/m^2 for 15 days. The pH, biomass, and antioxidant of microalgae were further detected. The pH values of microalgae were ranged from 7.3–9.0 within 15 days cultivation. No obvious difference was present among these plastic-bags. Under low global radiation with the average 4.8 MJ/m^2, the microalgal biomass produced by the vertical plastic-bag was 1.4-fold and 1.9-fold higher than that of the plastic-bag and horizontal plastic-bag, respectively. This result suggested that the vertical plastic-bag photobioreactor had a high mass transfer with low shear stress. Furthermore, the antioxidant activity was assayed by the total polyphenol and total reductive capability. The result revealed that the microalgae contained 360.4 μg/L-culture corresponding to gallic acid using vertical plastic-bag, followed by horizontal plastic-bag with 289.6 μg/L-culture and plastic-bag with 230.6 μg/L-culture by the total polyphenol determination after 15 days cultivation. The similar result was also observed in the total reductive capability by vertical plastic-bag with 413.4 μg/L-culture corresponding to ascorbic acid, followed by horizontal plastic-bag with 373.2 μg/L-culture and plastic-bag with 353.7 μg/L-culture. Consequently, the vertical plastic-bag had the potential to be applied in the industrial field.

1. Introduction

Comparing with the traditional agriculture for oil crop, microalgae grew rapidly and required a low area for cultivation [1]. They can cultivate in pollutant environment or non-agricultural land, and do not compete with other crops for food production. Microalgae accumulate large amounts of proteins, oils, and carbohydrates which can be used for healthy food, bioenergy, and pharmaceuticals [2]. Several essential amino acids are abundant in microalgae. Additionally, various carbohydrates such as starch, glucose, cellulose, hemicellulose, xylose, galactose, rhamnose, arabinose, and fucose were found, which can be applied in alcohol fermentation [3]. Microalgae usually contain essential fatty acids such as linolenic acid (C18:2), alpha-linolenic acid (C18:3) and arachidonic acid (C20:4) which provide an alternative to vegetable oils [4]. Moreover, ω-3 polyunsaturated fatty acids such as eicosapentaenoic acid (EPA; C20:5) and docosahexaenoic acid (DHA; C22:6) offered good effects on the brain, retinal development, and cardiovascular health are present in some algae. On the other hand, high-value products as β-carotenoid and astaxanthin revealing high antioxidant activity also can be obtained through microalgae cultivation. Thus, some species such as *Chlorella vulgaris*, *Dunaliella salina*, *Haematococcus pluvialis*, *Spirulina maxima*, *Nannochloropsis* and *Tetraselmis* has been commercialized for nutritional supplements and animal feeds [5-6].

Microalgae culture can be divided into an open pond and a closed photobioreactor [7]. Although the open pond system including a circular pond and a raceway pond can be mass-produced and cost-effective, climate and pollution problems are the disadvantages of this technology. Closed photobioreactor contains flat-plate, tubular and column. The advantages of closed photobioreactor are high yield, controllable, low CO_2 loss, high culture density and not easily contaminated; however, high equipment cost is necessary [7]. Plastic-bags combine the advantages of these two technologies with low cost and less contamination. In this study, three plastic-bags were designed for *Chlorella* sp. outdoor cultivation. The pH and biomass of *Chlorella* sp. production were determined during different sunshine and global radiation. The antioxidant properties of *Chlorella* sp. were further measured.

2. Materials and methods

2.1. Cultivation of Chlorella sp.
The medium (1.7 g urea, 0.8 g $MgSO_4$-$7H_2O$, 0.4 g KH_2PO_4, 0.005 g $FeSO_4$-$7H_2O$, 0.007 EDTA per liter containing 1 mL trace metal solution with 2.86 g H_3BO_3, 1.81 g $MnCl_2$, 0.22 g $ZnSO_4$, 0.08 g $CuSO_4$, 0.021 g $NaMoO_4$, and 10 μl H_2SO_4 per liter) was used for outdoor cultivation of *Chlorella* species.

2.2. Growth and biomass of Chlorella sp.
Plastic-bag was composed of polypropylene material. Three plastic-bags were designed by heat sealing machine for vertical and horizontal liquid circulation via gas introduction with 1.5 L/min. Two-liter medium was added into the plastic-bags for outdoor cultivation of *Chlorella* species. The microalgal growth was determined by a spectrometer at 682 nm. Biomass productivity (g/L) was calculated according a standard curve (OD value = 6.1354 × biomass (g/L) – 0.0233) [8].

2.3. Determination of total polyphenols
The total polyphenols assay was determined based on the Folin-Ciocalteu method. The *Chlorella* sp. was harvested, centrifuged at 10,000 xg for 10 min, and dried by freeze dryer. The microalgal sample was ground, extracted, and mixed with Folin-Ciocalteu reagent for 5 min reaction. Then, a 10% sodium carbonate solution with 0.5 mL was added to the mixture and left for 30 min in the dark. The sample was centrifuged at 10,000 xg for 10 min at room temperature. The supernatant was detected by a spectrometer at 760 nm. Various concentration of gallic acid (12.5, 25, and 50 μg/mL) was used as a positive control to calculate a standard curve.

2.4. Determination of total reductive capability
The total reductive activity was determined based on the method of Oyaizu [9]. The dried *Chlorella* sp. sample was added and mixed with 0.2 M sodium phosphate buffer (pH 6.6) and 1 % K_3FeCN_6 at 50°C for 20 min. After incubation, 10 % trichloroacetic acid was added, and the mixture was centrifuged at

10,000 ×g for 10 min. The supernatant was mixed with H_2O and 0.1% $FeCl_3$. The absorbance was determined by a spectrometer at 700 nm. Various concentration of ascorbic acid (12.5, 25, and 50 μg/mL) was used as a positive control to calculate a standard curve.

3. Results and discussion

3.1. Design of plastic-bag for Chlorella sp. cultivation

The hydrodynamic and mass transfer of photobioreactors were important for microalgal culture [7, 10]. A few studies on mass transfer were reported in phototrophic cultures of microalgae [7]. Because plastic-bag was relatively economical and good for microalgal cultivation, three plastic-bag types were designed to investigate the effect of mass transfer and gas holdup on microalgal productivity (Fig. 1). The heat sealing was used to shape the liquid circulation in plastic-bag; thus, the vertical and horizontal paths of plastic-bag were created.

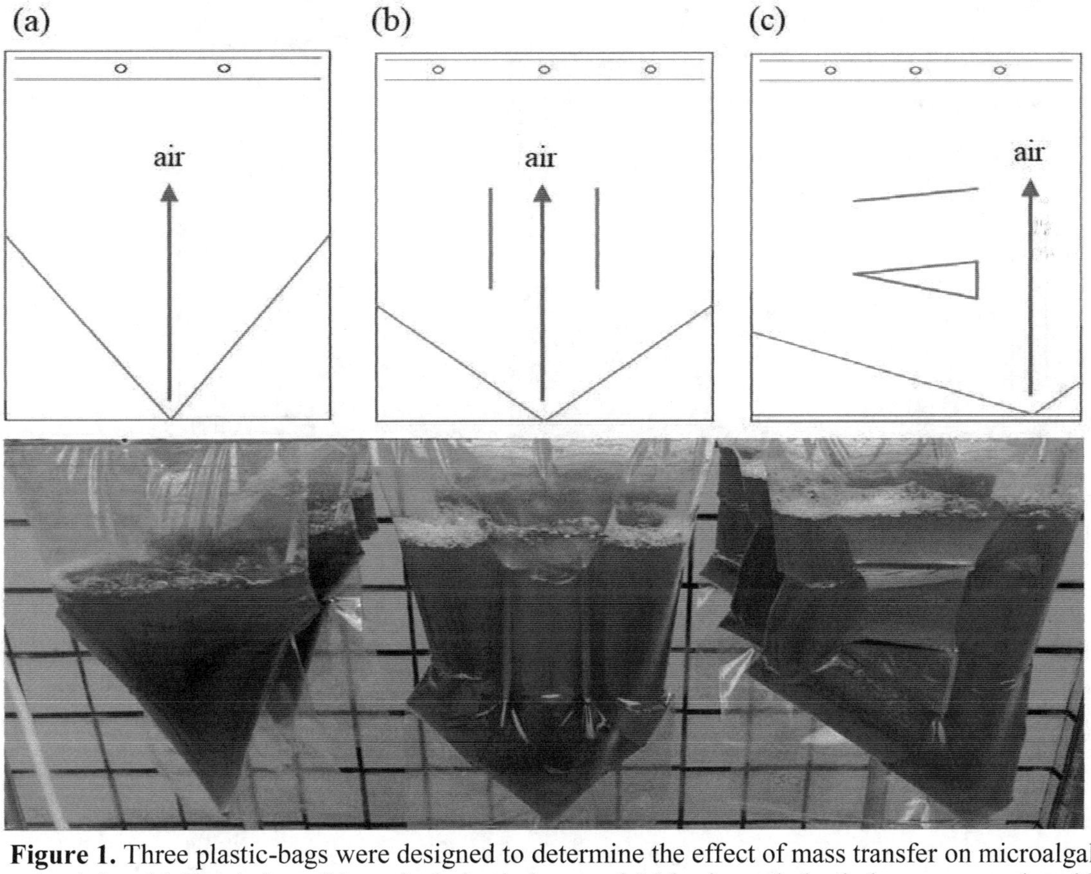

Figure 1. Three plastic-bags were designed to determine the effect of mass transfer on microalgal productivity. (a) Plastic-bag, (b) vertical plastic-bag, and (c) horizontal plastic-bag were produced by heat sealing machine.

When the air passed through the culture medium, the mass transfer from the gas to the liquid was conducted. Therefore, the bubble flow was simulated by the software of COMSOL Multiphysics (www.comsol.com). The volume fraction of bubbles was evaluated using the model of Bubbly Flow with a 2D axisymmetric component. The parameters of COMSOL Multiphysics were set up as follows: bubble diameter, 0.004 m; gas volume, 2.5E-5 m^3/s; inlet velocity, 1.9894 m/s; Slip-velocity proportionality constant rhog in 0.9727[$kg/(m^3)$]. Observing plots between t = 0.3 s and t = 1.2 s (Fig. 2), the gas bubbles were rising up and imparted momentum to the culture medium, recirculating and forming the vortex in the plastic-bag. Comparing the convection and diffusion of the gas bubbles, the predominance of convective transport was obvious in the vertical plastic-bag.

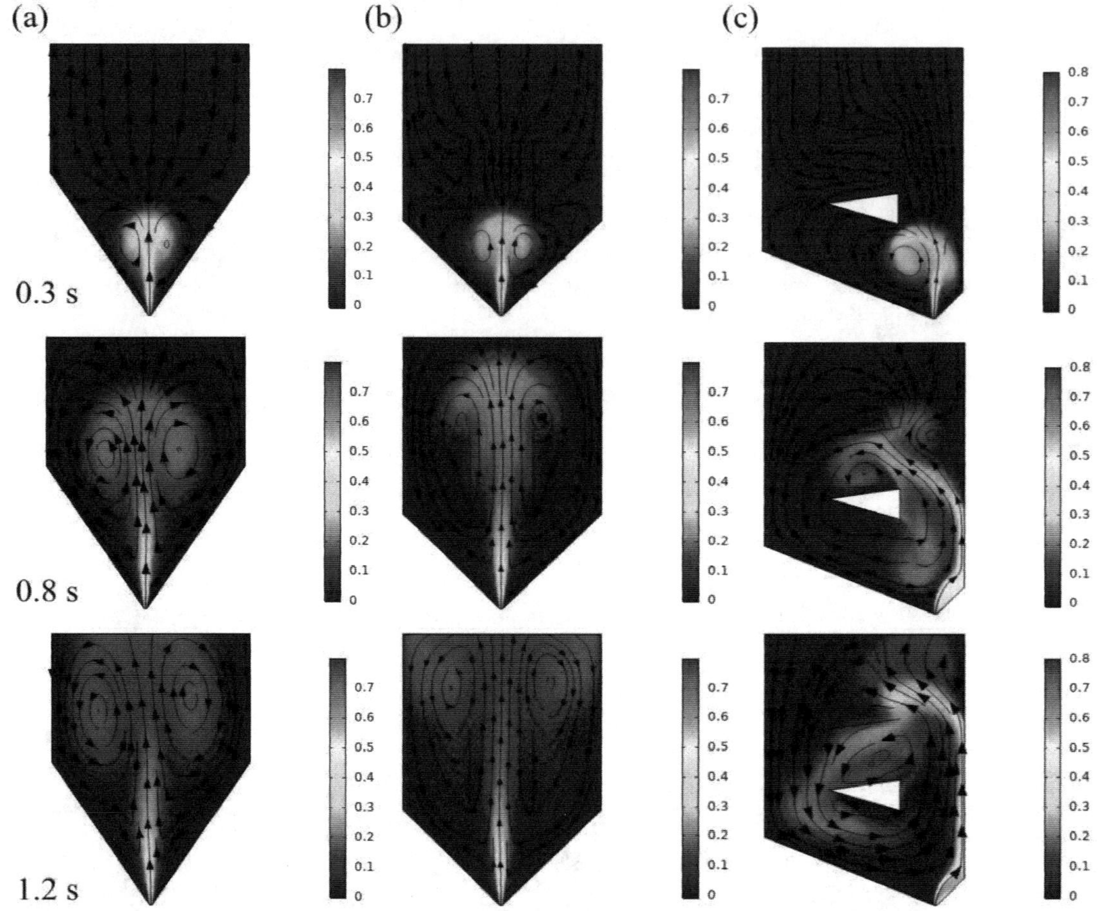

Figure 2. Time evolution of gas bubble volume fraction in the (a) plastic-bag, (b) vertical plastic-bag, and (c) horizontal plastic-bag at t = 0.3 s, t = 0.8 s, and t = 1.2 s. The right columns indicate the volume fraction of bubbles in the plastic-bag.

3.2. Outdoor cultivation of Chlorella sp. with different sunshine times and global radiation

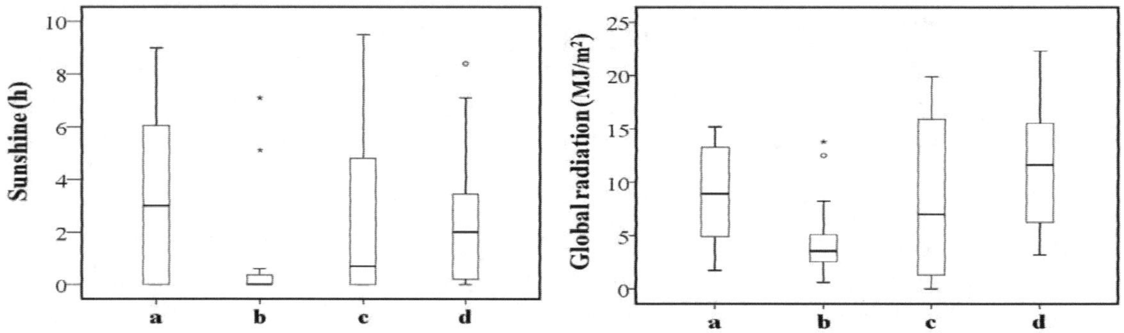

Figure 3. Microalgae were cultured outdoors with different sunshine times and global radiation. The sum of sunshine times and an average of global radiation were a, 46.7 h and 9.15 MJ/m^2; b, 13.7 h and 4.8 MJ/m^2; c, 38.2 h and 8.07 MJ/m^2, and d, 37.1 h and 11.25 MJ/m^2.

Artificial light can be applied in the microalgal culture system [11]; however, it could increase the cost of microalgal production. Comparing with the open pond, the plastic-bag had a good light path, which was similar to flat-plate photobioreactor with a large illumination surface area [12]. The light to biomass efficiency was an important factor for industrial productivity [13]. In this study, to detect microalgal production by using different plastic-bags, the microalgal culture was set up outdoors during different sunshine times and global radiation within 15 days. Four kinds of weathers are shown in Fig. 3.

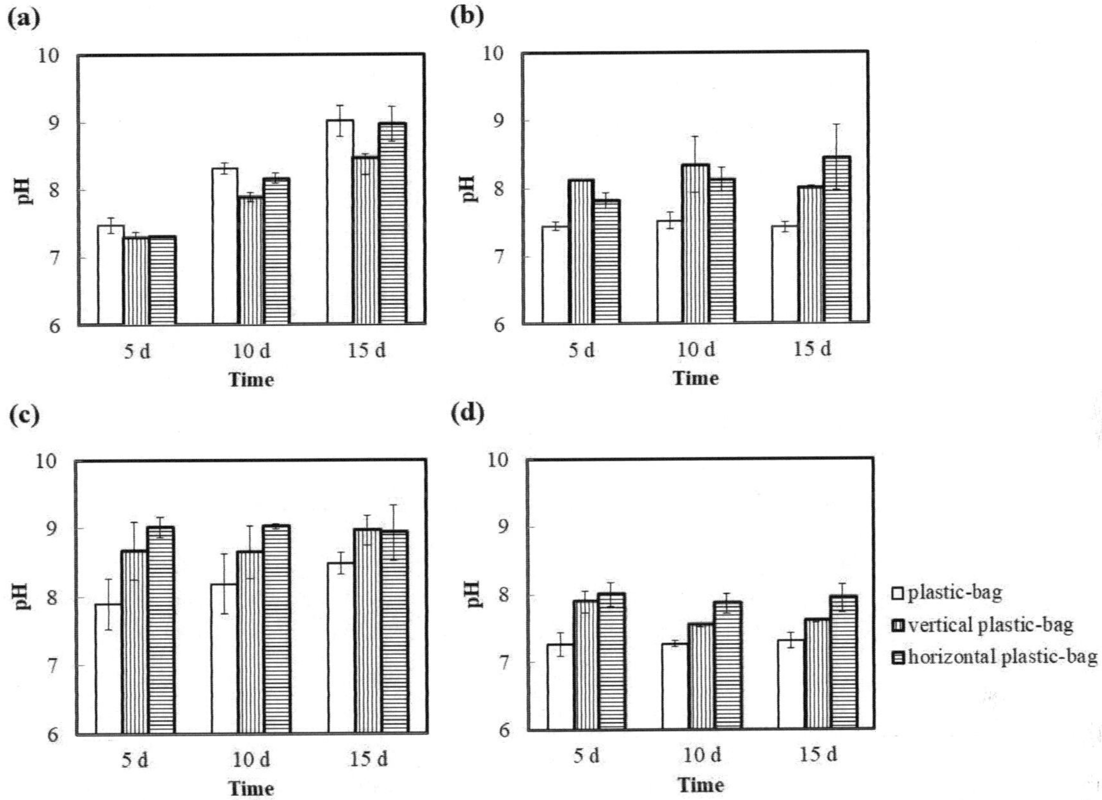

Figure 4. The pH value of microalgal culture within 15 days. The pH value was determined according to the weather of (a) 46.7 h and 9.15 MJ/m^2, (b) 13.7 h and 4.8 MJ/m^2, (c) 38.2 h and 8.07 MJ/m^2, and (d) 37.1 h and 11.25 MJ/m^2.

3.3. Determination of pH and biomass of Chlorella sp. cultivation

The pH value was also difficult to keep constant growth factor [13]. Therefore, the pH was monitored during microalgal growth (Fig. 4). The pH values were ranged from 7.3~9.0 within 15 days of growth. The pH change on vertical and horizontal plastic-bag photobioreactors was not obvious under different weathers. Nevertheless, the pH of horizontal plastic-bag was slightly higher than the others. In addition, the biomass of microalgae was determined (Fig. 5). The result revealed that the vertical plastic-bag had higher biomass than the other plastic-bag types in the various weathers. Under the high global radiation, no obvious difference was observed in all plastic-bags. However, the microalgal biomass produced by the vertical plastic-bag was 1.4-fold and 1.9-fold higher than that of the plastic-bag and horizontal plastic-bag, respectively, under the low global radiation of 4.8 MJ/m^2. This result suggested that the vertical plastic-bag can exert a great ability of mass transfer under poor global radiation. This design was similar to the vertical column photobioreactor which had a high mass transfer with low shear stress [7, 14]. When the weather is poor outdoors, the vertical plastic-bag can be effectively used.

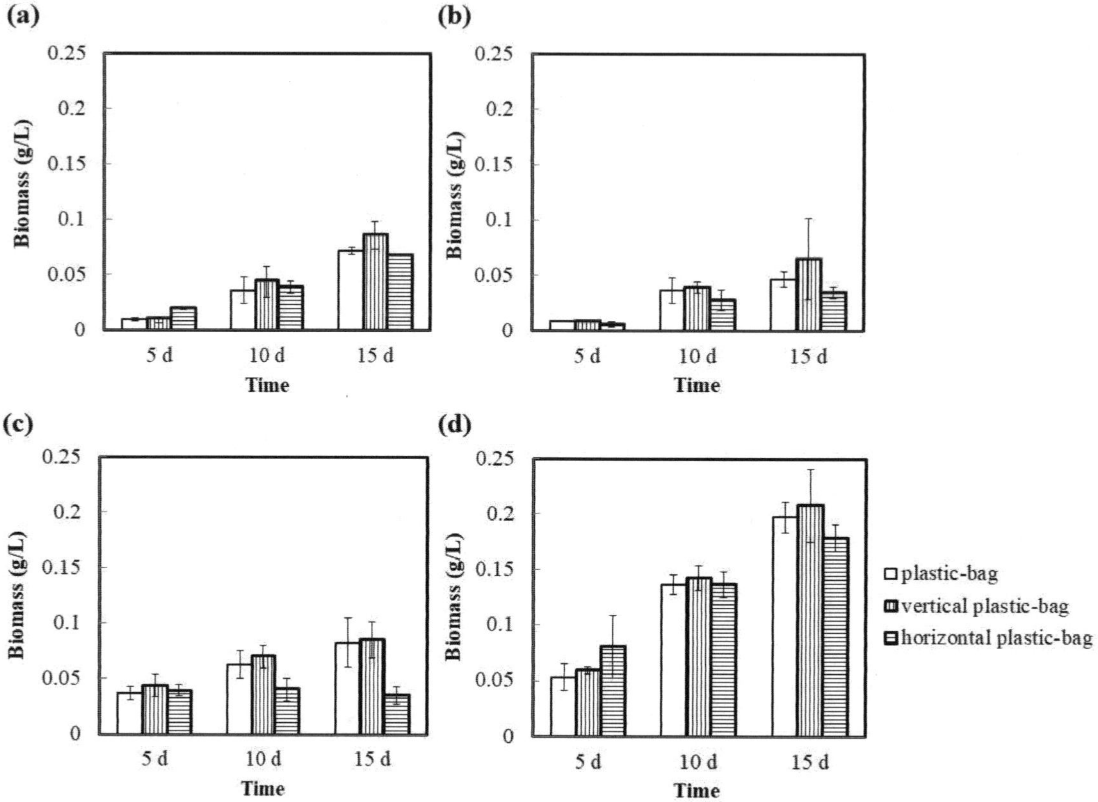

Figure 5. The biomass of microalgal culture within 15 days. The biomass was harvested according to the weather of (a) 46.7 h and 9.15 MJ/m^2, (b) 13.7 h and 4.8 MJ/m^2, (c) 38.2 h and 8.07 MJ/m^2, and (d) 37.1 h and 11.25 MJ/m^2.

3.4. Determination of antioxidant activity of Chlorella sp. cultivation
To comprehend the bioactive substance of microalgae, the antioxidant activity was determined by the total polyphenol and total reductive capability (Table 1). Polyphenolic compounds are recognized as the factor to alleviate oxidative stress [15]. Hence, a total polyphenol content of microalgae was measured using the Folin-Ciocalteu method. The result indicated that microalgae contained 360.4 μg/L-culture corresponding to gallic acid using vertical plastic-bag after 15 days cultivation, followed by horizontal plastic-bag with 289.6 μg/L-culture and plastic-bag with 230.6 μg/L-culture. On the other hand, the result of the reductive capability of microalgae was consistent with the total polyphenol. The microalgae of vertical plastic-bag had 413.4 μg/L-culture corresponding to ascorbic acid. Consequently, the microalgae cultivated by vertical plastic-bag had the highest total polyphenol and total reductive activity.

Table 1. The antioxidant detection of microalgal biomass under the condition of 46.7 h sunshine times and average 9.15 MJ/m^2.

	Corresponding to gallic acid[a] (μg/L-culture)	Corresponding to ascorbic acid[b] (μg/L-culture)
Plastic-bag	230.6	353.7
Vertical plastic-bag	360.4	413.4
Horizontal plastic-bag	289.6	373.2

[a] The total polyphenol of microalgae was corresponding to gallic acid.
[b] The total reductive capability of microalgae was corresponding to ascorbic acid.

4. Conclusions

Much attention has been directed toward the industrial production of algae. Several kinds of research have been reported on the application of algae for foods, biodiesel, bioalcohol, and pharmaceuticals. For bioenergy, cost-down and mass productions of algae are the focus of current research. In this study, three plastic-bag types were designed to explore the effect of mass transfer on microalgal production. Our result suggested that vertical plastic-bag presented higher biomass productivity than the other plastic types, especially when the weather and light were poor. Well growth of microalgae was also in response to better antioxidant activity detected by the total polyphenol and total reductive capability. Taking together, the vertical plastic-bag which was simply processed can be used to obtain better biomass productivity. This result had the potential to be applied in the industrial field.

Acknowledgments

This work was supported by the ELEZA-HYC Company, Limited, Taiwan, and the Health Research Talent Training Project of Fujian Provincial Health Commission (2019-ZQNB-20), Fujian, China.

References

[1] Spolaore P, Joannis-Cassan C, Duran E and Isambert A 2006 *J. Biosci. Bioeng.* **101** pp 87-96
[2] Zakaria S M, Kamal S M M, Harun M R, Omar R and Siajam S I 2017 *Molecules* **22** p 1105
[3] El-Dalatony M M, Kurade M B, Abou-Shanab R A I, Kim H, Salama E S and Jeon B H 2016 *Bioresour. Technol.* **219** pp 98-105
[4] Draaisma R B, Wijffels R H, Slegers P M, Brentner L B, Roy A and Barbosa M J 2013 *Curr. Opin. Biotechnol.* **24** pp 169-177
[5] Borowitzka M A 2013 *J. Appl. Phycol.* **25** pp 743-756
[6] Indira Priyadarshani B R 2012 *J. Algal Biomass Utln.* **3** pp 89-100
[7] Ugwu C U, Aoyagi H and Uchiyama H 2008 *Bioresour. Technol.* **99** pp 4021-4028
[8] Chiang Y L, Chen Y P, Yu M C, Hsieh S Y, Hwang I E, Liu Y J, Huang C Y, Tseng C P and Liaw L L 2018 *J. Renew. Sustain. Ener.* **10** p 033102
[9] Oyaizu M 1986 *Jpn. J. Nutr.* **44** pp 307-316
[10] Hall D O, Fernandez F G, Guerrero E C, Rao K K and Grima E M 2003 *Biotechnol. Bioeng.* **82** pp 62-73
[11] De Francisci D, Su Y, Iital A and Angelidaki I 2018 *Environ. Technol.* **39** pp 581-592
[12] Sun Y, Liao Q, Huang Y, Xia A, Fu Q, Zhu X and Zheng Y 2016 *Bioresour. Technol.* **220** pp 215-224
[13] Benedetti M, Vecchi V, Barera S and Dall'Osto L 2018 *Microb. Cell Fact.* **17** p 173
[14] Lopez M C, Sanchez Edel R, Lopez J L, Fernandez F G, Sevilla J M, Rivas J, Guerrero M G and Grima E M 2006 *J. Biotechnol.* **123** pp 329-342
[15] Zhang T J, Zheng J, Yu Z C, Huang X D, Zhang Q L, Tian X S and Peng C L 2018 *Tree Physiol.* **38** pp 1486-1501

Characterization and overexpression of esterases-encoding Lip900 and Lip3954 through metagenomic sequencing of paddy soil

Yu Pei Chen[1,2,3,4], Pei Chi Lee[4], Hong Tan Wu[1,2,3], Guey Horng Wang[1,2,3], Chai Fang Tsai[5] and Chiu Chung Young[5,6]

[1]Department of Medical Technology, Xiamen Medical College, Xiamen, China.
[2]Application Technique Engineering Center of Natural Cosmeceuticals College of Fujian Province, Xiamen Medical College, Xiamen, China
[3]Xiamen Key Laboratory of Traditional Chinese Medicine Bioengineering, Xiamen Medical College, Xiamen, China
[4]Department of Biological Science and Technology, China University of Science and Technology, Taipei, Taiwan
[5]Department of Soil and Environmental Sciences, National Chung Hsing University, Taichung, Taiwan
[6]Innovation and Development Center of Sustainable Agriculture, National Chung Hsing University, Taichung, Taiwan

E-mail:ccyoung@mail.nchu.edu.tw

Abstract. A lot of industrial genes can be explored from the metagenomic sequences. In this study, metagenome sequencing of paddy soil was carried out, and several putative open reading frames (ORFs) involved in the lipolytic activity can be identified. Lipolytic enzymes are widely used in different industrial applications, such as biodiesel production, bioremediation, and waste treatment. To verify the lipolytic enzymes of assembly ORFs, two putative genes encoding esterase, namely Lip900 and Lip3594, which shared 47.6% and 43.7% identities with the uncharacterized esterase proteins, were synthesized and constructed with pET-30a for *Escherichia coli* overexpression. Lip900 and Lip3594 belonging to VI and XII families were successfully obtained and characterized. The results of recombinant proteins indicated that Lip900 and Lip3594 preferred to hydrolyze short-length *p*-nitrophenyl (*p*-NP) esters such as *p*-NP butyrate (C4). The optimal temperature and pH for both Lip900 and Lip3594 were 30°C and pH 7, respectively. Nevertheless, Lip3594 had a higher relative activity than Lip900 when the temperature was over 40°C. The effect of various reagents on Lip900 and Lip3594 activities was determined. The inhibition of Lip900 and Lip3594 was observed in the presence of $MgSO_4$, $MnSO_4$, $NiSO_4$, and sodium dodecyl sulfate (SDS). However, the addition of ethylenediaminetetraacetic acid (EDTA) can improve the lipolytic activity, indicating these esterases without metal ions as the cofactor. Moreover, Lip900 and Lip3594 were resistant to methanol, ethanol, and butanol. Over 81.6% of the relative activity of Lip900 can be attained when these organic solvents of alcohol were added to 10%. These results revealed that Lip900 and Lip3594 have potential applications in various industries.

1. Introduction

Lipolytic enzymes can be applied in various industrial areas including biodiesel production, bioremediation, and waste treatment. They were grouped into eight families according to their amino acid sequences [1]. Furthermore, the new families were successively found to reach fourteen families, which family I was recognized as the true lipase and family II to XIV was esterase [2]. In the lipase engineering database, lipolytic enzymes were classified into three classes, GGGX, GX, and Y, on the basis of the oxyanion hole, which stabilizes the catalytic reaction [3]. The development of high-yield production has attracted considerable attention. Therefore, the heterologous expression of an esterase from various organisms in *E. coli*, *Bacillus subtilis*, *Trichoderma reesei*, *Pichia pastoris*, and *Saccharomyces cerevisiae* has been thoroughly investigated, and esterases have been successfully produced [4].

Several studies reported that metagenomics library construction is used to screen the functional genes [5-6]. However, the technique of DNA library construction and screening is time-consuming. Recently, metagenomic sequencing has been a powerful alternative and an already tool for cost reduction [7]. The novel genes can be explored from the metagenomic sequencing which shows a valuable source. In this study, the esterases were screened through the metagenomics sequencing of paddy soil. The novel esterases, Lip900, and Lip3954 which shared only 47.6% and 43.7% identities with the unpublished sequences, were successfully synthesized and overexpressed in *E. coli* for characterization. The recombinant esterases were purified, their structures were predicted, and the activities were studied.

2. Materials and methods

2.1. Metagenomic sequencing of the paddy soil

Soil DNA obtained from the Taiwan Agricultural Research Institute, Council of Agriculture, Executive Yuan (24°01'N, 120°41'E) was conducted through next-generation sequencing (NGS) [8]. The metagenomic sequencing and genome *de novo* analysis were performed by the Sequencing-Tech company (Taipei, Taiwan). The Kyoto Encyclopedia of Genes and Genomes (KEGG) was utilized for the analysis of metabolic pathways [9], and ORFs were annotated by MetaGeneMark and BLASTP of National Center for Biotechnology Information (NCBI). The phylogenetic tree was carried out by the neighbor-joining method in MEGA-X software with 1000 bootstrap replicates [10].

2.2. Overexpression plasmid construction of esterase

Two putative ORFs encoding esterase were designated as Lip900 and Lip3594. Lip900 and Lip3594 with *Eco*RV adaptors at 5' and 3' sequences were synthesized and provided by GeneDireX Inc. (Las Vegas, NV, USA). The pET-30a was treated with *Eco*RV restriction enzyme, and the linear pET-30a was ligated with the synthesized DNA. Two overexpression plasmids for *E. coli* were obtained, namely pLip900 and pLip3594. The two plasmids transferred into *E. coli* C43(DE3) which was induced by adding IPTG to 0.1 mM and incubated at 25°C overnight. The cells were disrupted with lysozyme, and the recombinant proteins were purified by a Ni-NTA purification system (Invitrogen, Carlsbad, CA, USA). Protein content was determined by the standard curve of BSA protein with different concentration on the SDS-PAGE.

2.3. Enzymatic activity assay of esterase protein

Lipolytic activity was detected through the detection of *p*-nitrophenyl (*p*-NP) esters as substrates by using a spectrophotometer at OD_{405}. The mixture in the enzymatic assay contained the purified esterase and various *p*-NP esters: *p*-NP butyrate (C4), *p*-NP valerate (C5), *p*-NP octanoate (C8), *p*-NP decanoate (C10), *p*-NP dodecanoate (C12), *p*-NP myristate (C14), and *p*-NP palmitate (C16). The enzymatic reaction included 0.25 mM *p*-NP esters (C4–C10), 2.5% ethanol, 50 mM Tris-HCl (pH 7), and the recombinant protein with a final volume of 800 μL. To detect the lipolytic activity with long-chain *p*-NP esters (C12-C16), 0.2 mM *p*-NP esters, 2.5% ethanol, 0.1% gum arabic, 0.1% deoxycholate, 50 mM Tris-HCl (pH 7), and recombinant protein were used. The enzymatic mixture was incubated at 30°C and pH 7 for 10 min. The optimal temperature and pH of enzymatic activity were measured between 20°C–70°C and pH 4–10 using *p*-NP butyrate as the substrate. The 50 mM sodium acetate/acetic acid buffer, 100 mM sodium phosphate, and glycine/NaOH buffers were used for the assays at pH 4.0–5.0, pH 6.0–8.0, and pH 9.0–10. The reaction was interrupted with 95°C incubation for 15 min. To analyze the effect of different reagents on esterases, various chemical compounds were used containing 10 mM $MgSO_4$, $MnSO_4$, $NiSO_4$, EDTA, SDS and organic solvents of alcohol such as methanol, ethanol, and butanol.

2.4. Nucleotide sequence accession number

The accession numbers of Lip900 and Lip3954 genes have been submitted to the NCBI. The accession numbers are MK641590 and MK641591.

3. Results and discussion

3.1 Esterases identification in metagenomic sequencing

High-throughput sequencing provides a large amount of gene information which can be utilized in the enzymatic engineering. Several metagenomes have been conducted in various fields, such as soil, animal gut, and marine to establish the microbial structure [11-13]. In this study, metagenomic sequencing of paddy soil was carried out with 55.68 Gb sequences by the Illumina Genome Analyze, and the metagenomic project had been deposited at DDBJ/ENA/GenBank under accession numbers BioProject PRJNA498754 and BioSample SAMN10334111 [14]. To explore the lipolytic enzyme in the metagenome, the KEGG pathway of lipid metabolism was analyzed (Fig. 1). The genes involved in fatty acid metabolism were predominant, followed by glycerolipid metabolism and fatty acid biosynthesis. Further observation of glycerolipid metabolism could reveal genes related to the esterase. The result indicated that the triacylglycerol acylhydrolases (lipases) (EC 3.1.1.3) could be found.

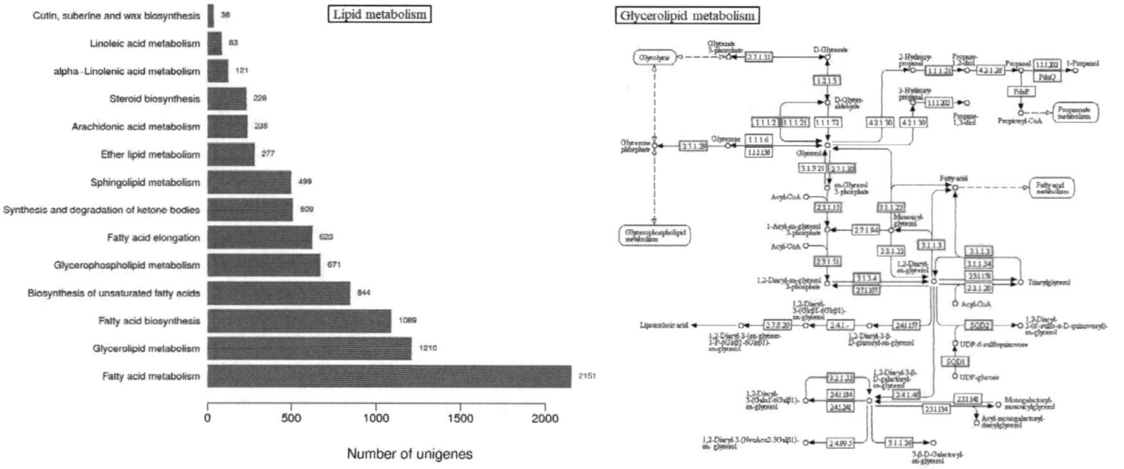

Figure 1. Gene classification of paddy soil metagenome by using KEGG analysis. The red box indicates the sequencing reads which can be identified in the metagenome of paddy soil.

In this study, the database search and ORF prediction further provide the information of putative esterase genes. Of the hypothetical esterases, two putative ORFs (L900 and L3594) encoding esterase with 481 and 308 amino acids with the low identical match were identified (Table 1). These two ORFs were designated as Lip900 and Lip3594; they had 47.6% and 43.7% identities with uncharacterized lipases from *Bradyrhizobium* sp. and uncultured bacterium, respectively.

Table 1. Lipase search from the data mining of putative open reading frame with over 250 amino acids in length based on the NCBI BLAST result.

No.	Closed sequence	Amino acid length	Identity (%)	Coverage (%)
L529	lipase [*Pseudomonas* sp. GM30]	276	83.7	100
L900	putative Lipase [*Bradyrhizobium* sp. STM 3843]	481	47.6	93
L1145	esterase/lipase [*Leptolyngbya* sp. PCC 7375]	288	67.9	99
L1620	lipase [*Tsukamurella paurometab*ola]	280	75.4	100
L1699	lysophospholipase [*Pyrinomonas methylaliphatogenes*]	300	63.5	66
L2145	phospholipase C [Candidatus *Koribacter versatilis*]	253	72.4	100
L2830	phospholipase [*Methylocystis* sp. SB2]	281	71.7	98
L3207	phospholipase C [Candidatus *Koribacter versatilis*]	284	52.5	99
L3594	esterase/lipase [uncultured bacterium]	308	43.7	87
L3766	phospholipase [Candidatus *Solibacter usitatus*]	277	65.8	100

3.2 Characterization of Lip900 and Lip3594

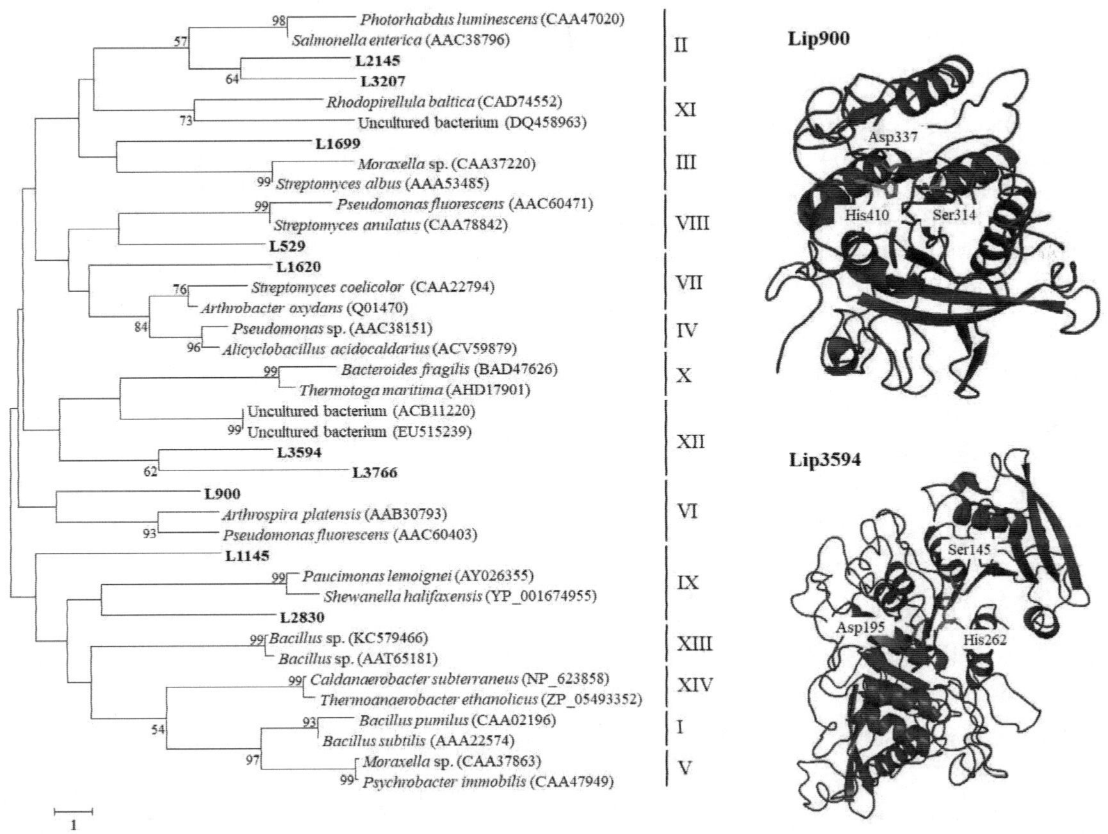

Figure 2. Phylogeny construction of esterase sequences, and prediction of Lip900 and Lip3594 protein structures by using SWISS-MODEL [15].

The lipolytic enzymes can be divided into 14 families by Rao et al. [2]. A phylogenetic tree was constructed by the deduced amino acids for the assessment of esterase relationship among the Lip900, Lip3594 and other esterases from various organisms (Fig. 2). The result suggested that the two genes were classified into two families, VI and XII. Furthermore, three-dimensional structures of Lip900 and Lip3594 were predicted based on the SWISS-MODEL [15]. A typical α/β-hydrolase fold of Lip900 and Lip3594 with a central β-sheet barrel surrounded by α-helices can be found [16]. In addition, the nucleophilic serine, acidic aspartate or glutamate, and histidine residues were conserved in the catalytic triad of esterase [16-17]. According to the alignment of deduced amino acids, the results revealed that the predicted catalytic residues were observed at Ser314, Asp337, and His410 for Lip900 and Ser145, Asp195, and His262 for Lip3594 (Fig. 3). The conserved G-X-S-X-G motif also can be perceived in both genes [1].

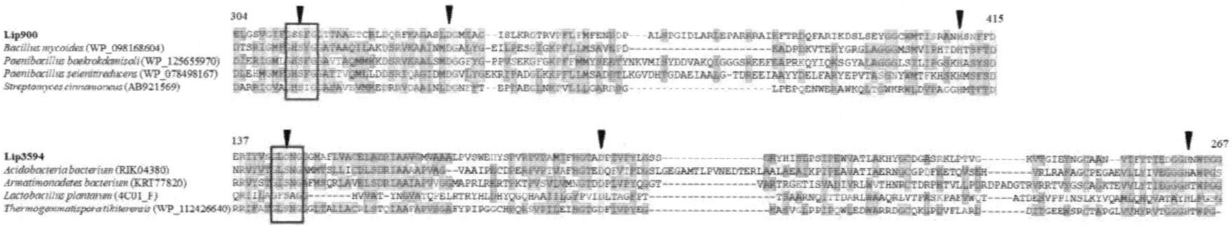

Figure 3. The alignment of deduced amino acids of esterases. The filled triangles indicated the assumed active site of serine, aspartate, and histidine.

3.3 Esterase activity assay

To confirm the property of Lip900 and Lip3594, *E. coli* was used as a heterologous host for esterase expression. Lip900 and Lip3594 were synthesized and introduced into the expression vector of pET-30a including a His-tag fusion of N-terminus. The recombinant proteins were purified, and various *p*-NP esters were utilized for demonstrating the esterase activity. The result indicated that Lip900 and Lip3594 tended to be the hydrolysis of short-length *p*-NP esters (C4)(Fig. 4). This was in agreement with other esterases from *Thalassomonas agarivorans*, and *Aeromonas* sp. [18]. The temperature and pH profiles of the recombinant Lip900 and Lip3594 were further determined. The optimal temperature of Lip900 and Lip3594 was 30°C. At 20°C, both relative activities were dropped below 30%. However, the residual activity of Lip3594 remained 72% at 50°C and was higher than that of Lip900 with 21%. The optimal pH of Lip900 and Lip3594 was pH 7. When pH was below 5, and above 9, a rare activity of Lip900 and Lip3594 was measured.

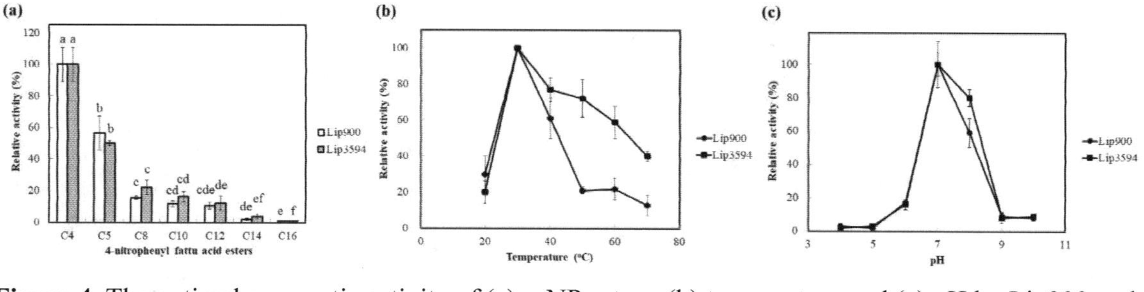

Figure 4. The optimal enzymatic activity of (a) *p*-NP esters, (b) temperature and (c) pH by Lip900 and Lip3594.

3.4 Effect of various reagents on Lip900 and Lip3594

The effect of different chemical reagents and metal ions on the recombinant esterase activities was shown in Fig. 5. Lip900 was obviously inhibited by $MgSO_4$, $MnSO_4$, $NiSO_4$, and SDS; besides, the remarkable inhibition of Lip3594 was observed by addition of $NiSO_4$, and SDS. By contrast, EDTA can improve a considerable stimulation of esterase activity. This implied that metal ions were not required as the enzymatic cofactor. The similar result was reported for the esterases from

Psychrobacter cryohalolentis and *Yarrowia lipolytica* [19-20]. High organic solvents usually result in lost enzymatic activity [21]. Rare effect of organic solvents on Lip900 was perceived as methanol, ethanol, and butanol below 5%. Additionally, no enzymatic activity loss on Lip3594 was found as ethanol below 10%. When these organic solvents were gradually increased, the enzymatic activity of Lip900 and Lip3594 was slowly decreased. These results suggested that Lip900 and Lip3594 had a high tolerance to organic solvent of methanol and ethanol, respectively. These enzymes were more stable than that of esterases from *Bacillus methylotrophicus* and *Yarrowia lipolytica* [20, 22].

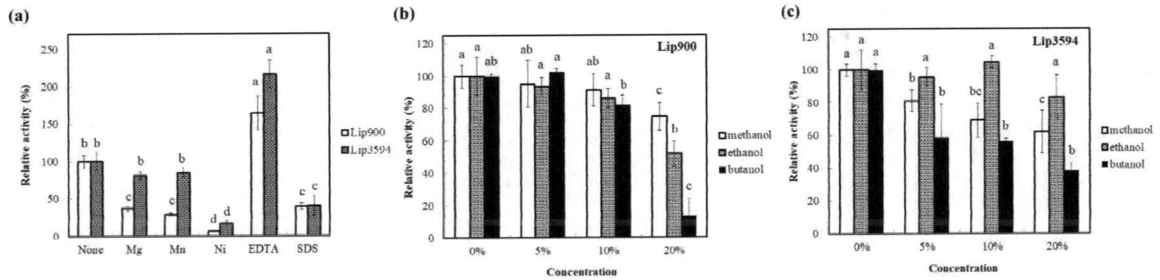

Figure 5. (a) Effect of various reagents on Lip900 and Lip3594. Different concentration of methanol, ethanol, and butanol was added into (b) Lip900 and (c) Lip3594 reaction with pH 7 at 30°C.

4. Conclusions

Esterases were the new trend for industrial application and account for 21% of the worldwide industrial enzyme market. They are the third largest commercialized enzymes after proteases and carbohydrases [23]. In this study, two genes, Lip900, and Lip3594, involved in lipolytic activity were identified through the paddy soil of metagenomic sequencing. According to the classification of phylogeny, Lip900 and Lip3594 were belonging to family VI and XII. These enzymes can be successfully overexpressed in *E. coli*, and the enzymatic profile can be characterized in detail. Lip900 and Lip3594 had the ability for hydrolyzing short-length *p*-nitrophenyl esters. In addition, they were both sensitive to the metal ion of nickel. Addition of EDTA into the recombinant esterases can enhance the enzymatic activity. On the other hand, Lip900 and Lip3594 were obviously resistant to methanol and ethanol, respectively. This property was useful for biodiesel application. Taking together, the metagenomic sequencing-based strategy had the potential to be applied in the various fields for mining the novel genes.

Acknowledgments

This work was supported by the Ministry of Science and Technology, Taiwan (Grant 102-2313-B-157-003-MY3), the Educational and Scientific Research Program for Young Scholar Sponsored by Educational Department of Fujian Province (JAT170715), Fujian, China, and the Innovation and Development Center of Sustainable Agriculture from The Featured Areas Research Center Program within the framework of the Higher Education Sprout Project by the Ministry of Education (MOE) in Taiwan.

References

[1] Arpigny J L and Jaeger K E 1999 *Biochem J* **343 Pt 1** pp 177-183
[2] Rao L, Xue Y, Zheng Y, Lu J R and Ma Y 2013 *PLoS One* **8** e60645
[3] Widmann M, Juhl P B and Pleiss J 2010 *BMC Genomics* **11** p 123
[4] Valero F 2012 *Methods Mol Biol* **861** pp 161-178
[5] Lu M, Dukunde A and Daniel R 2019 *Appl Microbiol Biotechnol* 10.1007/s00253-019-09695-1
[6] Istvan P, Souza A A, Garay A V, Dos Santos D F K, de Oliveira G M, Santana R H, Lopes F A C, de Freitas S M, Barbosa J and Kruger R H 2018 *Biotechnol Lett* **40** pp 1395-1406
[7] Bilal T, Malik B and Hakeem K R 2018 *J Microbiol Methods* **155** pp 65-69
[8] Chen Y P, Liaw L L, Kuo J T, Wu H T, Wang G H, Chen X Q, Tsai C F and Young C C 2019 *J*

Biosci Bioeng **128** pp 274-282
[9] Kanehisa M, Goto S, Hattori M, Aoki-Kinoshita K F, Itoh M, Kawashima S, Katayama T, Araki M and Hirakawa M 2006 *Nucleic Acids Res* **34** D354-357
[10] Kumar S, Stecher G and Tamura K 2016 *Mol Biol Evol* **33** pp 1870-1874
[11] Gastauer M, Vera M P O, de Souza K P, Pires E S, Alves R, Caldeira C F, Ramos S J and Oliveira G 2019 *Sci Data* **6** 190008
[12] Qin J, Li R, Raes J, Arumugam M, Burgdorf K S, Manichanh C, Nielsen T, Pons N, Levenez F and Yamada T, et al. 2010 *Nature* **464** pp 59-65
[13] Muangchinda C, Rungsihiranrut A, Prombutara P, Soonglerdsongpha S and Pinyakong O 2018 *J Hazard Mater* **357** pp 119-127
[14] Chen Y P, Liaw L L, Kuo J T, Wu H T, Wang G H, Chen X Q, Tsai C F and Young C C 2019 *J. Biosci. Bioeng.* **128** pp 274-282
[15] Biasini M, Bienert S, Waterhouse A, Arnold K, Studer G, Schmidt T, Kiefer F, Gallo Cassarino T, Bertoni M and Bordoli L, et al. 2014 *Nucleic Acids Res* **42** W252-258
[16] Casas-Godoy L, Duquesne S, Bordes F, Sandoval G and Marty A 2012 *Methods Mol Biol* **861** pp 3-30
[17] Sanishvili R, Yakunin A F, Laskowski R A, Skarina T, Evdokimova E, Doherty-Kirby A, Lajoie G A, Thornton J M, Arrowsmith C H and Savchenko A, et al. 2003 *J Biol Chem* **278** pp 26039-26045
[18] Chen Y P, Liang S S, Hwang I E, Chien M C, Wang G H and Liaw L L 2018 *MATEC Web of Conferences* **169** p 01018
[19] Novototskaya-Vlasova K A, Petrovskaya L E, Rivkina E M, Dolgikh D A and Kirpichnikov M P 2013 *Biochemistry (Mosc)* **78** pp 385-394
[20] Li T, Zhang W, Hao J, Sun M and Lin S X 2019 *Biotechnol Rep (Amst)* **21** e00295
[21] Gupta M N, Batra R, Tyagi R and Sharma A 1997 *Biotechnol. Prog.* **13** pp 284-288
[22] Sharma P, Sharma N, Pathania S and Handa S 2017 *J Genet Eng Biotechnol* **15** pp 369-377
[23] World enzymes 2014 *Industry study with forecasts for 2017 & 2022* (Cleveland: The Freedonia Group)

IOP Conf. Series: Materials Science and Engineering 1113 (2021) 012006 doi:10.1088/1757-899X/1113/1/012006

Generalized proportional integral disturbance observer-based fuzzy sliding mode control for active magnetic bearing system

Van Nam Giap and Shyh Chour Huang

Department of Mechanical Engineering, National Kaohsiung University of Science and Technology, Kaohsiung, Taiwan

E-mail: shuang@nkust.edu.tw

Abstract. This paper presents a generalized proportional integral disturbance observer-based a combination of a fuzzy logic control and sliding mode control for the suspension active magnetic bearing system. Firstly, a generalized disturbance observer based on proportional and integral is built to estimate the parameters variation and the outside disturbances. In order to guarantee the system-state will be stabilized on pre-defined states, a disturbance feedback is constructed with a low pass filter. Subsequently, the proportional integral derivative surface sliding mode control is designed. Finally, the Fuzzy logic controller is constructed to determine the hitting controller gain. The stability analysis is given following the Lyapunov law. The proposed control structure can guarantee the system transient response quite good, no overshoot value, and settling time is quite narrow. Two testing cases of the proposed controller with and without the generalized disturbance observe is compared.

1. Introduction

The active magnetic bearing system offer a noncontact working between the stator and rotor [1-4]. However, the active magnetic bearing system is a highly unstable system. In order to control this system previous published paper as our research [3] proposed the fuzzy sliding mode control for the suspension active magnetic bearing system. The fuzzy is used to regulate the controller current value, this research given good chattering rejection method, with small overshoot, narrow settling time, free-chattering value, but the research ignored the outside disturbance and uncertainty values. Chen et al. [5] proposed the neural network to approximate the uncertainty value of the active magnetic bearing system, their proposed method is very good at tracking with non-periodic trapezoidal and periodic trapezoidal signals. Zad et al. [6] presented a design and adaptive sliding-mode control of hybrid magnetic bearings, the results of their paper are good but the settling time still high. Due to the sensitive with the output disturbance of the suspension active magnetic bearing system, this paper uses a general proportional integral to estimate the output disturbance and inside uncertainty values. The convergence of the system state is guaranteed by a low-pass-filter. The concept of disturbance observer was presented and has been developed by many researchers. Kim et al. [7] said that the disturbance is estimated by a general disturbance observer, the original work of their paper is based on the concept of the friction observer. Chen et al. [8] applied the general proportional integral observer-based composition control method for robotic thermal tactile sensor with disturbances, their paper is well achieved with the regulation temperature of the robotic thermal tactile under unknown time varying disturbances. In [9] Wang et al. present the same method to estimate the output disturbance for the induction motor. The results of their control method are very good at reject the output disturbance and parameter uncertainty. This paper proposes the sliding mode control with a proportional-integral-derivative surface is used to construct the controller for the active magnetic bearing system.

The concept of the sliding mode was initially developed in mid-1950s. In [10] Utkin presented the concept and the stability of a sliding mode controller. Sliding mode control is a type of the variable structure controller [11]. The stability of system should be satisfied the Lyapunov condition. The control signals of sliding mode controller are consisted the equivalent value and the switching value. The switching control value is used to force the system-state converge on the pre-defined surface, and the equivalent control signal is used to stabilize the system-state on that surface. In order to reduce the chattering value of the sliding mode control, this paper use the saturation function to replace the sign function of the switching control part. The fuzzy output is used to approximate the hitting control gain of the switching control part. The archived results will be shown in the subsequent part.

2. Mathematical modeling of the SAMB system

The control structure of the suspension active magnetic bearing system is included two opposite poles, one rotor, and one embedded thrust disk, one computer includes a multi channels analog to digital card with 16 bits revolution. Multi channels digital to analog card with 16 bits revolution, an eddy-current position sensor, and an amplifiers 0.5 A/V. The details of the system are presented as Fig. 1 below.

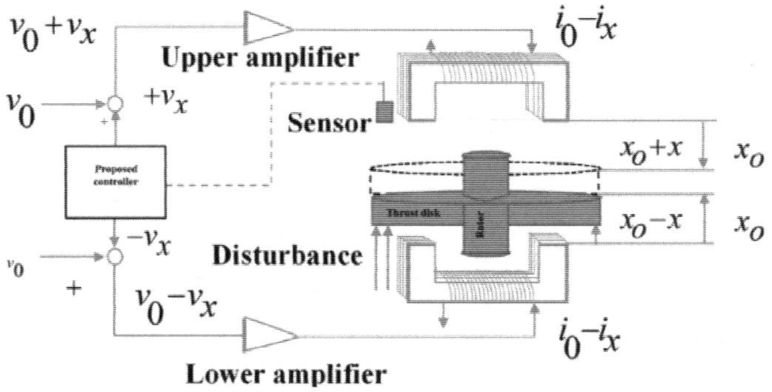

Figure 1. Active magnetic bearing control system

The mathematical model of the SAMB is built from the literature [4] as

$$F_x = F_+ - F_- = K\left(\frac{(i_0+i_x)^2}{(x_0-x)^2} - \frac{(i_0-i_x)^2}{(x_0+x)^2}\right) \tag{1}$$

where F_+ and F_- are the upper and lower forces of upper and lower the magnet coils, respectively. K is the coefficient of the force values. i_0 and i_x are the initial current and controlled current values. Using Taylor expansion of the Eq. (1) leads to

$$F_x = k_p \cdot x + k_i \cdot i_x \tag{2}$$

With k_p, k_i are the amplification factors of the rotor position and the magnet coil currents, respectively. Where

$$k_p = \left.\frac{\partial F_x(x,i_x)}{\partial x_x}\right|_{x=0,i_x=0} = \left(\frac{2k(i_0+i_x)^2}{(x_0-x)^3} - \frac{-2k(i_0+i_x)^2}{(x_0-x)^3}\right) = 4k\frac{i^2}{x^3}$$

$$k_i = \left.\frac{\partial F_x(x,i_x)}{\partial i_x}\right|_{x=0,i_x=0} = \left(\frac{2k(i_0+i_x)}{(x_0-x)^2} - \frac{-2k(i_0+i_x)}{(x_0-x)^2}\right) = 4k\frac{i}{x^2}$$

Follows the Newton II law:

$$m\ddot{x} = F_{x(t)} - f_{dt_{x(t)}} \tag{3}$$

where m is the mass of the inside rotor and an embedded thrust disk, F_x is the Lorentz force, and $f_{dt_{x(t)}}$ is the unexpected output disturbance. The Eq. (3) can be written as

$$\ddot{x}(t) = \frac{1}{m}(-c\dot{x} + k_p.x(t) + k_i.i_{x(t)} - f_{dt_{x(t)}}) \tag{4}$$

denotes, $A_n, = -c/m, B_n = k_p/m, C_n = k_i/m, \gamma = 1/m.$ Then the Eq. (4) can be re-written as

$$\ddot{x}(t) = (A_n + \Delta A)\cdot\dot{x}(t) + (B_n + \Delta B)\cdot x(t) + (C_n + \Delta B)\cdot i_x(t) - \gamma.f_{dx(t)} \tag{5}$$

where A_n, B_n, and C_n are known system state matrixs. The values of $\Delta A \in [\Delta A_l, \Delta A_h]$, $\Delta B \in [\Delta B_l, \Delta B_h]$, and $\Delta C \in [\Delta C_l, \Delta C_h]$ are unknown parameters, it represents the uncertainties values of the system and parameter variation. the system model is represented as

$$\ddot{x}(t) = (A_n)\cdot\dot{x}(t) + (B_n)\cdot x(t) + (C_n)\cdot i_x(t) + d \tag{6}$$

The Eq. (6) can be written as

$$\dot{X}(\mathrm{t}) = \begin{bmatrix} 0 & I \\ B_n & A_n \end{bmatrix}\cdot X(\mathrm{t}) + \begin{bmatrix} 0 \\ C_n \end{bmatrix}\cdot i_x(t) + \begin{bmatrix} 0 \\ I \end{bmatrix}\cdot d \tag{7}$$

where $d = (\Delta A\cdot\dot{x}(t) + \Delta B\cdot z(t) + \Delta B\cdot i_x(t) - \gamma.f_{dx(t)}) < \beta$, and $X(\mathrm{t}) = [x(\mathrm{t}), \dot{x}(\mathrm{t})]^T$ β is a positive constant upper boundary of the unknown disturbance of the system. The system parameters are shown in Table 1 below.

Table 1. System parameters

Parameters	Description	Value	Unit
m	Mass of rotor	2.565	kg
k_{ai}	Current stiffness of electromagnetic force of thrust disk AMB	40	N/A
k_{ap}	Position stiffness of electromagnetic force of thrust disk AMB	25200	N/M
z_0	Nominal air gap where thrust disk is centered	1	mm
T	Thrust disk mass	0.38	kg
c	Damping constant	0.001	N.s/m
v_0	Reference voltage	1.4	V
i_0	Amplifier range	0.5	A/V

3. Proposed approach

A highly unstable like the suspension active magnetic bearing system could be provided a robust controller. The disturbance observer can be used to reject the outside disturbance and system parameters variation. A generalized proportional integral disturbance is applied to reject the unexpected values and guarantee the asymptotic convergence of the rotor position to the pre-defined position.

3.1. Generalized proportional integral disturbance observer

In facts, the disturbance rejection is one of the most importance thing to estimate the controller performance. This paper presents a GPIDO as follows

$$d = \sum_{m=0}^{n} a_m t \tag{8}$$

the a_m are unknown constant values. The observer system mathematical model is described as

$$\begin{aligned} \dot{\xi} &= \begin{bmatrix} 0 & I \\ B_n & A_n \end{bmatrix} \cdot X(\mathrm{t}) + \begin{bmatrix} 0 \\ C_n \end{bmatrix} \cdot i_x(t) + \begin{bmatrix} 0 \\ \delta_i \end{bmatrix} \cdot g_i \\ \hat{d} &= \begin{bmatrix} 0 \\ \delta_i \end{bmatrix} \cdot g_i \end{aligned} \tag{9}$$

where $i = 1 \div n$, and

$$g_i = \begin{cases} X - \xi \\ \int_0^t g k-1(\tau) d\tau \end{cases} \tag{10}$$

The estimate disturbance value is asymptomatically tracking the lump of disturbance and uncertainty when the disturbance gain is resulted the $p(\mathrm{s})$ are Hurwitz stable, where

$$p(\mathrm{s}) = \mathrm{s}^{n+1} + \delta_0 s^n + .. + \delta_{n-1} s + \delta_n a \tag{11}$$

For reducing the noise of high frequency term, this study used the low-pass-filter gain to estimate disturbance value. The low-pass-filter is

$$G(\mathrm{s}) = \frac{1}{Ts + 1} \tag{12}$$

A suitable low-pass-filter time constant need to be chosen.

3.1.1. The stability of proposed observer

In this section the detail analysis of the stability of the disturbance estimation will be revealed. By denoting $\tilde{d} = \hat{d} - d$, then the derivative of the estimation disturbance error is

$$\dot{\tilde{d}} = \dot{\hat{d}} - \dot{d} = (\delta_0 \dot{g}_0 + \delta_1 \dot{g}_1 + .. + \delta_n \dot{g}_n) - \dot{d} = (-(\tilde{d}) + \delta_1 \dot{g}_1 + .. + \delta_n \dot{g}_n) - \dot{d} \tag{13}$$

After $n+1$ steps of derivation the estimated disturbance error value will leads to

$$\tilde{d}^{n+1} = -(\delta_0 \tilde{d}^n + ... + \delta_n \tilde{d}) - d^{n+1} \tag{14}$$

or

$$\tilde{d}^{n+1} + (\delta_0 \tilde{d}^n + ... + \delta_n \tilde{d}) = d^{n+1} \tag{15}$$

From Eq. (8) after $n+1$ steps of derivation, the term d^{n+1} goes to zero. The Eq. (15) could be written as

$$\tilde{d}^{n+1} + (\delta_0 \tilde{d}^n + ... + \delta_n \tilde{d}) = 0 \tag{16}$$

The derivative error of Eq. (8) is asymptotically stable, and the estimated disturbance can well track the disturbance and uncertainty only when $p(s) = s^{n+1} + \delta_0 s^n + ... + \delta_{n-1} s + \delta_n$ are Hurwitz. A filter with approximate bandwidth as a low-pass-filter is used to guarantee the disturbance error value goes to zero. The stability is proved as

$$\tilde{d} = d - L^{-1}(\mathrm{G(s)}) \times \hat{d} \tag{17}$$

The disturbance and uncertainty are accurately approximated by

$$d \times L^{-1}(\mathrm{G(s)}) = \hat{d} \tag{18}$$

The Eq. (17) then could be

$$\tilde{d} = L^{-1}(1 - \mathrm{G(s)}) \times d \tag{19}$$

The transformation of the term

$$L^{-1}(1 - \mathrm{G(s)}) = 1 - \frac{1}{T} e^{-\frac{t}{T}} \to 0 \tag{20}$$

3.2. *Sliding mode control*

The sliding mode control are included an equivalent control value, and a switching control value. Every system-state will slide and converge on a pre-defined state. This paper proposes the sliding surface by

$$s(t) = k_d \cdot \dot{e}(t) + k_p \cdot e(t) + k_i \cdot \int_0^t e(\tau) \cdot d\tau \tag{21}$$

refer $e(t) = x_r - x_m$ where x_r is the reference distance value, and x_m is the measured distance. k_p, k_i, k_d, are positive constant values and it should be chosen such that the real parts of the roots then sliding surface is satisfied Lyapunov law as

$$\dot{V}(t) = \dot{s}(t) \cdot s(t) < 0 \tag{22}$$

Taking the derivative of the Eq. (21)

$$\dot{s}(t) = k_d \cdot \ddot{e}(t) + k_p \cdot \dot{e}(t) + k_i \cdot e(t) \tag{23}$$

Combining Eq. (10) and Eq. (20), the current is calculated as

$$I_{cref}(t) = \frac{1}{k_d \cdot C}\left[k_d \cdot \ddot{x}_r(\mathrm{t}) - k_d \cdot (A_n \cdot \dot{x}_m(t) + B_n \cdot x_m(t) + d) + k_p \cdot \dot{e}(t) + k_i \cdot e(t) + k \cdot sign(s(t))\right] \tag{24}$$

where the derivative surface value is chosen by term of switching control value as $\dot{s}(t) = -k \cdot sign(s(t))$. Then the control value is divided by two terms of values as following

$$I_{cref} = I_{eq} + I_{sw} \tag{25}$$

In term of reduction chattering value from switching control part, the saturation is used to replace the sign function. Where $sat(s) = sign(s) \cdot \min\{1, |s|\}$ Or

$$sat(s) = \begin{cases} 1 \ \ if \ \ s > \varepsilon \\ \frac{1}{\varepsilon} \cdot s \ \ if \ \ s \in [-\varepsilon; \varepsilon] \\ -1 \ \ if \ \ s < -\varepsilon \end{cases} \tag{26}$$

Then Eq. (21) could be written as

$$I_{cref}(t) = \frac{1}{k_d \cdot C}((k_d \cdot \ddot{x}_r(t) - k_d \cdot (A \cdot \dot{x}_m(t) + B \cdot x_m(t) + L) + k_p \cdot \dot{e}(t) + k_i \cdot e(t) + k \cdot sat(s(t))) \tag{27}$$

The Eq. (22) can be written to obtain the disturbance and uncertainty value as

$$s(\mathrm{t}) \cdot \dot{s}(t) = s \cdot \left(k_d \cdot \ddot{e}(t) + k_p \cdot \dot{e}(t) + k_i \cdot e(t)\right) \tag{28}$$

or

$$\begin{aligned} s(\mathrm{t}) \cdot \dot{s}(t) &= s \cdot (k_d \cdot \ddot{x}_r(t) - ((A_n + \Delta A) \cdot \dot{x}_m(t) + (B_n + \Delta B) \cdot x_m + L + (C_n + \Delta C)(I_{eq} + I_{sw})) \\ &+ k_p \cdot \dot{e}(t) + k_i \cdot e(t)) \\ &= s \cdot (k_d \cdot \ddot{x}_r(t) - ((A_n + \Delta A) \cdot \dot{x}_m(t) + (B_n + \Delta B) \cdot x_m(t) + L \\ &+ (C_n + \Delta C)(\frac{1}{k_d \cdot C}[k_d \cdot \ddot{x}_r(t) - k_d \cdot (A_n \cdot \dot{x}_m(\mathrm{t}) + B_n \cdot x_m(t)) + k_i e(t) + k_p \cdot \dot{e}(t) + I_{sw})) \\ &+ k_p \cdot \dot{e}(t) + k_i \cdot e(t)) \\ &= s \cdot (k_d(-\Delta A + C_n^{-1} \cdot A_n \cdot \Delta C) \cdot \dot{x}_m(t) + (-\Delta B + L - (C_n^{-1} \cdot \Delta C) \cdot (k_d \cdot \ddot{x}_r(t) \\ &+ k_i e(t) + k_p \cdot \dot{e}(t))) - s \cdot k_d(C_n + \Delta C) \cdot I_{sw} \end{aligned} \tag{29}$$

or

$$\begin{aligned} &\le |s| \cdot (\left|k_d\right|(\left|-\Delta A + C_n^{-1} \cdot A_n \cdot \Delta C\right|) \cdot \left|\dot{x}_m(t)\right| + (\left|-\Delta B + C_n^{-1} \cdot B_n \cdot \Delta C\right|) \cdot \left|x_m(t)\right| \\ &-|L| - (\left|C_n^{-1} \cdot \Delta C\right|) \cdot ([k_d \cdot \left|\ddot{x}_r(t)\right| + k_i \left|e(t)\right| + k_p \cdot \left|\dot{e}(t)\right|)) - s \cdot k_d(C_n + \Delta C) \cdot I_{sw} < 0 \end{aligned} \tag{30}$$

The inequality (34) is satisfied when the hitting control gain should be defined as

$$\begin{aligned} k = (C_n + \Delta C_l)^{-1} \cdot (\left|-\Delta A_h + C_n^{-1} \cdot A_n \cdot \Delta C_l\right|) \cdot \left|\dot{x}_m(t)\right| + (\left|-\Delta B_h + C_n^{-1} \cdot B_n \cdot \Delta C_l\right|) \cdot \left|x_m(t)\right| - |L| - \\ (\left|C_n^{-1} \cdot \Delta C_h\right|) \cdot ([k_d \cdot \left|\ddot{x}_r(t)\right| + k_i \left|e(t)\right| + k_p \cdot \left|\dot{e}(t)\right|)) \end{aligned} \tag{31}$$

Then the inequality 28 could be written as

$$s(\mathrm{t})\dot{s}(\mathrm{t}) \le -k\left|s(\mathrm{t})\right| \tag{32}$$

The hitting control gain should by approximate for the chattering as small as possible. The paper provides the fuzzy logic controller to implement the work.

3.3. Fuzzy logic control

Fuzzy logic is another practical mathematical addition to classic Boolean logic [12]. The Fuzzy rules were chosen somehow the output of the fuzzy is satisfied the Lyapunouv condition. The hitting control gain is output of the fuzzy system. The sliding mode stability provides $s(\mathrm{t}) \cdot \dot{s}(\mathrm{t}) \leq -k_f |s(\mathrm{t})|$, with k_f is the fuzzy output. In case $s(\mathrm{t}) > 0$ the k_f increase will lead to $\dot{s}(\mathrm{t})$ decrease and vice versa. Otherwise, in case $s(\mathrm{t}) < 0$ the k_f decrease will leads to $\dot{s}(\mathrm{t})$ increase and vice versa. The fuzzy membership function includes five distinguished partitions as NB (Negative Big), NM (Negative Medium), ZO (Zero), PM (Positive Medium), and PB (Positive Big). The fuzzy rules are constructed as Table 2 and Fig. 2 below.

Table 2. The Fuzzy rules

$s(\mathrm{t})$						
	kf	NB	NM	ZO	PM	PB
	NB	VH	H	M	H	VH
	NM	H	H	M	H	H
$\dot{s}(\mathrm{t})$	ZO	M	M	L&H	M	M
	PM	L	L	M	L	L
	PB	ZO	L	M	L	ZO

Fuzzy logic membership function are determined as

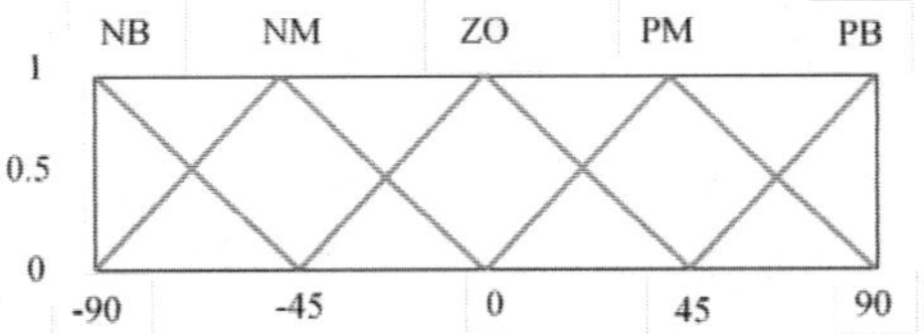

(a) sliding mode control surface

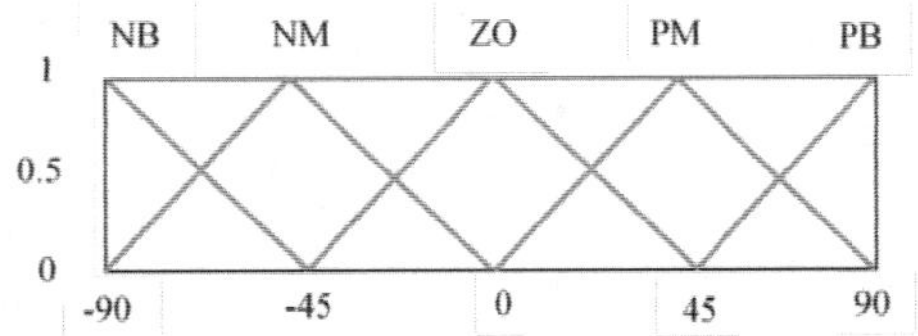

(b) derivative of the sliding mode surface

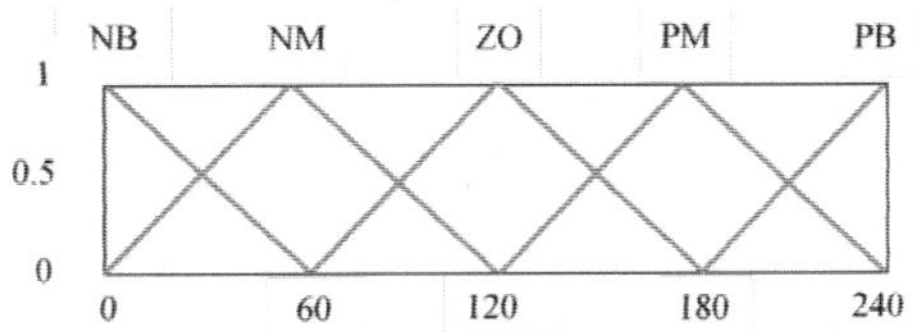

(c) fuzzy output signal Sliding mode surface

Figure 2. Fuzzy rules

The control system is constructed as Fig. 3 below.

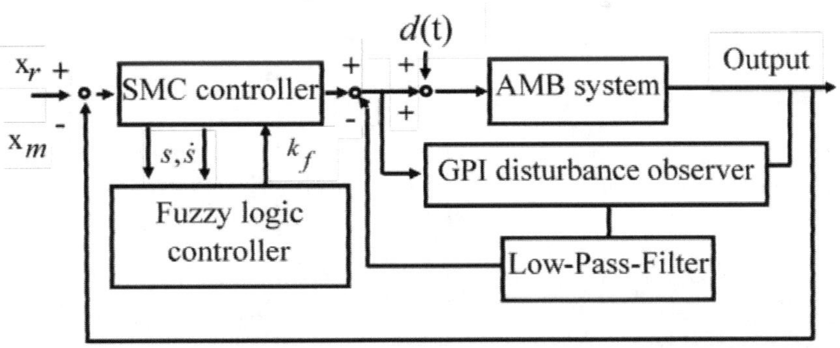

Figure 3. GPI disturbance observer based fuzzy sliding mode control for AMB system

The proposed method results are given by the next part. The control parameters are in Table 3.

Table 3. The controller parameters are

Parameter	Description	Value
k_p	Proportional coefficient	3000
k_i	Integral coefficient	3000
k_d	Derivative coefficient	1
k	Positive gain	k_f
ε	Saturation coefficient	0.8
T	Low-pass-filter time	0.001
δ_0	Disturbance coefficient	50
δ_1	Disturbance coefficient	100

4. Proposed approach

The approximation fuzzy system used centroi-weight defuzzification to operate the fuzzy system. The stability of the system now is combination of Eqs. (24) and (36) with a hitting control gain is determined by k_f. With a case of testing the outside disturbance. This section brief describes the effectiveness of the proposed control method by these performances. An illustrative example of distance tracking response is most important impact, their properties is including the settling time, overshot value, and steady state, the disturbance response is important also.

4.1. The Fuzzy sliding mode control performance

This section gives some result of the case without generalized proportional integral disturbance observer, a disturbance signal is given from outside to test the performance of the proposed method was implemented. The given output signal determined that the proposed control method is good at tracking the flexible signal. The uncertainty and disturbance are rejected, but not completely rejected the output disturbance signal. The given output are shown in Fig. 4 below.

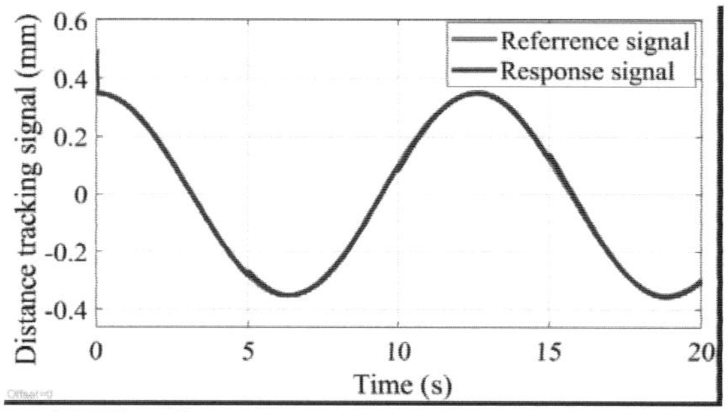

(a) The distance response signal at first 20s

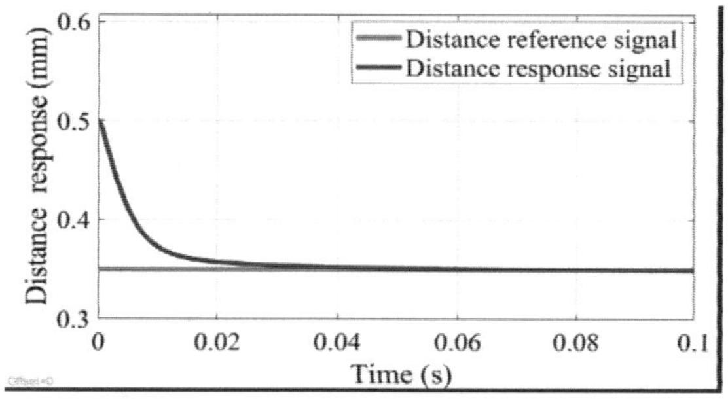

(b) The distance response signal at first 0.1s

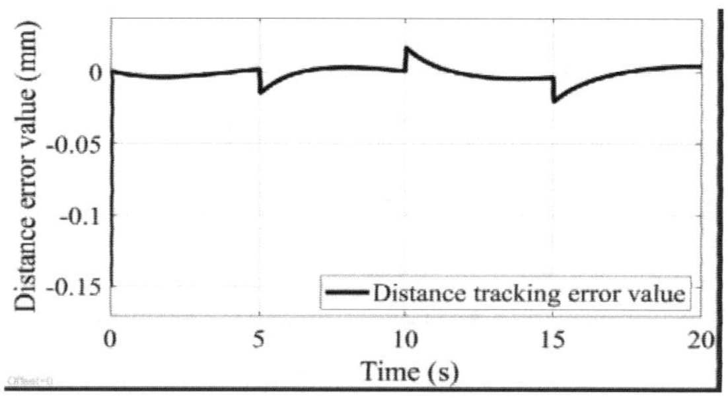

(c) The distance tracking error value

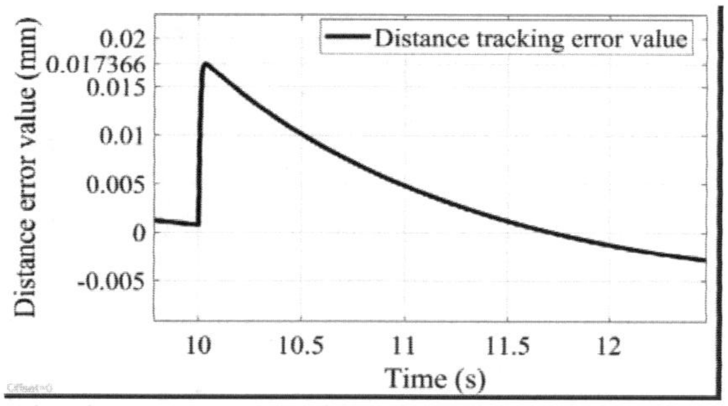

(d) The maximum value of the distance tracking error value

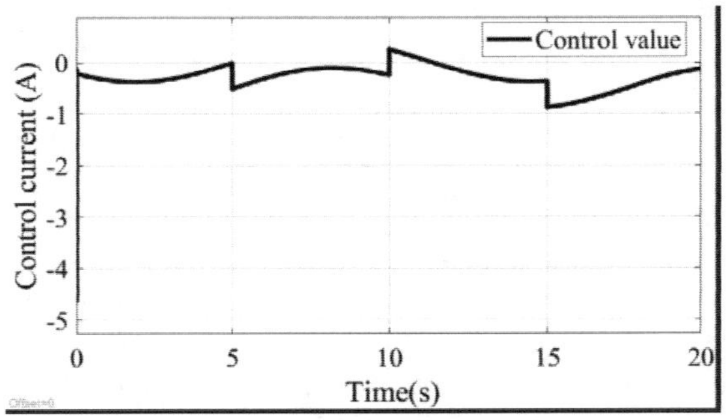

(e) The control current value

Figure 4. Fuzzy sliding mode control for AMB system given output signals

4.2. The GPI disturbance observer based fuzzy sliding mode control performance

The given output signal are shown in Fig. 5 below.

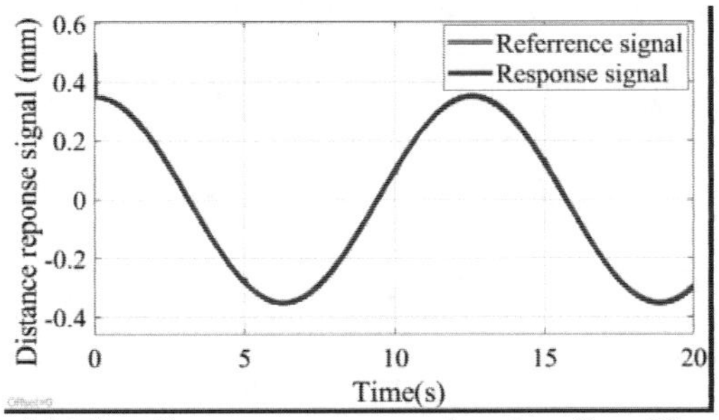

(a) The distance response signal at first 20s

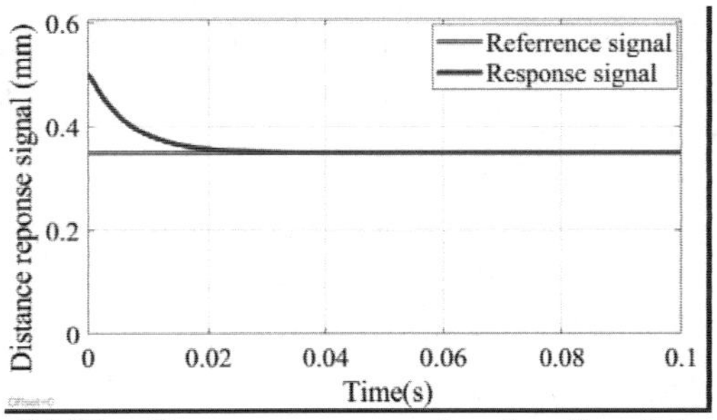

(b) The distance response signal at first 0.1s

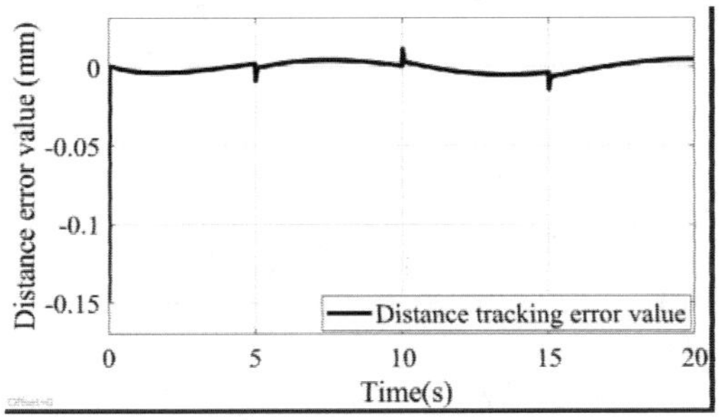

(c) The distance tracking error value

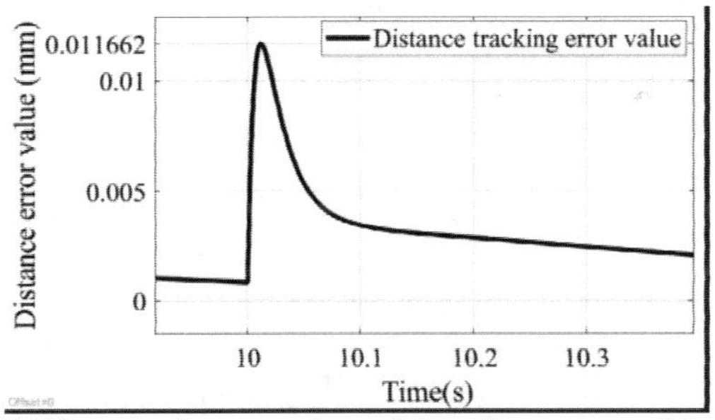

(d) The maximum value of the distance tracking error value

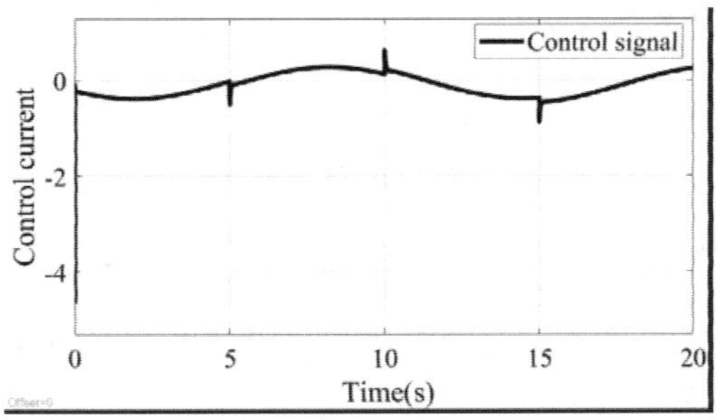

(e) The control current value

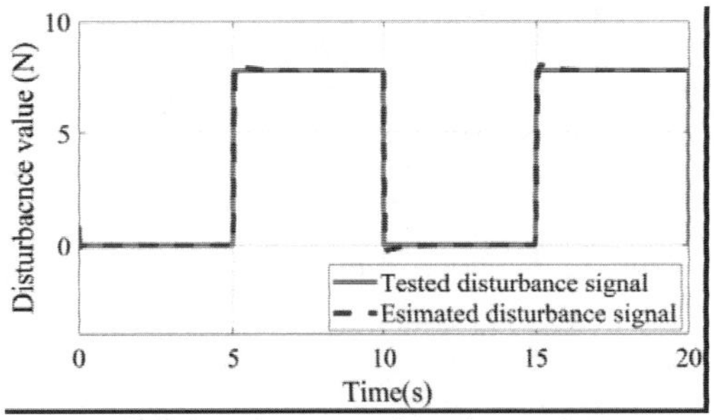

(f) Disturbance estimation value

Figure 5. GPI Disturbance observer based fuzzy sliding mode control for AMB system output signals

This section will present the performance of the GPI disturbance observer-based fuzzy sliding mode control for the active magnetic bearing system. The outside disturbance is completely rejected by a feedback gain. The results are represented as

The given output signals figure out that the performance of the proposed method is quite good at tracking the sinusoidal reference signal, the settling time just approximately 3ms in case of no disturbance observer, and 2ms in case of disturbance observer is equipped, respectively. There are no overshoot values in both case, and the average distance tracking error values are equal to 1.344um. The maximum of the distance tracking error value is approximately equalled 17.366mm and 11.662mm in case 1 and case 2, respectively. The testing disturbance is responded by a lump of disturbance observer, which means the exogenous disturbance is automatically rejected.

5. Proposed approach

This study used a proportional integral derivative to construct the sliding mode control surface, a fuzzy logic control is used to construct the hitting control gain. There are somehow the Lyapunov is guaranteed for system stability. A general proportional integral disturbance observer is equipped to estimate the unknown outside disturbance and unknown parameters variation. The design methodology is utilized to control the suspension active magnetic bearing system. The archived result are figured out that proposed methodology is good at tracking the dynamic input signal, the proposed controller can reject the output disturbance and the variation of the parameters. The main advantages are, the settling time of proposed controller is very small, no overshot value, and the average of the distance tracking error value is quite small. The energy of control value is strongly small. The proposed fuzzy controller is suitable for building the hitting control gain. In comparison of two cases is aimed to figure out that the proposed controller is good at rejecting the outside disturbance and system parameters variation.

Acknowledgments

The authors acknowledge and thank the Ministry of Science and Technology of the Republic of China for their partial financial support of this study under Contract Number MOST 108-2622-E-992-009 - CC3.

References

[1] Schweitzer G and Maslen E H 2009 *Magnetic bearings-theory, design, and application to rotating machinery* (Springer, Dordrecht)
[2] Lin J, Chen S Y, and Huang M S 2011 *ITE Control Theory and Apply* **5**(11) pp 1287-1303
[3] Su T J, Li T Y, Tsou T Y, Giap V N, and Nguyen Q D 2017 *J. Advances in mechanical engineering* **9**(12) pp 1-8
[4] TSai J S H, Su T J, Cheng J C, Lin Y Y, Giap V N, Guo S M, and Shieh L S 2018 *J. International Journal of Systems Science* **49**(6) pp 1273-1299
[5] Chen S Y and Lin F J 2011 *J. IEEE Transaction on control systems technology* **19**(3) pp 636-643
[6] Zad H S, Khan T I and Lazoglu I 2018 *J. IEEE Transactions on Industrial Electronics* **65**(3) pp 2537-2547
[7] Kim K S, Rew K H and Kim Sh 2010 *J. IEEE Transactions on Automatic Control* **55**(8) pp 1905-1911
[8] Chen C Y, Wang Y D and Li S H 2017 *J. International Journal of Advanced Robotic Systems* **14**(3) pp 1-10
[9] Wang J X, Wang F X, Wang G L, Li Sh H and Yu L 2018 *J. IEEE Transactions on Industrial Informatics* **14**(9) pp 4159-4168
[10] Utkin V 1997 *J. IEEE Transactions on Automatic Control* **22**(2) pp 212-222
[11] Ablay G 2015 *J. Journal of Process Control* **32** pp 10-15
[12] Ross T J 2010 *Fuzzy logic with engineering applications, (third edition)* (John Wiley & Sons, Ltd)

IOP Conf. Series: Materials Science and Engineering 1113 (2021) 012007 doi:10.1088/1757-899X/1113/1/012007

Artificial neural network base on grey relational analysis estimate displacement of bridge-type amplifier

Quang Phuoc Tran[1], Ngoc Thai Huynh[1,2] and Shyh Chour Huang[1]

[1]Department of Mechanical Engineering, National Kaohsiung University of Science and Technology, Kaohsiung, Taiwan, ROC
[2]Faculty of Mechanical Engineering of Industrial University of Ho Chi Minh City, No. 12 Nguyen Van Bao St., Go Vap Dist., Ho Chi Minh City, Vietnam

E-mail: shuang@nkust.edu.tw

Abstract. The power of complaint mechanisms is able for a dramatic reduction in the number of parts required to compass a specified task. It may reduce manufacturing and assembly time and cost. Reduce the number of joints can also increase mechanism precision. The output displacement must be larger than input displacement with lower stress. Thus, Optimize geometry-shape for besting parameter design of the structure was investigated in this paper. The part was designed by Autodesk Inventor and determined displacement amplifier and stress by FEM in ANSYS. We estimate the displacement of a bridge-type amplifier using Artificial Neural Network (ANN) and regression method (RM) base on Grey relational analysis (GRA). The variability of the thickness, the incline angle, the length and the width of the flexure hinges (FH) affects the variability of target displacement (DI) and stress (ST). The influence of the width and the area of the location force is very small and not clear. The result of the simulation is verified by ANOVA and also compared with the predicted value of ANN, GRA, and RM. The displacement of a bridge-type amplifier is reached 0.95 mm with displacement input 0.01 mm. The ratio amplifier is got 95 times.

1. Introduction

In recent years, many researchers have designed many types of flexure hinge to replace traditional joints. For example, Yong and Lu proposed the kinetostatic modeling for 3-RRR compliant mechanisms [1]. These joints were used as the rotation joints for a 3-DOF paralleled mechanical systems for smooth and high precision motion in Micro/ Nano scale, designed by Tian et al. [2]. Qi et al. [3] use the kinematics theory and elastic beam theory (EBT) to calculate DAR of a bridge-type FH. The equivalent formula and FEM were applied to analyze failure for Triple-LET and LET FHs by Qiu, Yin and Xie [4]. Tian et al. [5] applied closed-form compliance equation to compute deformation of filleted V-shaped FHs and was verified by FEM. Yang et al. [6] applied super elastic materials for a FH and their numerical computations were able to accurately forecast the displacement and effectively reduce the computation cost compared by FEA in ANSYS and was con-firmed by the experiments. Ling et al. [7] applied Lagrange's equation to design semi-analytical FEM for a flexure hinge in complex compliant mechanisms. Dao and Huang [8-13] designed and applied multi-objective optimization to optimize design parameters of compliant mechanisms.

The objective of this paper analyzed and optimized effects of design parameters of a bridge-type amplifier displacement employing FHs by using FEM and ANN base on GRA. The model of the mechanism and finite element method was presented in section 2. After that optimization approaches were description in section 3. The FEM and optimization results was analyzed and discussed in section 4. The section 5 is the conclusions.

2. Modelling and methodology

2.1. Modelling

The mechanism was presented in Figure 1 with dimension of 70mm x 25mm x 10mm. Figure 1b is the solid part, Figure 1a is the drawing 2D. The mechanism has 8 FHs with variable length (L), thickness (T), width (W) include the incline angle (Ø) of 2 FHs.

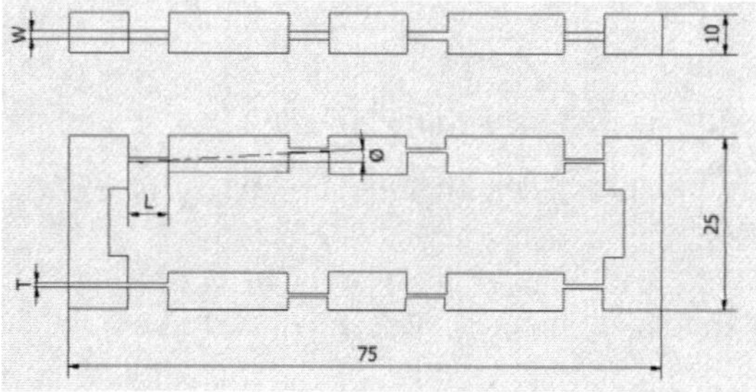

(a)

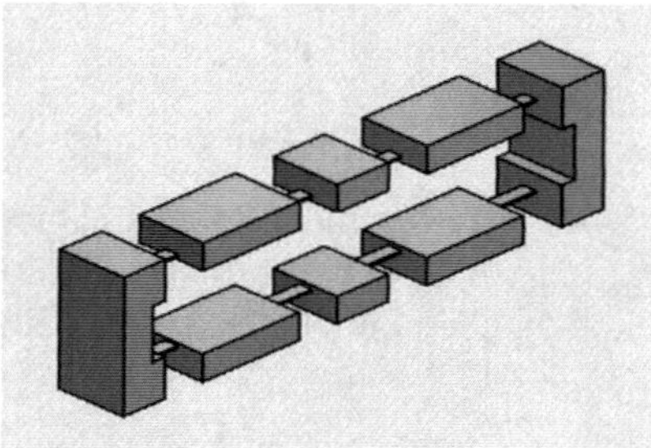

(b)

Figure 1. Active magnetic bearing control system

2.2. Finite element analysis

The aluminum alloy was used for the mechanism as listed in Table 1. The first, the mechanism is drawn in Autodesk Inventor and then imported to ANSYS. The alloy was aluminum AL-7075. In Figure 2a, the meshed mechanism was created by automatic mesh with adaptive size. The mesh convergence as h-refinement function is used to reduce the element sizes at the each FHs. The mesh type is used as triangular mesh. The number of nodes and the number of elements is equal to 130524 nodes, 73352 elements, respectively.

Table 1. Material mechanical properties

Alloy	E (GPa)	ν (ratio)	σ_Y(MPa)
AL7075	72	0.33	503

In Figure 2b, the static structural function in Ansys is used to analyze the output displacement and stress of the structure. The fixed support is chosen at the surface A. Two input displacements which magnitude is 0.01mm are located at the surface B and C. The direction of movements is in the opposite direction. The output displacement and output stress are observed.

(a)

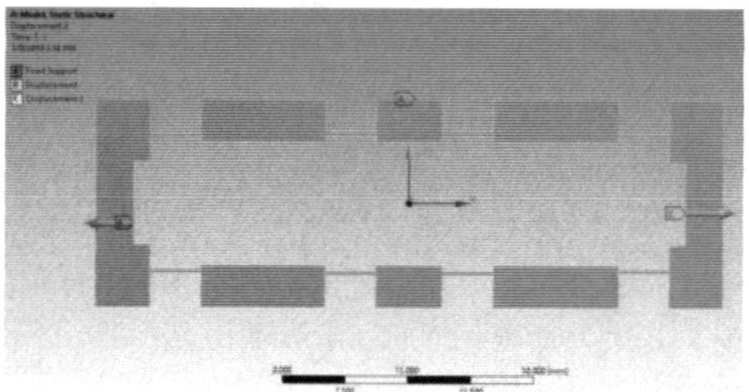

(b)

Figure 2. (a) The grid model, (b) Location of Fixture and DI for the model

3. Optimization

3.1 Selection of simulation parameter

In this study, the geometry parameters were designed by the orthogonal array table. The Taguchi method can effectuate optimally and highly operated in diverse environments with a powerful capacity. The geometry parameter are showed as the length of the flexure hinge (LOFH), the width of the flexure hinge (WOFH), the thickness of the flexure hinge (TOFH) and incline angle. The parameters for the simulation are shown in Table 2.

Table 2. System parameters

Factor		Unit	Level		
			1	2	3
Incline angle	Ø	Degree	0.6	0.75	0.9
LOFH	L	mm	4	5	6
WOFH	W	mm	2	4	6
TOFH	T	mm	0.3	0.4	0.5

3.2 Grey 's method

Normalization: Rewrite each sequence between 0 and 1 as follows [8, 11].

The larger the better formula:

$$D_i^* = \frac{D_i^{(0)}(k) - \min D_i^0(k)}{\max D_i^{(0)}(k) - \min D_i^{(0)}(k)} \quad (1)$$

The smaller the better formula:

$$D_i^* = \frac{\max D_i^{(0)}(k) - D_i^0(k)}{\max D_i^{(0)}(k) - \min D_i^{(0)}(k)} \quad (2)$$

Where $D_i^*$is the normalized S/N for the kth response (k= 1,2,...n). Max $D_i^{(0)}(k)$ and min $D_i^{(0)}(k)$ are the maximum and minimum values of $D_i^{(0)}(k)$, respectively.

Grey relational coefficient (GRC)(γ) : is quantified in the Grey relational space. GRC is as requirement before solving for the Grey relational grade (GRG).

$$\Delta_{0i} = \left\| D_0^*(k) - D_i^* k) \right\| \quad (3)$$

$$\Delta_{\min} = \max_{\forall j \in i} \min_{\forall k} \left\| D_0^*(k) - D_j^*(k) \right\| \quad (4)$$

$$\Delta_{\max} = \max_{\forall j \in i} \max_{\forall k} \left\| D_0^*(k) - D_j^*(k) \right\| \quad (5)$$

Where Δ_{0i} is deviation between $D_0^*(k)$ and $D_i^*(k)$. Calculate grey relational coefficient (GRC).

$$\gamma_i(k) = \frac{\Delta_{\min} + \xi\Delta_{\max}}{\Delta_{0i} + \xi.\Delta_{\max}} \quad (6)$$

Here, ζ ϵ [0,1] is the distinguishing coefficient. Determine the weight

$$\omega_e(x) = x.e^{(1-x)} + (1-x)e^x - 1 \quad (7)$$

Where $\omega_e(x)$ is the mapping function in the entropy measurement. This function obtains the maximum value when x=0.5 and $e^{0.5}$-1=0.6487 and the mapping value in [0, 1] is obtained as follows:

$$w \equiv \frac{1}{(e^{0.5}-1)} \sum_{i=1}^{m} \omega_e(x) \tag{8}$$

$\in = \{\gamma_i(1), \gamma_i(2), ..., \gamma_i(n)\}$. Note that $i = 1, 2, ..., n$. Determine the total of the grey relational coefficient:

$$D_j \equiv \sum_{i=1}^{m} \gamma_i(j), j = 1, 2, ..., n \tag{9}$$

Estimate the normalized coefficient:

$$k = \frac{1}{(e^{0.5}-1) \times m} = \frac{1}{0.6487 \times m} \tag{10}$$

Determine the entropy:

$$e_j = k \sum_{i=1}^{m} \omega_e(\frac{\gamma_i(j)}{D_j}), j = 1, 2, ..., n \tag{11}$$

Here, $\omega_e(x)$ uses Eq. (10). Compute total of entropy:

$$E = \sum_{j=1}^{n} e_j \tag{12}$$

Determine the weight:

$$\omega_j = \frac{1}{n-E} \cdot \frac{[1-e_j]}{\sum_{j=1}^{n} \frac{1}{n-E} \cdot [1-e_j]}, \tag{13}$$

here, $j = 1, 2, ..., n.$

GRG (ψ_i) is written as follows:

$$\psi_i = \sum_{k=1}^{n} \omega_k \gamma_i(k) \tag{14}$$

Predict output characteristics and GRG: output the acceleration, contact force and GRG can be estimated as:

$$\mu_D = D_m + \sum_{i=1}^{q} (D_0 - D_m) \tag{15}$$

$$\mu_\sigma = \sigma_m + \sum_{i=1}^{q} (\sigma_0 - \sigma_m) \tag{16}$$

$$\mu_G = G_m + \sum_{i=1}^{q} (G_0 - G_m) \tag{17}$$

Where μ_D, μ_σ are the optimization values of the predicted mean of displacement and stress.

The (CI) value for the optimal parameter level combination was estimated at a 95% confidence level as follows:

$$CI_{CE} = \pm\sqrt{F_{\alpha}(1, fe)V_{e}(\frac{1}{n_{eff}} + \frac{1}{R_{e}})} \quad (18)$$

where $\alpha = 0.05$; $F_{\alpha}(1, f_e) = F_{0.05}(1, f_e)$; f_e is the error of freedom degrees; V_e is the error of the adj MS value; $n_{eff} = n/(1+R)$; n is the quantity testing for the orthogonal array; R is the sum of the effect parameters and R_e is the simulation times.

3.3 Artificial neural network

This study used 3 classes, the input class with five input parameters, the hidden layer with eleven neurons. The network is trained using the Levenberg-Marquardt hybrid (trainlm) [7, 14-15]. The input L, Ø, T, W, fillet radius of the FH were taken as the input factors, and displacement or stress were taken as the output factors. The simulation values in Table 3 were utilized for training.

Table 3. Design of experiment and simulation result

No.	Ø	L	W	T	Displacement (mm)	Stress (MPa)
1	0.6	4	2	0.3	0.9806	102.9614
2	0.6	4	4	0.4	0.6536	106.8779
3	0.6	4	6	0.5	0.4446	99.8693
4	0.6	5	2	0.4	0.7687	97.6351
5	0.6	5	4	0.5	0.5312	95.6649
6	0.6	5	6	0.3	0.9962	91.0395
7	0.6	6	2	0.5	0.6236	82.5444
8	0.6	6	4	0.3	1.0819	75.0702
9	0.6	6	6	0.4	0.7761	81.9193
10	0.75	4	2	0.3	0.9519	93.9727
11	0.75	4	4	0.4	0.6851	105.3196
12	0.75	4	6	0.5	0.4928	101.1267
13	0.75	5	2	0.4	0.7835	91.6975
14	0.75	5	4	0.5	0.5744	93.9629
15	0.75	5	6	0.3	0.9639	81.9564
16	0.75	6	2	0.5	0.6590	79.8422
17	0.75	6	4	0.3	1.0259	66.7393
18	0.75	6	6	0.4	0.7906	76.8598
19	0.9	4	2	0.3	0.8967	92.8028
20	0.9	4	4	0.4	0.6870	95.9143
21	0.9	4	6	0.5	0.5182	89.5852
22	0.9	5	2	0.4	0.7676	84.2787
23	0.9	5	4	0.5	0.5920	90.5757
24	0.9	5	6	0.3	0.9059	73.1690
25	0.9	6	2	0.5	0.6646	75.1992
26	0.9	6	4	0.3	0.9507	58.8989
27	0.9	6	6	0.4	0.7733	71.0449

3.4 Statistical analysis

The precision model is assessed by four error standards as follows. The root mean squared error (RMSE) is the difference between forecast values and the simulation values or observed actual values:

$$RMSE = \sqrt{\frac{1}{m}\sum_{i=1}^{m}(x_i - y_i)^2} \tag{19}$$

The mean square error (MSE):

$$MSE = \frac{1}{m}\sum_{i=1}^{m}(x_i - y_i)^2 \tag{20}$$

The mean absolute percentage error (MAPE):

$$MAPE = \frac{100\%}{m}\sum_{i=1}^{m}\left|\frac{x_i - y_i}{x_i}\right| \tag{21}$$

The coefficient of determination (R^2) must be at least 0.8 for forecast models to be accepted:

$$R^2 = 1 - \frac{\sum_{i=1}^{m}(x_i - y_i)^2}{\sum_{i=1}^{m}(x_i - \overline{y})^2} \tag{22}$$

where m is number of simulation of experiments, x_i and y_i represent the simulation and predicting value, respectively, and $\bar{y}$ is the mean simulation value.

4. Results and discussion

4.1. Influence of thickness

Figure 3 presents the response of the displacement and stress of the mechanism. It was pointed out that the displacement output is decreased by 1.058mm, 0.748mm, 0.538mm when the thickness T is changed 0.3 mm, 0.4mm, 0.5mm, respectively. The stress is relatively decreased by 76.50MPa, 74.943MPa, 74.50MPa. The incline angle of FHs is 0.6 degree. The width is 2mm.The length is 4mm.

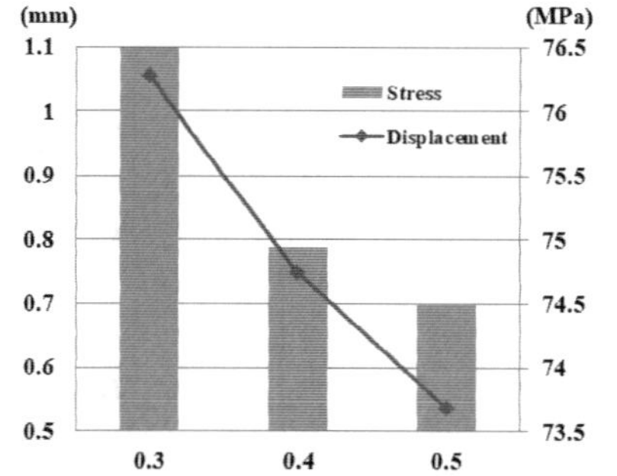

Figure 3. FEM result for displacement and stress with different TOFH (0.3mm-0.4mm-0.5mm)

4.2. Influence of width

In Figure 4, the incline angle between two FHs is 0.6 degree. The thickness of FHs is 0.3mm. The length of FHs is 4mm.When the width of the FHs is increased by 2mm, 4mm, 6mm, the output displacement is decreased by 1,084mm, 1,058mm, 1,03mm, while the output stress is increased 74,436MPa, 74.7MPa, 75.347MPa, respectively.

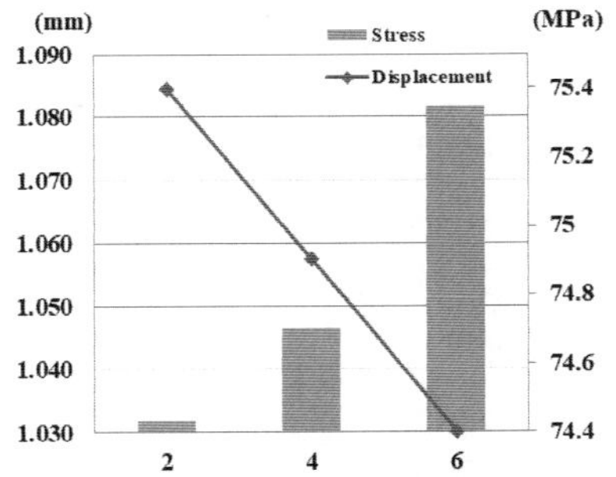

Figure 4. FEM result for displacement and stress with different WOFH (2mm-4mm-6mm)

4.3. Influence of incline angle

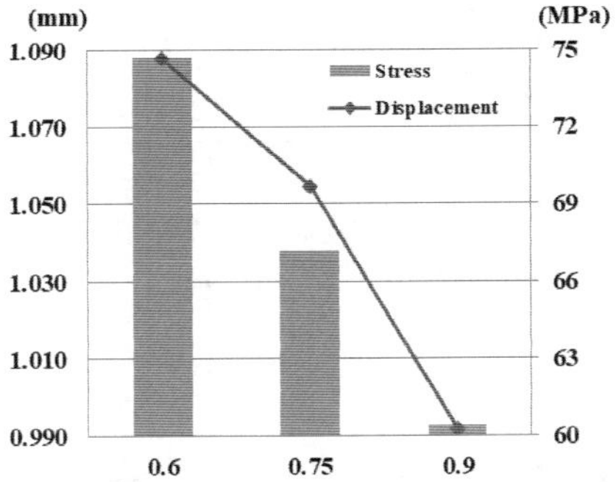

Figure 5. FEM result for displacement and stress with different Incline angle (0.6-0.75-0.9 (degree))

In Figure 5, the thickness, the length and the width of FHs are kept at 0,3mm, 4mm, 2mm, respectively. When the incline angle between two FHs is increased by 0.6 degrees, 0.75 degrees, 0.9 degrees, the displacement of output is decreased by 1.088mm, 1,055mm, 0.992mm, the stress is also decreased by 74.7MPa, 67,19MPa, 60.41MPa, respectively.

4.4. Influence of incline length

In Figure 6, the width of FHs is 2mm. the thickness of FHs is 0.3mm.The incline angle between two Fhs is 0.6degree. When the length L is changed by 4mm, 5mm, 6mm, the value of the displacement is increased by 0.995mm, 1.0621mm, 1.121mm and the stress is decreased by 110.68MPa, 88.317MPa, 71.93MPa, respectively.

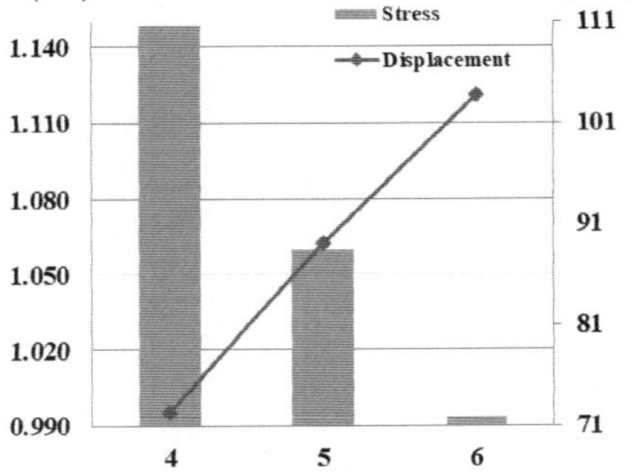

Figure 6. FEM result for displacement and stress with different LOFH (4mm–5mm–6mm)

4.5. The Grey relation analysis

Eqs. (1)-(3) is used to calculate the deviation. Level max-min of deviation values was determined by Eqs. (4)-(5). Solve the GRG value by Eqs. (8), (16). The result is displayed in Table 4. The maximum of GRG in Figure 7 shows the optimal value of the Grey relational analysis. The combination parameter obtained for DI and ST is Ø3L2W3T1.

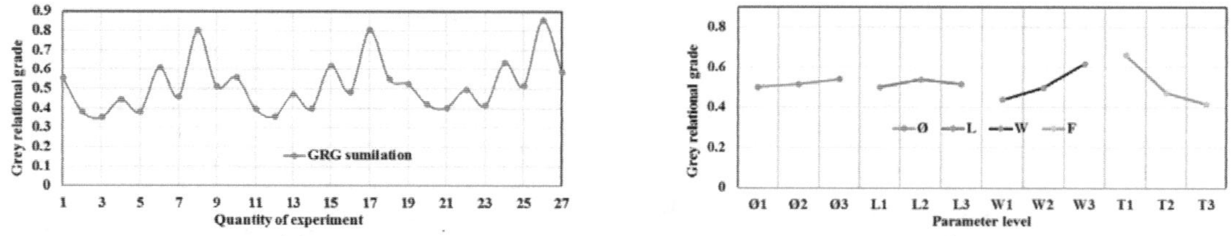

(a) (b)

Figure 7. Plot GRG values for simulation (a) and respone graph of GRG (b)

Table 4. The parameter value of Grey method

No.	$D_i^*(1)$	$D_i^*(2)$	$\Delta_{oi}(1)$	$\Delta_{oi}(2)$	$\gamma_i(1)$	$\gamma_i(2)$	GRG	Rank
1	0.841	0.0816	0.159	0.9184	0.7587	0.3525	0.5558	9
2	0.328	0	0.672	1	0.4266	0.3333	0.3800	25
3	0	0.1461	1	0.8539	0.3333	0.3693	0.3513	27
4	0.5086	0.1926	0.4914	0.8074	0.5043	0.3824	0.4434	18
5	0.1359	0.2337	0.8641	0.7663	0.3665	0.3949	0.3807	24
6	0.8655	0.3301	0.1345	0.6699	0.788	0.4274	0.6078	6
7	0.2809	0.5072	0.7191	0.4928	0.4101	0.5036	0.4568	17
8	1	0.663	0	0.337	1	0.5974	0.7989	3
9	0.5202	0.5202	0.4798	0.4798	0.5103	0.5103	0.5103	13
10	0.796	0.269	0.204	0.731	0.7102	0.4062	0.5583	8
11	0.3775	0.0325	0.6225	0.9675	0.4454	0.3407	0.3931	23
12	0.0757	0.1199	0.9243	0.8801	0.351	0.3623	0.3566	26
13	0.5319	0.3164	0.4681	0.6836	0.5165	0.4224	0.4695	16
14	0.2037	0.2692	0.7963	0.7308	0.3857	0.4062	0.3959	22
15	0.8148	0.5194	0.1852	0.4806	0.7297	0.5099	0.6199	5
16	0.3365	0.5635	0.6635	0.4365	0.4297	0.5339	0.4818	15
17	0.9121	0.8366	0.0879	0.1634	0.8505	0.7537	0.8021	2
18	0.5429	0.6257	0.4571	0.3743	0.5224	0.5719	0.5471	10
19	0.7094	0.2934	0.2906	0.7066	0.6324	0.4144	0.5235	11
20	0.3804	0.2285	0.6196	0.7715	0.4466	0.3932	0.4199	19
21	0.1156	0.3604	0.8844	0.6396	0.3612	0.4388	0.4000	21
22	0.5069	0.471	0.4931	0.529	0.5035	0.4859	0.4947	14
23	0.2314	0.3398	0.7686	0.6602	0.3941	0.431	0.4125	20
24	0.7238	0.7026	0.2762	0.2974	0.6442	0.627	0.6356	4
25	0.3453	0.6603	0.6547	0.3397	0.433	0.5955	0.5142	12
26	0.7941	1	0.2059	0	0.7083	1	0.8540	1
27	0.5158	0.7468	0.4842	0.2532	0.508	0.6638	0.5858	7

4.6. Analysis of variance (ANOVA)

The design factor have greatly impacted on GRG with P-values are less than 0.05. The width variable is 0.137. It is insignificant impact factor than another parameter. The distribution of the length and the thickness is greatly with 30.72% and 57.88%. The R-square value is equal 91.39 %. The result is as Table 5.

Table 5. The variance of GRG

Source	DF	Seq SS	Contribution	Adj SS	Adj MS	F-Value	P-Value
Regression	5	0.196357	91.39%	0.196357	0.039271	44.56	0.000
Ø	1	0.003534	1.64%	0.003534	0.003534	4.01	0.058
L	1	0.066006	30.72%	0.054147	0.054147	61.44	0.000
W	1	0.000129	0.06%	0.002106	0.002106	2.39	0.137
T	1	0.124354	57.88%	0.023805	0.023805	27.01	0.000
W*T	1	0.002334	1.09%	0.002334	0.002334	2.65	0.119
Error	21	0.018508	8.61%	0.018508	0.000881		
Total	26	0.214865	100.00%				

S	R-sq	R-sq(adj)	PRESS	R-sq(pred)
0.0296874	91.39%	89.34%	0.0313926	85.39%

The RE of GRG was reached by Minitab as:

$$GRG = (0.7645 + 0.0934Ø + 0.06938L - 0.0339W - 1.184T + 0.0882W*T)^2$$

4.7. The ANN Models

The results of training are shown in Figure 8. The best validation performance is 9.645×10^{-12}. The gradient is 1.0019×10^{-5}.

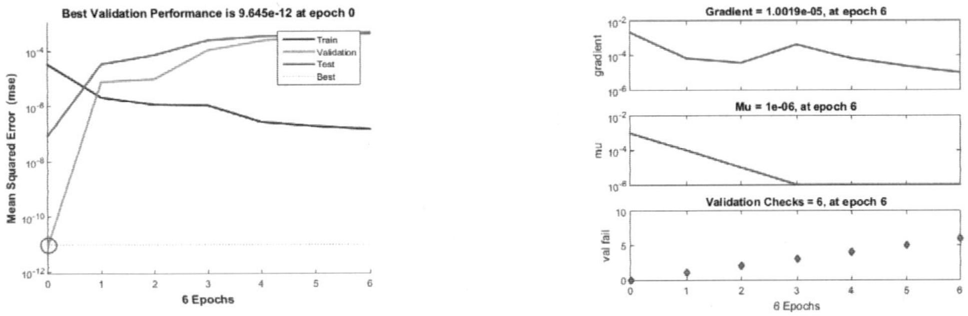

(a) (b)

Figure 8. Performance plot for GRG (a) and Training state of ANN for GRG(b)

Table 6 shown that RE and ANN were significant for optimal design. R square = 99.77%. RMSE is less than 0.05.

Table 6. Analysis of variance of GRG

Criteria	Predicted model			
	ANN for GRG		RE for GRG	
	Training	Testing	Training	Testing
RMSE	0.01	0.0062838	0.04	0.022377
MSE	0	0.000118	0.0016	0.000501
MAPE	0.42	0.36	5.93	1.92
R quare(%)	99.77	99.45	91.15	99.43

As the level of Rank, the optimal value of GRG by Eqs. (15)-(17). The predicted value of DI, ST, GRG are 0.5403mm, 75.3451MPa and 0.6578.

5. Conclusions

The target of this study is to reach the maximum response displacement achieved when optimizing the geometric design parameters of the amplifier structure. L27 orthogonal array was investigated with four levels of inline angle, length, width, and thickness of bridge-type flexure hinges. The FEA values were taken to the Grey model to approximate the best parameter geometry and determine optimal by the ANN method. The outcome obtained high displacement with the design parameter as following: Ø3L2W3T1. The dis-placement is 0.9507mm. In the future, compare the optimal result to other optimal methods to achieve the best value and the optimal results will be verified by measurement data on the actual model.

Acknowledgments

The authors acknowledge and thank the Ministry of Science and Technology of the Republic of China for their partial financial support of this study under Contract Number MOST 108-2622-E-992-009 - CC3.

References

[1] Y K Yong and T F Lu 2009 *Mechanism and Machine Theory* **44**(6) pp. 1156-1175
[2] Y Tian, B Shirinzadeh and D Zhang 2010 *Microelectronic Engineering* **87**(2) pp. 230-241
[3] K Q Qi, Y Xiang, C Fang, Y Zhang and C S Yu, 2015 *Mechanism and Machine Theory* **87** pp. 45-56
[4] L Qiu, S Yin and Z Xie 2016 *Engineering Failure Analysis* **66** pp. 35-43
[5] Y Tian, B Shirinzadeh and D Zhang 2010 *Precision Engineering* **34**(3) pp. 408-418
[6] M Yang, Z Du and W Dong 2016 *Precision Engineering* **46** pp. 177-183
[7] M S Sukumar, P V Ramaiah and A Nagarjuna 2014 *Procedia Engineering* **97** pp. 365-371
[8] T. P. Dao 2016 *Mathematical Problems in Engineering* **2016** pp. 1-17
[9] T P Dao and S C Huang 2015 *Journal of Mechanical Science and Technology* **29**(8) pp. 3309-3320
[10] T P Dao and S C Huang 2016 *Microsystem Technologies* **23**(2) pp. 441-456
[11] T P Dao and S C Huang 2017 *Microsystem Technologies* **23**(10) pp. 4815-4830
[12] T P Dao and S C Huang 2017 *Sādhanā* **42**(9) pp. 1527-1542
[13] T P Dao and S C Huang 2017 *Journal of Mechanical Science and Technology* **31**(3) pp. 1293-1303
[14] S Kumar, A Batish, R Singh and T P Singh 2014 *Journal of Mechanical Science and Technology* **28**(7) pp. 2831-2844
[15] G Kant and K S Sangwan 2015 *Procedia CIRP* **31** pp. 453-458

IOP Conf. Series: Materials Science and Engineering 1113 (2021) 012008 doi:10.1088/1757-899X/1113/1/012008

Developing programmable robot for K12 STEAM education

Cheng Tiao Hsieh

Department of Industrial Design, Ming Chi University of Technology, New Taipei City, Taiwan

E-mail: cthsieh@mail.mcut.edu.tw

Abstract. STEAM is the acronym that is a group of academic disciplines, Science, Technology, Engineering, Art and Mathematics. In recent years, the STEAM has been considered as the core parts of K12 education system in many countries. Regarding this need, this paper attempts to develop a cheap, interactive and programmable robot to assist K12 students to learn STEAM effectively and friendly. The robot was developed based the open source – Ottobot. Its primary external structure was made by 3D printers, so its shape can be customized by users. In addition, its electrical and electronic devices are compatible to Arduino. The development kits of Arduino are friendly and free, so it is very suitable to new users like K12 and the students with non-engineering background. The control board applied in this study is a low-cost Arduino Nano, which includes 14 digital I/O and 8 analog I/O. Associated with ESP8266 or HC-06, the robot can be controlled remotely via Wifi or Bluetooth. In this paper, we demonstrated the graphical programming tool, called mBlock, to code the firmware of the robot. Its friendly programming interface is similar to the Scratch, which provides K12 students a friendly and useful tool to explore programming world. This paper does not only present the detail information of building a programmable robot, but also demonstrate a friendly tool to assist K12 students to learn the programming ability for developing a smart robot or AI related products.

1. Introduction

STEAM stands for science, technology, engineering, art, and math. It can be seen as the term integrating "Art" into the STEM. There are many papers illustrating the importance of the STEM education. The STEM has been developed more than three decades. There are many success stories occurred in the world. In early of 1990s, the Center for the Advancement of Hispanics in Science and Engineering Education (CAHSEE) opened a summer program for the talented students at the STEM institute. Their success programs have really captured people's attention as well as encouraged educators and researchers to further explore the STEM activities in various level schools. In 2001, the acronym STEM was first adopted by the National Science Foundation (NSF), and many related NSF projects were proposed in various aspects, e.g. strengthening teachers' teaching skills for STEM, developing STEM courses, creating good STEM learning circumstance for kids, and so on. The goal of the STEM education is to train talented students to become eligible people for future high-tech society. In the traditional education system, science, math, and engineering were taught separately. Students may understand a specific discipline, but they may difficult to connect the disciplines they had learned together to furtherly solve specific high-tech problems or projects. In recent years, the "Maker Movement" brought a novel thought about educating people to a "maker". "Makers" is a group of talented people who owns abundant knowledge and skills in various fields. Under traditional education system, students seem hard to be educated to become a "Maker". It is the reason that many countries like to introduce the STEM into their education system. STEM puts all of needed disciplines together, and then to create better learning environment to offer students a completed education package. According to this fact, many countries like USA, Australia, Canada, and Taiwan have started to adopt the STEM in their K12 education system. In 2019, the Taiwan education administrator concentrated on two disciplines, programming and multiply languages. The "language" is the tool to assist students to link high-tech sources for exploring their desired knowledge and technologies. As for "programming", it can be seen as the important tool to let people communicate with computers. In current age, computer had deeply involved human's life. If students can learn programming in their early learning stage, it is helpful to let them to get familiar with the current digital world. In order to enhance Taiwan students' competitiveness, those two disciplines have been required in K12 school system. Regarding the "programming" discipline, many papers claim that robots are a good tool to assist young students to learn programming skills easily and happily. Therefore, this paper attempts to develop a useful tool, programmable robot for students, to assist young students, especially K12 students.

2. Related works

The importance of STEM education has been discussed in many papers [1-2]. Unlike the traditional education system, it focuses on not only science and mathematics, but also engineering and technology. Under traditional education system settings, students seem not able to obtain a completed knowledge to solve their practical high-tech projects which involve the knowledge and technologies in various disciplines. In 2011, US President Obama had recognized the importance of STEM for their K12 education, and he announced to increase amount of teachers in STEM. In recent years, some of educators think STEM education seems not contain enough disciplines for children. Therefore, STEM was extended to STEAM [3]. The contents of STEM are usually hard to let children understand. In 1988, Patrick McShane used robotics as learning tools to teach children efficiently. As we know, robotics plays an important role in current and future industry. Its contents are perfect to match the contents of STEAM. There are many wonderful stories about applying robotics in STEAM [4-7]. They really prove that STEAM does support children to learn the "T", "E", and "A" happily and efficiently. In addition, friendly technologies like Arduino and Visual Programming Language(VPL) are also the key to make STEAM success [8]. By this education, a completed knowledge and skills can transform students to a high-tech person to satisfy the needs of future industry.

3. Hardware of the developed robot

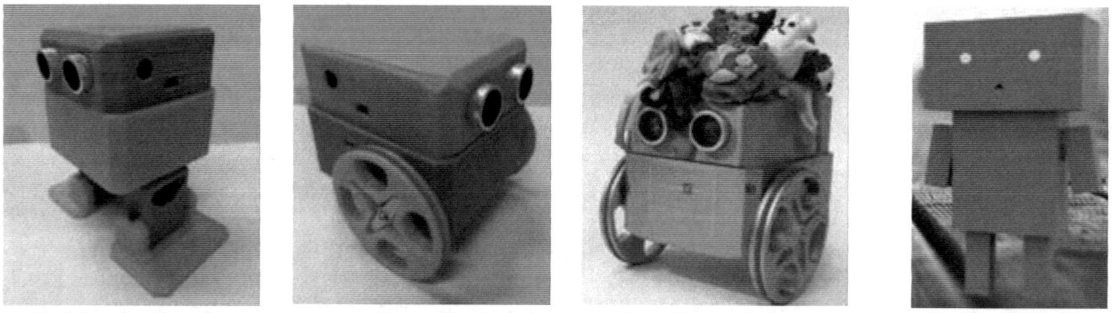

(a) Bipedal Ottobot (b) Wheel Ottobot (c) Clay Ottobot (d) Paper Ottobot

Figure 1. Different types of low cost programmable robots

This study referred to the "Ottobot" to develop a low-cost robot and used the developed robot to assist K12 students to learn programming courses effectively. There are two types of Ottobot, wheel Ottobot and bipedal Ottobot as shown in Figure 1(a) and 1(b). In 2019, author opened a course for the university students with design background. Most of deign students don't have any experience about programming. The course was used to train students to design and build Ottobot-like robot. Figure 1(c) and (d) are students' works and they all are functional. The students utilized their creative talents to design robots' shapes as well as applied their programming skills to make robots work functionally and meaningfully. Currently, 3D printer like FDM are the most common machine used by students, because its filaments are cheap and its fast print rate can support them to get their desired parts within a short time. It is their perfect tool to make their designs.

The original components of Ottobot consists of 3D printed parts, Arduino, Arduino extension board, servo motors, ultrasound, and buzzer. Figure 2(a) and 2(b) show the 3D printed parts of the wheel Ottobot and bipedal Ottobot. The wheel Ottobot requires two 360° servo motors, but bipedal Ottobot requires four 180° servo motors. For considering to build a cheaper robot, this paper adopted Arduino Nano, as shown in the left side of Figure 2(c), associated with an extension as the core control board as shown in the right side of Figure 2(c). The extension has 14 I/O pins, 8 analog pins, 6 PWM pins, and 12C extension pins. It can control up to 14 servo motors or sensors. Many sensors can be installed in the extension to enhance the capability of the robot. For example, the ultrasound can make the robot to know obstacles. buzzers, the robot is capable of singing or making a sound. Students are able to design multi-functional robots based on the sensors they have. It can also be seen as smart robots.

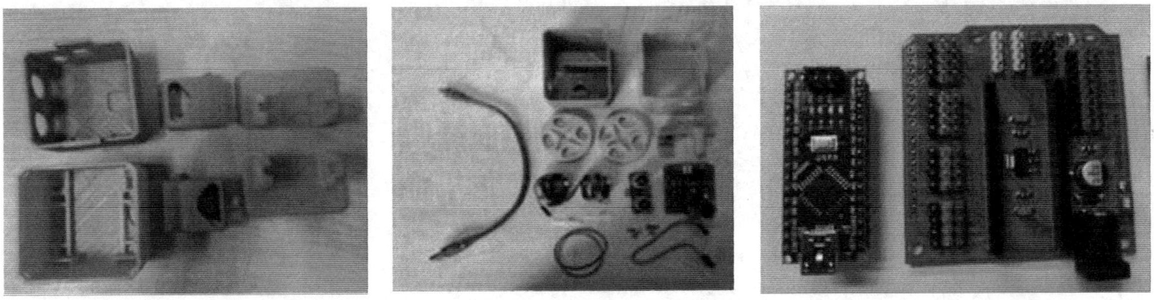

(a) 3D printed parts of Bipedal Ottobot (b) All components of wheel Ottobot (c) Arduino Nano and its extension

Figure 2. Parts of the developed robot

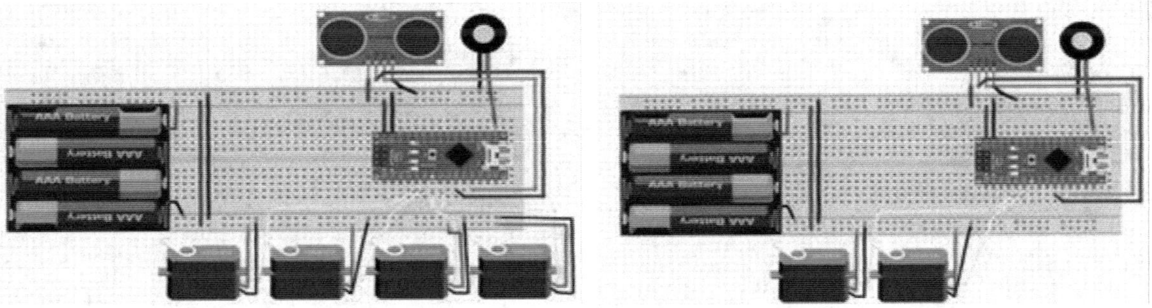

(a) Wire diagram of Bipedal Ottobot (b) Wire diagram of Wheel Ottobot

Figure 3. Wiring diagrams of developed robots

As previous description, bipedal Ottobot utilizes four 180° servo motors to simulate human walking and the wheel Ottobot depends on two 360° servo motors to move like car moving. The motion of each leg of Biped Ottobot has two degrees of freedom(DOF). Comparing to real human, it has less DOF, so it cannot perform complicated motions. However, if the four motors are controlled appropriately, the robot can move forward, backward and perform simple dancing. As for the motion of wheel Ottobot, it is much simpler than that of the bipedal Ottobot. It can only move forward/backward and turn right/left. A Ottobot usually consists of Arduino Nano, ultrasound sensor, buzzer and servo motors.

Figure 3 shows their wiring diagrams. Each of servo motors is installed on specific pin, and then firmware is capable of communicating and controlling motors via the pins.

If users like to control the robot via cell phone or the internet, the devices like HC06 and ESP8266 can support us to develop the remotely controllable robots. The HC06 as shown in Figure 4(a) is a Bluetooth transceiver module, and it can support the device to transit or receive data via the Bluetooth wireless communication. About the ESP8266 as shown in Figure 4(b), it is a Wi-Fi transceiver module, and it can support the device to send and receive data via Wi-Fi. Those two modules are compatible to Arduino Nano. With their assistances, a remotely controllable robot can be developed. Figure 4(c) is the example wiring diagram of ESP8266 associated with Arduino Nano. The total cost of the bipedal robot is less than 500 NTD.

(a) HC-06 Bluetooth transceiver module

(b) ESP8266 Wi-Fi transceiver module

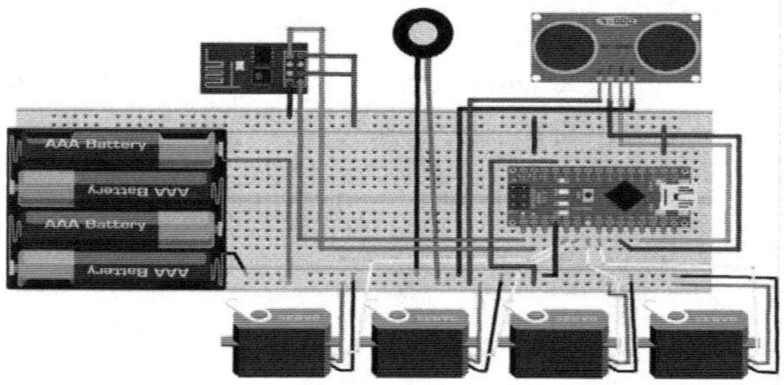

(c) Wiring diagram for ESP8266

Figure 4. Wiring diagram for remoted control

4. Coding software of developed robot

This part is the key to make the developed robot to move with specific tasks. Coding courses are always difficult to K12 students. This paper attempts to offer a better learning module to allow K12 students to accomplish their coding projects easily. By this module, students are able to learn coding skills as well as to test the code by the robots simultaneously. The core control board of the robot is Arduino Nano. The Arduino IDE is the most common application used to code Arduino firmware. However, its programming interface is text-based. It seems not suitable for K12 students. According to this event, VPL was applied in this study. It allows users to code a program by simply dragging graphical code elements, e.g. Scratch. The Scratch is a famous VPL for children. Most of elementary schools promotes Scratch courses for the students whom are interested in coding. There are several available VPL for coding Arduino firmware, such as Scratch for Adruino(S4A), Ardublock, and mBlock. The above VPL are quite similar. This paper introduced mBlock to assist design students to program their robot motions as shown in Figure 5. The mBlock was developed by makeblock, but it is free to anyone. The makeblock is a commercial robotics company. The company focuses on developing and selling programmable robots for kids, its goal is to help kids to learn coding via its robots. In order to let children easy to handle coding events, they developed a block-based coding platform called mBlock. Its platform features are similar to Scratch as shown in Figure 4(a). Users can switch the coding platform to either the Scratch mode or the Arduino mode. The Arduino mode allows users to generate the Arduino code from the original block-based code. In addition, it supports many compatible Arduino boards. Its operation is convenient and friendly to instructors and new learners.

The mBlock provides a useful extension manager that allows users to create their own command blocks for their specific usages. For example, the extension with the name "Robot" collects several command blocks for handling robot motions. In this project, author had developed a group of command blocks to handle the developed Ottobot. The educators who are interested in developing STEAM courses are able to create their own blocks for their specific courses. The mBlock does provide necessary and useful functions to create better code learning environment. Figure 6 shows the example codes demonstrating specific robot motions. Every robot motion can be generated based on specific robot hardware features. The code as shown in Figure 6(a) demonstrates that a bipedal robot preforms a specific gesture like "confused" when it meets an obstacle. The "gesture" block contains several "organized" robot gestures to offer students easy to develop a specific robot motion task. With the assistance of the blocks, students are able to easily program robot codes as well as to debug whether their codes and program logic are correct or not via real robot gestures. Figure 6(b) shows a series of blocks to enable a wheel robot to move forward/backward by controlling two 360° servo motors. Unlike 180° servo motor, 360° servo motor always rotates continuously. It is not capable of rotating with specific angle. Two servo motors with wheels were installed on both sides of robot, so they must be rotated in opposite direction to obtain one direction movement. In this project, students can learn not only coding skills but also robot mechanism.

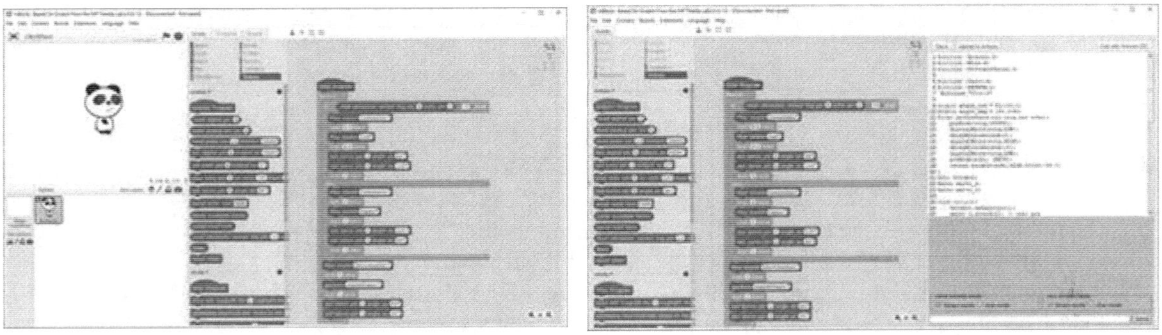

(a) mBlock with Scratch mode (b) mBlock with Arduino mode

Figure 5. Block-based programming interface

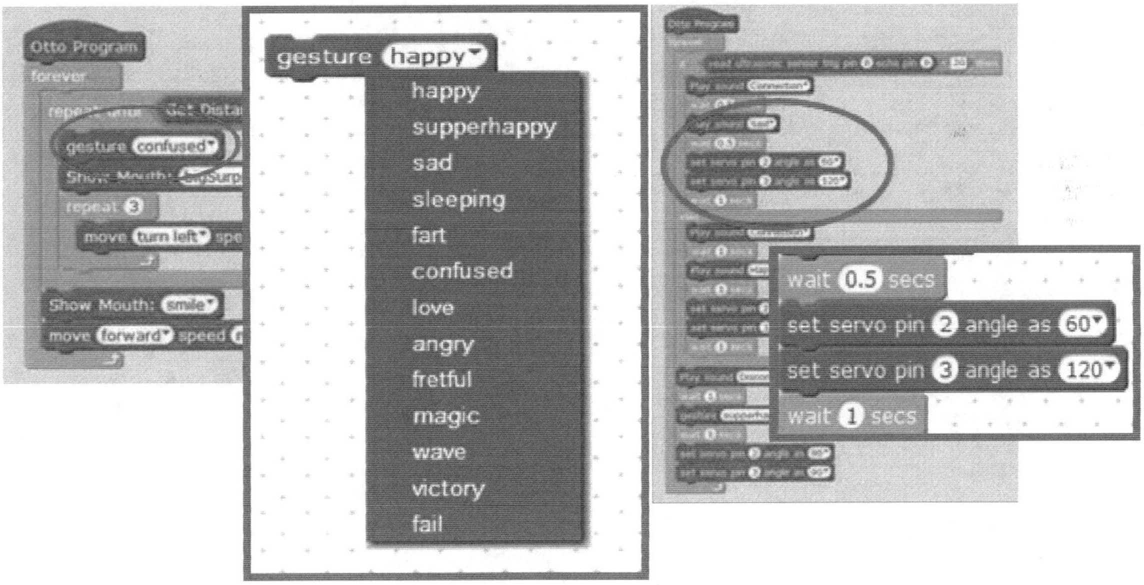

(a) Block-based codes for bipedal robot (b) Block-based codes for wheel robot

Figure 6. Example block codes for the developed robots

5. Conclusion

That people are educated to own a completed trainings and disciplines is a requirement for satisfying the needs of Industry 4.0 age. Fortunately, STEAM education is the best solution to fit the need. For considering to create better learning curriculums for K12 students. This paper presents the low cost robot that was developed to assist kids in learning STEAM courses efficiently and happily. The original development of the robot is the open source - Ottobot. The main structures of the robot are 3D printable. In STEAM course, students follow instructors' guideline and their own creativities to design robots' outlooks as well as utilize 3D printer to build the robot parts cheaply. The control unit of the robot is compatible to Arduino. Thus, this robot can also be used to learn about Arduino controlling and its related electronics. Fortunately, the mBlock is compatible with the robot. The mBlock is VPL-based platform, so it is very easy-to-use. Of course, it is definitely suitable for K12 students. With the assistance of the mBlock, instructors can also develop their own command blocks to teach students more efficiently. This paper had demonstrated the motion codes for two different types of robots, bipedal robot and wheel robot. Under STEAM programs, students follow instructors' teaching to build up a robot step by step. They can learn the knowledge of related engineering and technology. After the robot is completed, students can apply their knowledge of computer language to make robots move meaningfully. Robotics is important to current industry. If students can get familiar with it early, it will good the development of high-tech country.

References

[1] White D W 2014 *Florida Association of Teacher Educators Journal* **1**(14) pp. 1-8
[2] Ismail Z 2018 *Benefits of STEM Education," in "K4D Helpdesk Report* (International Development Department / University of Birmingham)
[3] Hamner E and Cross J 2013 *3rd IEEE Integrated STEM Education Conference* (Princeton, NJ, March 9)
[4] Ziaeefard S, Miller M H, Rastgaar M and Mahmoudian N 2017 *Robotics and Autonomous Systems* **97** pp. 40-50
[5] Sullivan A and Bers M U 2018 *Int J Technol Des Ed* **28**(2) pp. 325-346
[6] Eguchi A2016 *Robotics and Autonomous Systems* **75** pp. 692-699
[7] Eguchi A, 2014 *Proceedings of 4th International Workshop Teaching Robotics, Teaching with Robotics & 5th International Conference Robotics in Education* (Padova (Italy), pp. 27-34)
[8] Hughes J M, Arduino A Technical Reference. O'REILLY, 2016

IOP Conf. Series: Materials Science and Engineering 1113 (2021) 012009 doi:10.1088/1757-899X/1113/1/012009

FEM design of proportional solenoid for hydraulic proportional suspension damper

Chin Yi Cheng[1], Jyh Chyang Renn[1], Yi Zhe Xie[1], Yi Kai Huang[1] and Chun Bin Yang[2]

[1]Department of Mechanical Engineering, National Yunlin University of Science & Technology, Yunlin , Taiwan
[2]Metal Industries Research & Development Center, Kaohsiung, Taiwan

E-mail: zezheng@yuntech.edu.tw

Abstract. In this paper, the FEM simulation package FLUX2D is utilized as a tool to design a proportional solenoid for the hydraulic proportional suspension damper. The hydraulic proportional suspension damper can be used in vehicle semi-active suspension system due to its adjustable output damping force. In addition, the proportional solenoid plays the most significant role to achieve such a variable output damping force. From previous literatures, three different structures of proportional solenoid will be discussed in this paper. The first one utilizes the stepped conical armature design. Its advantages include low cost, easy manufacturing. However, the limited available linear stroke as well as the poor linearity are its major faults. The second structure uses a special tube consisting of three metal rings that are welded together. Between two magnetic steel rings, there is a non-magnetic copper ring serving to guide the magnetic flux and to produce a linear force/stroke relation. Clearly, higher cost and complex manufacturing process are its major disadvantages. However, larger available linear stroke and good linearity are two promising advantages. In this paper, the third design of proportional solenoid is adopted. The construction of the proportional solenoid is divided into several parts. After assembling these parts together, a ring of air gap with a well-designed geometrical shape exists which can be used to replace the non-magnetic copper ring. Without the copper ring, the expensive and complex welding process can be omitted. Meanwhile, the large available linear stroke as well as good linearity are both preserved. Finally, an experimental test device for measuring the force/stroke relation of proportional solenoid is implemented in this study. It is observed that the simulation results agree quite well with the experimental results.

1. Background and Introduction

As shown in Figure 1, the suspension system is classified as a passive, semi-active, and active suspension system, according to its ability to add or extract energy [1]. Among these three systems, the passive suspension is perhaps the most commonly used one and may be found in most vehicles. However, the passive suspension has no means of adding external energy to a system because it contains only passive elements such as a damper and a spring. For the active suspension, the obvious advantage is that it can supply energy from an external source and generate force to achieve the optimal desired performance. However, the inevitable high cost is its major fault. By contrast, for the semi-active suspension, it is possible to continuously vary the rate of energy dissipation using a controllable damper. Though it can only provide moderate performance, the low-cost configuration is its major advantage. For these reasons, the semi-active suspension using the hydraulic proportional suspension damper is investigated in this study. Figure 2 shows the 2-stage hydraulic proportional suspension damper discussed in this paper. It is observed that a proportional solenoid at the pilot-stage is used to determine the damping force output of the hydraulic proportional damper. Two significant features of the proportional solenoid are the quite linear force/stroke characteristic and the proportionality between the input current and the output armature force. Thus, it is possible to vary the damping force output by changing the input excitation current. In this paper, the FEM simulation package FLUX2D is chosen as a tool to design the proportional solenoid for the hydraulic proportional suspension damper. By surveying some previous literatures [2-4].

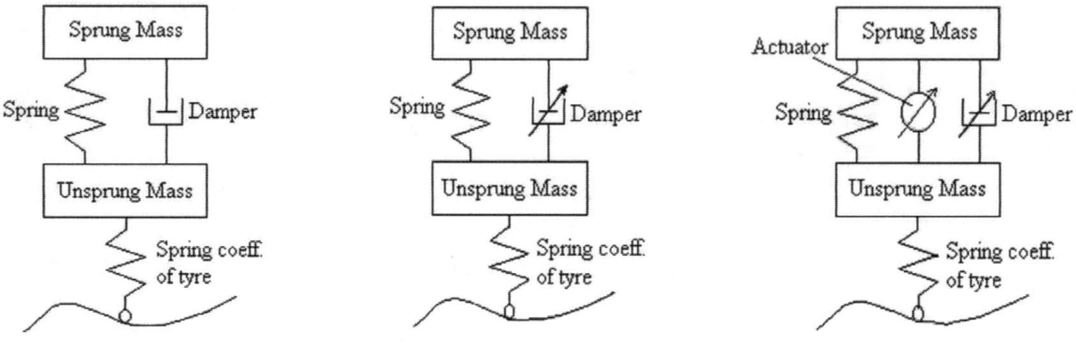

(a) Passive suspension (b) Semi-active suspension (c) Active suspension

Figure 1. Classification of vehicle suspension system

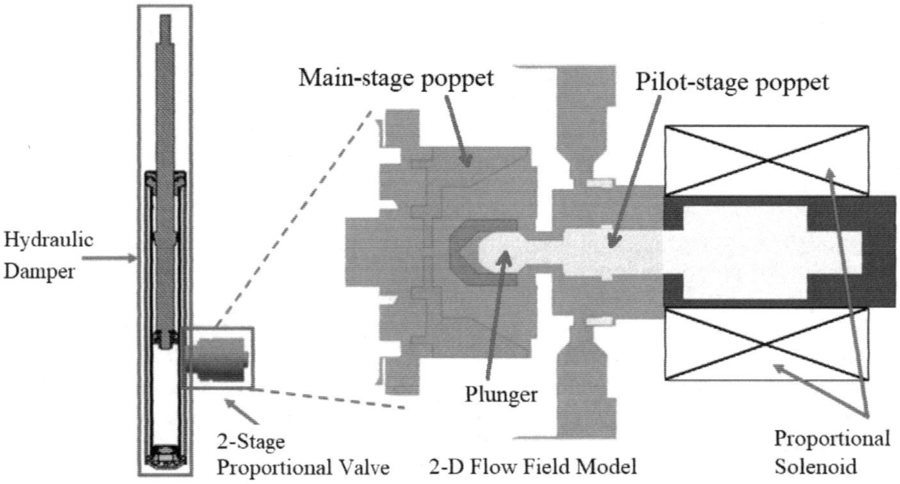

Figure 2. Scheme of the 2-stage hydraulic proportional damper

2. Three Different Kinds of Proportional Solenoid

Three different structures of proportional solenoid will be discussed in this paper. The first one utilizes the stepped conical armature design as shown in Figure 3. Its advantages include low cost, easy manufacturing. However, the limited available linear stroke as well as the poor linearity are its major faults. The second structure uses a special tube consisting of three metal rings that are welded together as shown in Figure 4. Between two magnetic steel rings (No.: 2, 6), there is a non-magnetic copper ring (No.: 5) serving to guide the magnetic flux and to produce a linear force/stroke relation. As the armature (plunger) moving, axial space increase as well as radial space decrease, this will make resultant force keep in same value, Figure 5 shows the relation between axial force, radial force and resultant force of proportional solenoid [5]. The axial force (Fa) of the base and the radial force (Fr) of the flange work with each other in specific section so that the solenoid force is equal to their sum. Therefore, the proportional electromagnet can be directly controlled, and the solenoid force depends on the operating current in the linear zone. Clearly, higher cost and complex manufacturing process are its major disadvantages. However, larger available linear stroke and good linearity are two promising advantages. In this paper, the third design of proportional solenoid is adopted as shown in Figure 6. The construction of the proportional solenoid is divided into several parts. After assembling these parts together, a ring of air gap with a well-designed geometrical shape exists which can be used to replace the non-magnetic copper ring. Without the copper ring, the expensive and complex welding process can be omitted. Meanwhile, the large available linear stroke as well as good linearity are both preserved. In the following, the FEM design of the proportional solenoid using FLUX 2D will firstly be outlined.

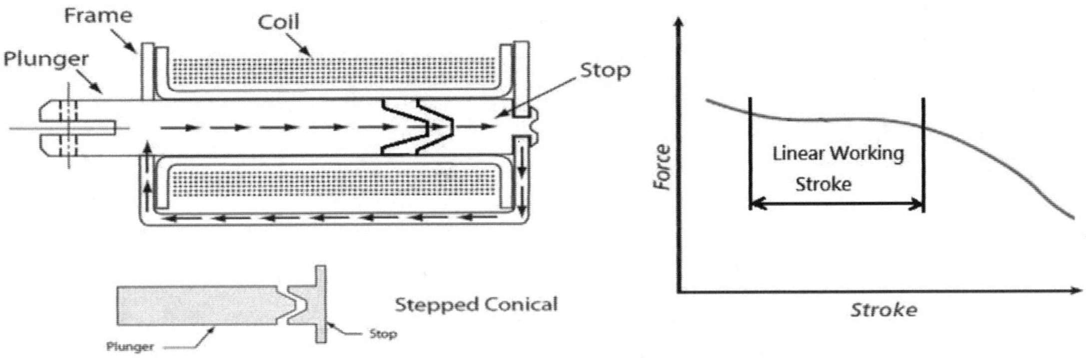

Figure 3. Proportional solenoid with the stepped conical armature design [2]

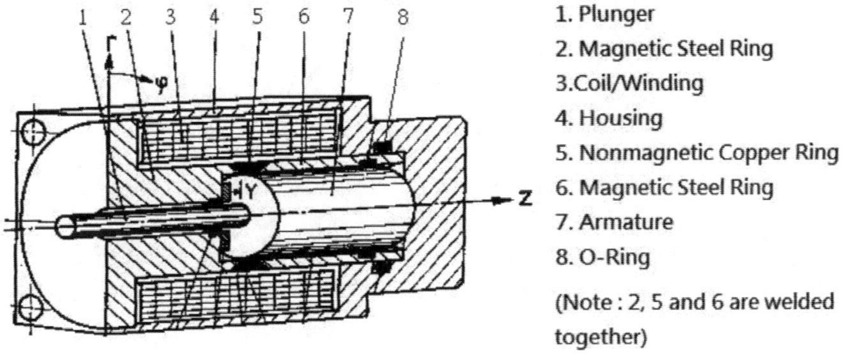

Figure 4. Proportional solenoid with a tube consisting of three metal rings welded together [3]

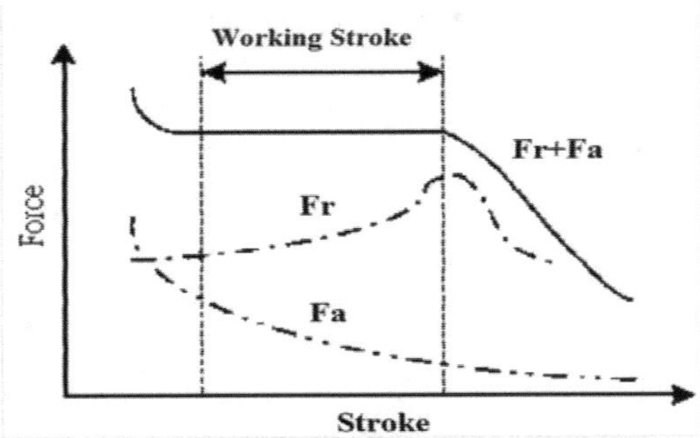

Fr: Radial Fore Fa: Axial Force Fr+Fa: Resultant Force

Figure 5. Radial Force and Axial Force of Proportional Solenoid [6]

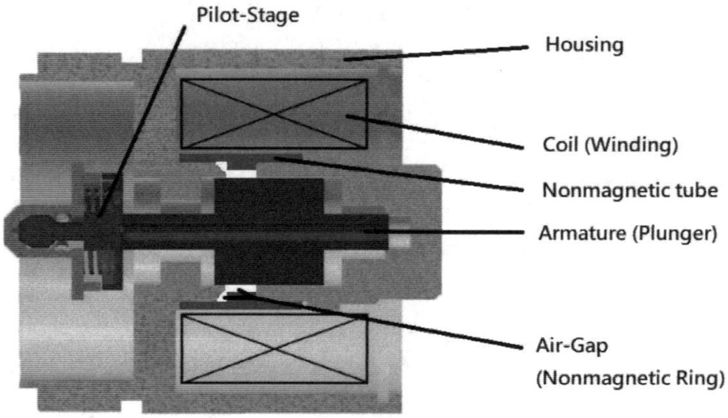

Figure 6. Proportional solenoid with the ring of air gap [4]

3. Designing the Proportional Solenoid Using Flux2D

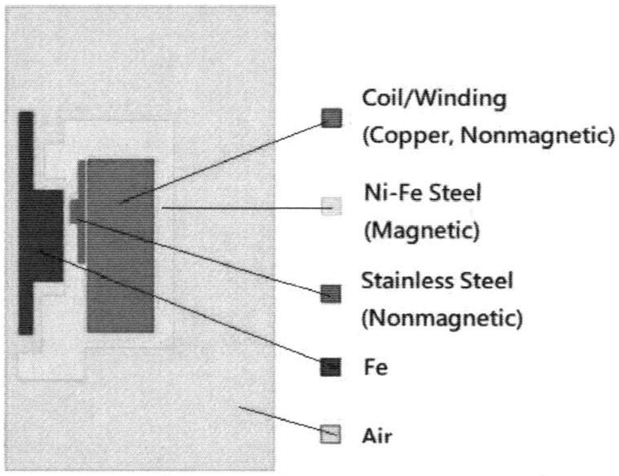

Figure 7. Construction of the study domain showing different materials

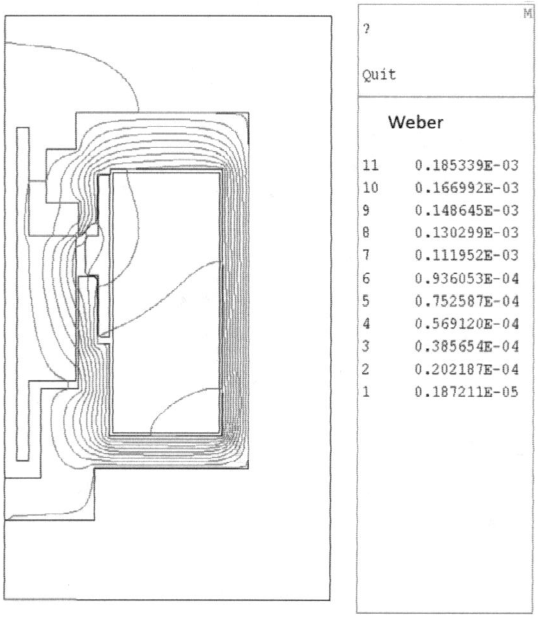

Figure 8. Flux distribution diagram of the proportional solenoid
(Input current: 2.0 A, Stroke: 0.5 mm)

In this paper, the commercial FEM software package FLUX 2D is used to design the electro-mechanical proportional solenoid. Generally speaking, conventional design approach based on equivalent circuit method and trial-and-error are time-consuming [6]. However, the introduction of FLUX 2D helps to test new ideas quickly and design a successful prototype rapidly. First of all, according to the design dimension of the proportional solenoid shown in Figure 6 the construction of the study domain for the proportional solenoid by showing different materials is shown in Figure 7. Since the geometry of the proportional solenoid is axis-symmetric, a two-dimensional cylindrical coordinate model of the magnetic field is established for the quantitative analysis of the armature's output force. Then, the FLUX 2D's automatic mesh points generator accomplishes the meshing in finite elements for the proportional solenoid. Before the simulations, the value of the excitation current as well as the magnetization (B-H) curves for different magnetic or nonmagnetic materials must be given in advance. The former is the known input signal and the latter can be found in the data sheets of corresponding magnetic or nonmagnetic materials. In addition, the boundary condition assuming all the field lines are perpendicular to the boundary is applied throughout the simulations. For the proportional solenoid, the maximal saturation flux density of the Ni-Fe steel permanent magnet is set to be 2.9 Tesla. The number of coil turns is 500 and the maximal external excitation current is set to be 2.0 A. After simulations, as an example, the flux distribution as well as the flux density of the proportional solenoid for the input current of 2.0 A and at the stroke of 0.5 mm can be derived as shown in Figure 8 and Figure 9, respectively. From the numerical table shown in Figure 9, the average flux density can be determined. Consequently, the static force/stroke relation for a given excitation current can be numerically derived. Similar simulations may be repeated for different input currents and at different strokes of the armature (plunger). Figure 10 shows the numerical static force/stroke characteristic curve for the proportional solenoid using four different excitation currents. It is observed that the static force/stroke relation is quite linear within the stroke range from 0.4 mm to 1.4 mm under the error of 15%.

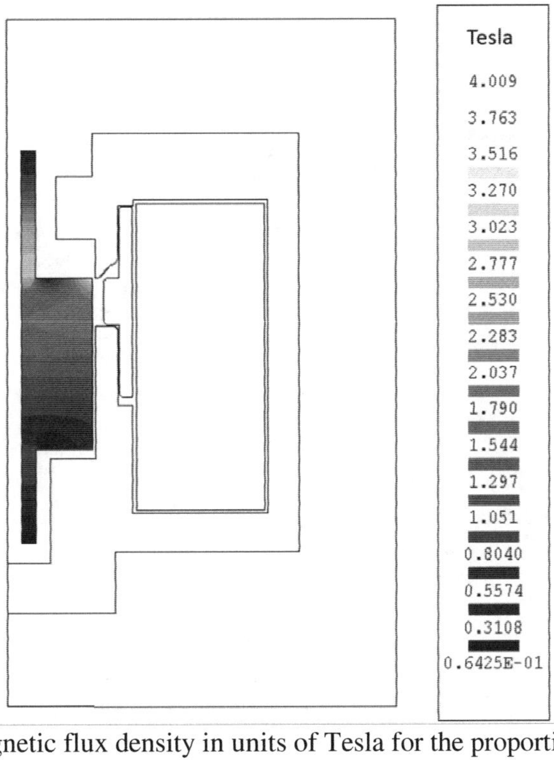

Figure 9. Magnetic flux density in units of Tesla for the proportional solenoid (Input Current; 2.0 A, Stroke: 0.5 mm)

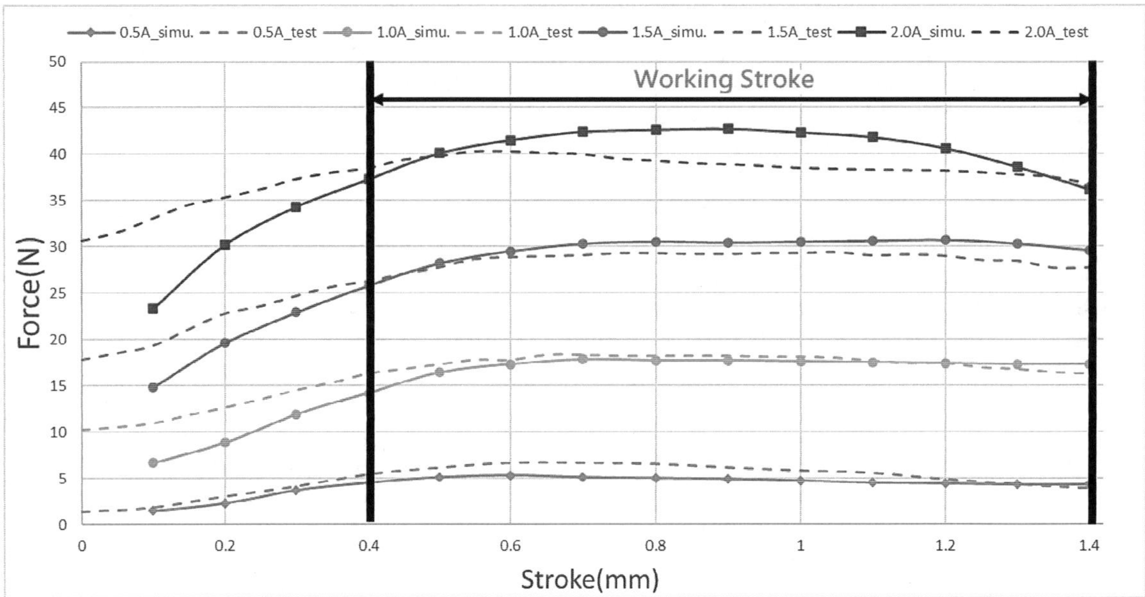

Figure 10. Static force/stroke relation using 4 different excitation currents

The spool movement in a proportional valve is small; a few mm stroke is typical. The valves are therefore very vulnerable to static friction. In order to make sure that the output flow or pressure is in direct proportion to the input current, we normally take the approximate straight-line part on the characteristic curve as the working range for pressure & current or flow & current. This working range is known as linear range for a proportional valve. When choosing a proportional valve, you should choose the one with narrower linear range.

The corresponding static force/current relation is also shown in Figure 11, in which the good linearity between the magnetic output force and the input excitation current is noticeable.

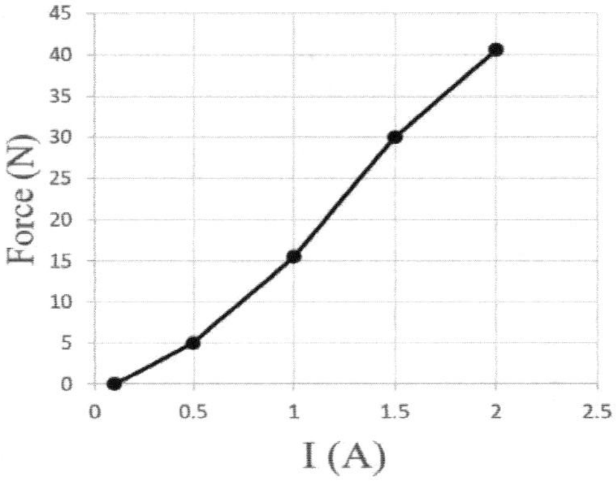

Figure 11. Numerical static force/current relation

4. Experimental Test Device

In this study, an experimental test device is developed for evaluating the performances of the proposed proportional solenoid. The scheme is depicted as shown in Figure 12. An open-loop controlled micro-stepping motor is utilized to control the position of armature (plunger) of the tested linear motor. The rotational angle and speed output of the micro-stepping motor can be derived directly from the number of pulses and the frequency of the generated pulse signal sent to the driver, respectively. In addition, the direction of rotation can be easily controlled by sending a Hi- (5 Volt) or Lo- (0 Volt) signal to one

input port of the driver. Besides, this test device provides a position sensor (LVDT) as well as a load cell for the measurement of the stroke and the output force of the armature (plunger).
In Figure 12, we set tested proportional solenoid at position No. 6 and let the micro-stepping motor goes right after proportional solenoid excitation. The load cell and LVDT device will measure the force and stroke value and send it to computer. Thus, the most important force/stroke relation can be obtained. Finally, the control of the unit as well as the acquisition and processing of the measured data are all integrated in a PC-based LabVIEW software controller, Figure 13 shows the LabVIEW User Interface of experimental device in this study. After the measuring the sample proportional solenoid with the excitation current of 0.5A to 2.0A, same with simulation, the result is shown in Figure 10 and can compare with the result of simulation.

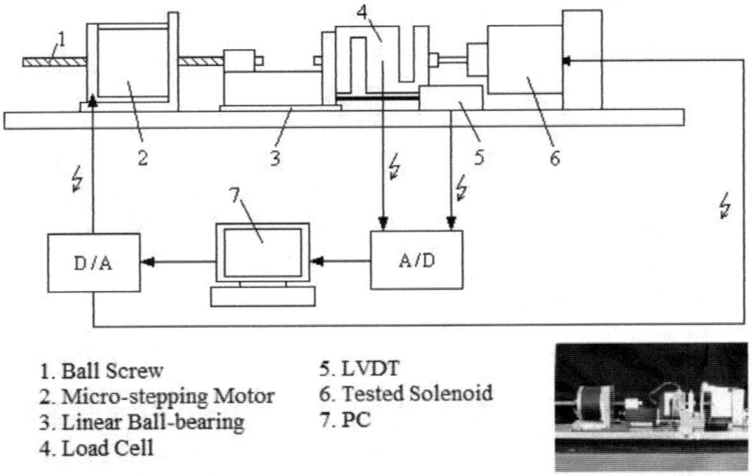

Figure 12. The static force/stroke test device for the proportional solenoid

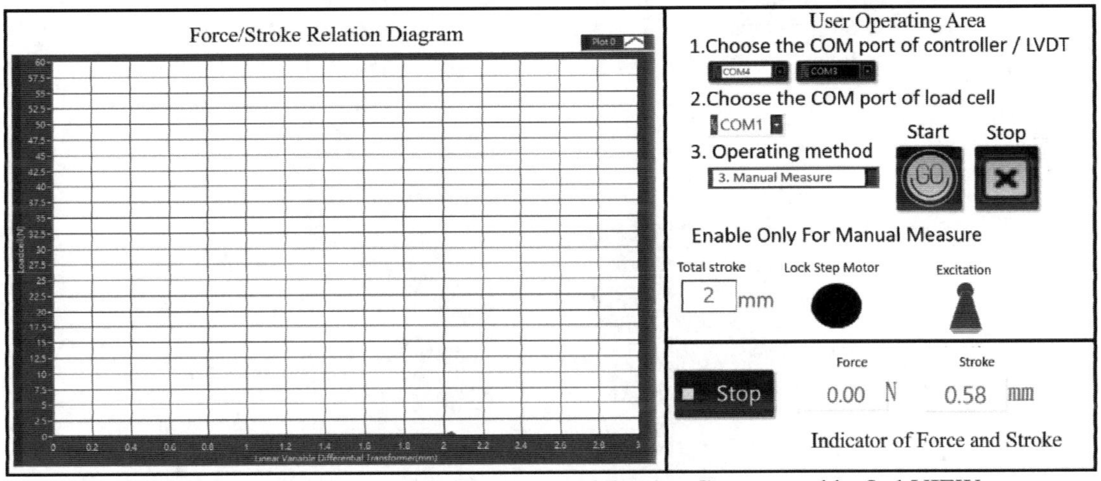

Figure 13. User Interface of Experimental Device Constructed by LabVIEW

5. Experimental Results and Discussion

Using the test device shown in Figure 12, the experimental force/stroke relation can be obtained. It is also observed that the range of linear working stroke is from 0.5 mm to 1.4 mm. Therefore, the average output force of the proportional solenoid within this stroke range can be calculated. Table 1. Comparisons between experimental and simulation shows the comparisons between experimental and simulation average forces for 4 different excitation currents. The minimum error percentage is only 3 % while the maximal error percentage is around 12 %. From engineering point of view, these small deviations are still acceptable.

Table 1. Comparisons between experimental and simulation

Input Current	Measurement(N)	Simulation(N)	Error(%)
0.5(A)	7	6.2	12.9
1.0(A)	18	17.5	2.8
1.5(A)	28	29.1	-3.9
2.0(A)	38	40.4	-6.3

6. Conclusion

In this paper, a proportional solenoid for hydraulic proportional suspension damper is proposed and successfully developed. Three conclusions may be drawn from this research.

1. It is observed that the simulation results agree quite well with the experimental results. The overall percentage deviation between simulation and experimental result is only approximately 6 %.
2. The linear working stroke of the proportional solenoid designed in this paper amounts 0.9 mm. In the real application of pilot-stage design shown in Figure 6, a linear working stroke of 0.5 mm would be sufficient. In addition, the maximal magnetic output force reaches 40 N which meets the design requirement of the pilot-stage. Therefore, it can be concluded that the design of the proportional solenoid proposed in this paper is quite successful.
3. From Figure 10, currently, this study noticeable that the force/stroke relation is not fully satisfactory at the higher excitation current of 2.0A. In the future, however, an optimization study has to be conducted to improve the linearity for the whole range of excitation current. Figure 14 suggests the three most important geometrical parameters for the triangle type control cone. That has to be studied, including the width, W, the length, L, and the angle θ are cone length, con thickness, cone depth, cone width, and the gap between plunger and cone, respectably. The attraction force characteristics of the manufactured solenoid actuator were simulation and experimentally calculated. After optimized the geometric shape of the proportional solenoid, the force/stroke relation may become better. [7-8]

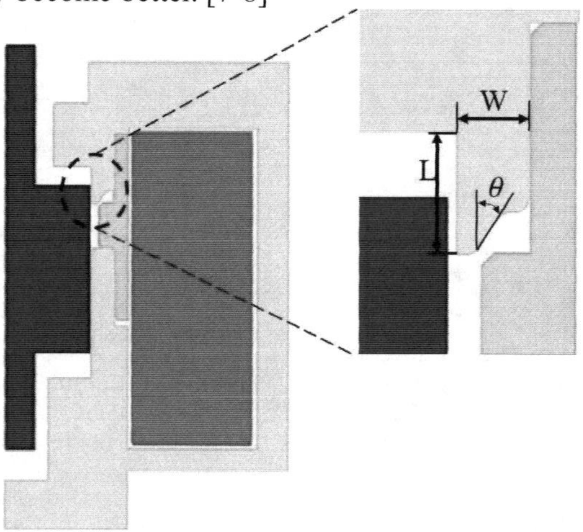

Figure 14. Three chosen geometrical parameters for future optimization study

Acknowledgments

The financial supports from the Metal Industries Research & Development Center, Taiwan and the Ministry of Science and Technology under grant number are both greatly appreciated.

IOP Conf. Series: Materials Science and Engineering 1113 (2021) 012009 doi:10.1088/1757-899X/1113/1/012009

References

[1] Renn, J. C. and Chen, H. M., "Design of a Novel Semi-active Suspension for Motorcycles with Fuzzy-Sliding Mode Controller," J. of the CSME, 26(3), pp. 287-296, 2005.

[2] Backe, W., Servohydraulik, Umdruck zur Vorlesung, RWTH Aachen, 6. Auflage, Germany, 1992.

[3] Backe, W., Steuerungs- und Schaltungstechnik II, Umdruck zur Vorlesung, RWTH Aachen, 4. Auflage, Germany, 1993.

[4] Renn, Jyh-Chyang, Jhan, Cian-Wei, "A Full-Digital Quasi-Proportional Force Output Solenoid," Proceedings ohe 14th International Conference on Automation Technology (Automation 2017), Paper ID: 1006, Kaohsiung, Taiwan, 2017.

[5] He Jen-Yu, Designing the Shape of Poppet for a Proportional Switching Solenoid, National Yunlin University, Master Thesis, Taiwan, 2016.

[6] Renn, Jyh-Chyang, Xu, Zheng-Liang, State-space Modeling and FEM Design of Moving-coil Transducer for Fluid-technical Proportional Valves," J. of the CSME, Vol. 24, No. 2, pp. 119-125, 2003.

[7] So-Nam Yun, Young-Bog Ham, Jung-Ho Park, Attraction force improvement strategy of a proportional solenoid actuator for hydraulic pressure control valve, 12th International Conference on Control, Automation and Systems, Engineering, 2012.

[8] So-Nam Yun, Young_Bog Ham, and Jung-Ho Park, New Approach to Design Control Cone for Electro-Magnetic Proportional Solenoid Actuator, IEEE/ASME International Conference on Advanced Intelligent Mechatronics, Taiwan, 2012.

IOP Conf. Series: Materials Science and Engineering 1113 (2021) 012010 doi:10.1088/1757-899X/1113/1/012010

Flow characteristics and heat transfer performance of a multi-layer winding hose system: theory and experimentation

Haiyan Qiang[1], Wanli Li[1], Yanran Wang[1], Guanyuan Li[1] and Junqi Xu[2]

[1]School of Mechanical Engineering,Tongji University, Shanghai, China
[2]National Maglev Transportation Engineering R&D Center,Tongji University, Shanghai, China

E-mail: cnlwl@tongji.edu.cn

Abstract. This paper investigates the flow characteristics and heat transfer performance of a multi-layer winding hose system (MLWH) that is widely used in double-wheel trench cutters. Numerical analysis with Fluent 15.0 was performed to investigate the laminar fluid flow characteristics for the multi-layer winding hose system, especially for the study the effect of the Reynolds number and radius of curvature on the fluid flow. The radius of curvature of the multi-layer winding hydraulic hose was gradually increased in the range of 1 to 3.The Reynolds number varied between 167 and 1510. According to numerical analysis, the curvature-induced centrifugal force is found to have a significant influence on pressure drop and heat transfer. An experimental platform was presented to simulate the working conditions for the multi-layer winding hose system and collect experimental data for comparison with numerical data. The results show that the flow characteristics and heat transfer performance of the multi-layer winding hose system is strongly influenced by the Reynolds number and radius of curvature.

1. Introduction

A curved piping structure, with a compact structure and better heat transfer performance, is widely used in industrial engineering systems such as refrigerating systems[1], solar energy systems[2], processing-industry systems[3-4], and engineering machinery systems[5]. In engineering machinery systems, high requirements exist for the length and curvature of the piping system; hoses which are commonly utilized a curved piping structure by winding the hoses onto a reel owing to their flexibility. Compared with straight piping structures, secondary flow will occur in curved structures and a curvature induced centrifugal will complex the kinetic energy, heat transfer and mass transfer problems of the working fluid in the curved structures.

In this paper, the hydraulic system of a double-wheel trench cutter is taken as an example. To accomplish the formed operations, multi-layer winding hoses are presented in practice. The MLWH is a kind of common curved pipe widely used in various industrial and engineering applications. As we all know, in the stream-line motion of a fluid in a curved pipe, the primary motion along the line of the pipe is accompanied by a secondary motion in the plane of the cross-section[6]. One of the special features of multi-layer winding flow can be recognized as secondary flow, which is formed with centrifugal force due to the curvature effect[7].

Ciofalo *et al.*[8] numerically investigated the gravitational and centrifugal buoyancy in helical coils. Akbaridoust *et al.*[9] studied the laminar, steady state flow in helical coils at constant wall temperature. Sheeba *et al.*[10] analyzed the heat transfer and flow characteristics in a helical coil and presented a correlation to predict the Nusselt number. Saleh *et al.*[11] analyzed the geometrical properties (different coil pitches and coil diameters) of helical wire turbulators and carried out a related experiment for Dean numbers more than 100. Dastmalchi *et al.*[12] numerically investigated the geometrical parameters and compared the predicted values to the experimental results. Wen *et al.*[13] studied the heat transfer performance of aviation kerosene in a vertical helical tube experimentally at supercritical pressure. The centrifugal secondary flow was proved to be the key factor for heat transfer enhancement. Kurnia *et al.*[14] evaluated heat transfer performances of helical coiled tube with different cross section (square, ellipse and circular). Naghibzadeh *et al.*[15] quantitatively studied the heat transfer performance and pressure drop problems for a helical pipe with a flattened cross section in several cases (with varying cross section geometry and Reynolds numbers. Omidi *et al.*[16] investigated the effects of a helical coil with a lobe cross section, by analysing the physical and geometrical parameters in a helical coil heat exchanger.

However, with the development of engineering machinery, the demands for an inconstant radius of curvature have increased.

Yoo *et al.*[6] performed a numerical analysis for a spiral coiled tube heat exchanger. The curvature effects on the pressure drop and heat transfer were found and the relationship among the friction factor, Nusselt number and Dean number were also discussed. Patil *et al.*[17] introduced a new dimensionless number R to describe the heat transfer phenomenon in helical pipes. Correlations were developed for thermal systems with spiral coils. Kurnia *et al.*[18] performed a numerical investigation of the heat transfer performance of a laminar non-Newtonian fluid with different configurations of square cross section. In addition, more attention was given to in-plane spiral type and conical spiral type structures. Khoshvaght *et al.*[19] investigated the flow characteristics and thermal performance of spirally-coiled twisted tube, which also proved that the twist-pitch and coil-pitch can influence the thermo-fluidic transport characteristics of the research objects. Sreedhara *et al.*[20] investigated the heat transfer performance of corrugated plate heat exchangers. At the same time, experiments were presented as well.

Although previous literatures for the spiral-coiled type of structure exist, more detailed investigation is required for a multi-layer winding hydraulic hose. Reference[5, 21-22] has already analyzed the heat transfer and flow characteristic of MLWH systems under turbulent flows. However, further research is needed. In this paper, simulations on the flow characteristics of the MLWH system in laminar flow are presented. An experimental platform is presented. Experimental data were obtained to compare the simulation data. The Reynolds numbers and curvatures are proved to effect the flow characteristics and the heat transfer performance.

The rest of this paper is organized as follows. In section 2, the numerical analysis of the MLWH is established. Section 3 designs an experimental analysis of the MLWH system. Simulation results are provided in Section 4. Finally, the conclusions and future works are shown in Section 5.

2. Numerical analysis

2.1. Modelling of a multi-layer winding hydraulic hose

As shown in Fig. 1, a single layer hydraulic hose (marked 1) is winded onto a reel (marked 2) with an Archimedean spiral. The hydraulic hose is controlled by rolling the reel clockwise and anticlockwise. If the reel is rolled clockwise, the free end of the hydraulic hose will be longer as marked by the red arrows in Fig. 1. Otherwise, the free end of the hydraulic hose will be shorter (marked by the green arrows in Fig. 1), if the reel is rolled anticlockwise.

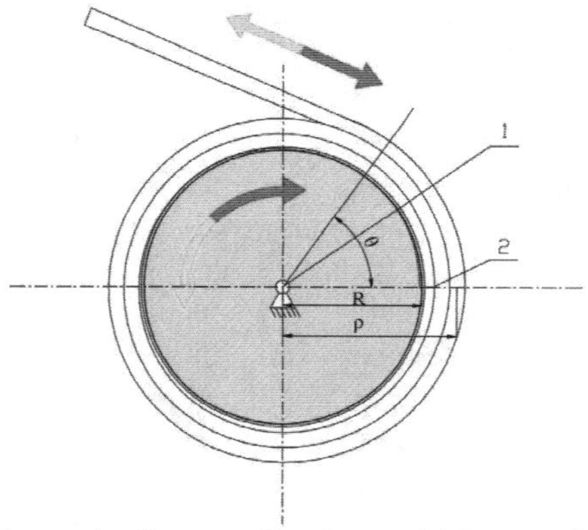

Figure 1. Schematic diagram for the multi-layer winding hoses

According to the mathematical definition of Archimedean spirals, the mathematical model of the MLWH is established under polar coordinates in Equ.1. As shown in Fig. 1, the main parameters in Equ.1 are listed.

$$\rho = a\theta + R \tag{1}$$

where, θ denotes the rotation angle of the reel. d_0 represents the outer diameter of the hose, $a = d_0/2\pi$. R denotes the minimum radius of the reel, where $\theta = 0$.

The radius of curvature is defined as follows:

$$R_i = \frac{1}{k} = \frac{\left(a^2\theta^2 + 600a\theta + 90000 + a^2\right)^{\frac{3}{2}}}{\left|a^2\theta^2 + 600a\theta + 2a^2 + 90000\right|} \tag{2}$$

where k denotes the curvature and $k = \frac{(\rho^2 + 2\rho'\rho' - \rho\rho'')}{(\rho^2 + \rho'\rho')^{\frac{3}{2}}}$.

2.2. *Governing equations*

Owing to the geometric effect of the MLWH systems, the flow characteristics in the MLWH vary considerably with different configurations. The Reynolds numbers are considered as the critical parameters and are set to 1600~2000 in this paper which is determined by the material of the hose. The chosen Reynolds numbers were substituted into Eq. (3) which was presented by Dravid *et al.* [22] and the curvature effects were taken into consideration in the calculation of the critical Reynolds numbers.

$$R_{e_{cr}} = 1300\left[1 + 8.6\left(\frac{d}{2R_i}\right)^{0.45}\right] \tag{3}$$

The critical Reynolds numbers were set within the field of $[2950, 3260]$. Correspondingly, the Reynolds number in the MLWH were set within $[167, 1510]$. Hence, the control equations in laminar flow are listed as follows:

Continuity equation:

$$\frac{\partial \rho}{\partial t} + \frac{\partial u}{\partial x} + \frac{\partial v}{\partial y} + \frac{\partial w}{\partial z} = 0 \tag{4}$$

where u , v and w represent the flow velocities in the axial, radius, and azimuthal directions respectively.

Assuming steady flow for the working fluid in the MLWH and $\frac{\partial \rho}{\partial t}=0$, the continuity equation can be rewritten in the vector form and listed as follows:

$$div(\rho \mathbf{v})=0 \tag{5}$$

where $\mathbf{v}$ denotes the flow velocity. The matrix $\mathbf{v}$ can be rewritten as $\mathbf{v}=(u \quad v \quad w)$.

Momentum equation:

$$\frac{\partial \mathbf{v}}{\partial t}+(v\cdot\nabla)\mathbf{v}=-\frac{1}{\rho}\nabla p+\upsilon\nabla^2\mathbf{v} \tag{6}$$

where $\upsilon=\frac{\mu}{\rho}$, μ denotes the viscosity of the working fluid. ρ represents the density of the working fluid.

2.3. Data reductions

Eq. (7) is presented to describe the relationship between the inlet velocity and the Reynolds numbers.

$$\mathrm{Re}=\frac{\rho u_{in} d_h}{\mu} \tag{7}$$

where ρ represents the density; u_{in} denotes the oil inlet velocity; μ represents the dynamic viscosity; d_h is the hydraulic diameter which is defined as below:

$$d_h=\frac{4A_c l}{A_t} \tag{8}$$

where A_c and A_t denote the frontal surface area and the total surface area of the MLWH, respectively; l denotes the length of the MLWH.

The Nusselt number is defined as below:

$$Nu=\frac{hd_h}{\kappa} \tag{9}$$

where κ is the thermal conductivity; h is the heat transfer coefficient.

$$h=\frac{Q}{A_t(T_w-T_b)^{\odot}} \tag{10}$$

where Q and $(T_w-T_b)^{\odot}$ represent the heat transfer rate and the logarithmic mean temperature difference, respectively, which are written as below:

$$Q=mC_p(T_{out}-T_{in}) \tag{11}$$

where T_{out} represents fluid temperature at outlet; T_{in} represents fluid temperature at inlet

$$(T_w-T_b)^{\odot}=\frac{\Delta T_{w-in}-\Delta T_{w-out}}{\log\left(\frac{\Delta T_{w-in}}{\Delta T_{w-out}}\right)} \tag{12}$$

In the equations above, m is the mass flow rate; c_p denotes the specific heat capacity; T_w represents the average wall temperature for the MLWH; $T_{w\text{-}in}$ represents the inlet wall temperature for the MLWH ; $T_{w\text{-}out}$ represents the outlet wall temperature for the MLWH.

The friction factor is obtained as below:

$$f = \frac{2\Delta p d_h}{\rho u_{in}^2 l} \tag{13}$$

where Δp denotes the pressure drop between the inlet and outlet parts of the MLWH system and is listed as follows:

$$\Delta p = p_{in} - p_{out} \tag{14}$$

2.4. Numerical simulation

The configurations for the MLWH are presented which the numbers of layers winded onto the reel are 1 and 3. The inlet velocities for the MLWH winded onto the reel considered in the current work were set as $0.63m/s$, $1.88m/s$, $3.14m/s$, $4.36m/s$ and $5.65m/s$. The Reynolds numbers for these inlet velocities were 167, 503, 838, 1170 and 1510 correspondingly. To investigate the flow characteristic and thermal performance for the considered cases, numerical simulations were performed. The boundary conditions were composed of the constant mass flow rate at the inlet, no-slip boundary wall conditions and a constant wall heat flux. A second-order upwind scheme was employed to discretize the governing equations. The momentum and continuity equations were connected by the SIMPLE algorithm. The discretized equations were solved by Fluent 15.0 software. Oil was used as the working fluid inside the MLWH.

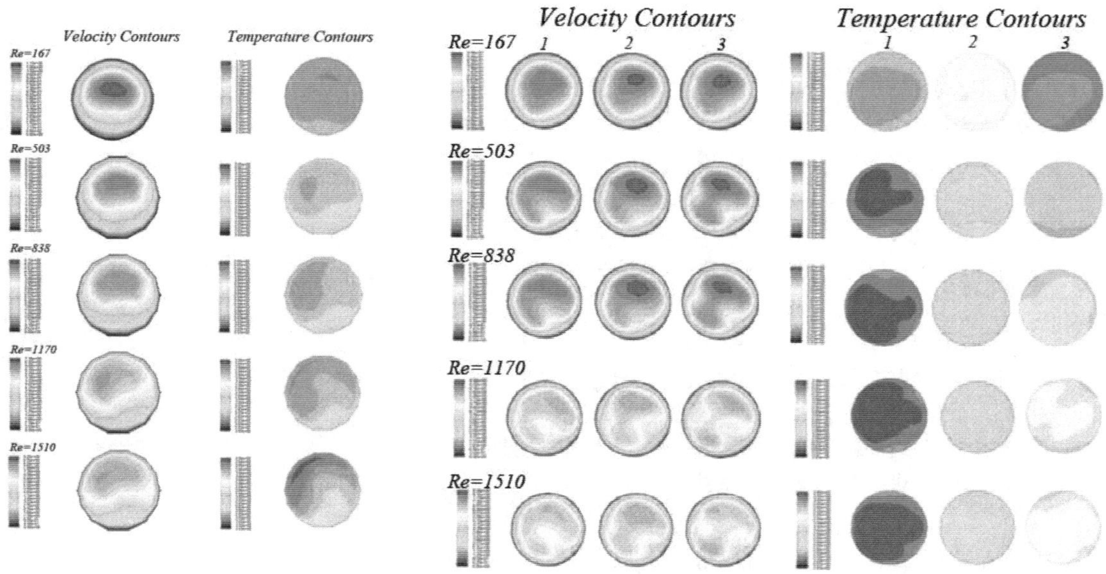

(a) Case1 (b) Case 2

Figure 2. Simulation of the velocity and temperature contours for MLWHs with different configurations and Reynolds numbers

Fig. 2 shows the simulation results for the velocity and temperature contours for MLWHs with different configurations and Reynolds numbers.

Fig. 2(a)-(b) shows the velocity and temperature contours for the MLWH in the cross section with different configurations. It is common that the highest velocity and centrifugal force for the working fluid occurs in the center part of the MLWH at the entrance. However, one can observe from the velocity contour that at a certain distance away from the entrance, the working fluid is pushed away from the center of the MLWH by the centrifugal force which results in a shift of the maximum velocity from the center to the outer area of the hose wall. This is of great importance for enhancement of heat transfer and the shift in temperature towards the outer curvatures.

The velocity contours show that with the maximum velocity shifting to the inner bend of the MLWH, the centrifugal force also points to the inner bend under a constant Reynolds number. When the constant Reynolds numbers are 167, 503 and 838, the directions of the centrifugal force and maximum velocity are pointed towards the inner bend of the MLWH in Case 1-2. However, for constant Reynolds numbers of 1170 and 1510, the flow state in the most of the cases becomes turbulent, which makes the flow characteristics much more complex; these are not studied further in this paper.

As shown in the temperature contours for each case, the temperature of the working fluid in the MLWH varies little in Case 1. In Case 2, the temperature of the working fluid varies in different layers of the MLWH. At the innermost part of the MLWH for each case, the temperature of the working fluid varies lightly, while temperature at the outermost part varies widely. When the Reynolds number increases, the temperature increases correspondingly, especially at the outermost part.

3. Experimental analysis

3.1. Experimental principles

A multi-layer winding hose testing system is proposed according to a double-wheel trench cutter by the similarity theory. The scheme for the experiment is shown in Fig. 3.

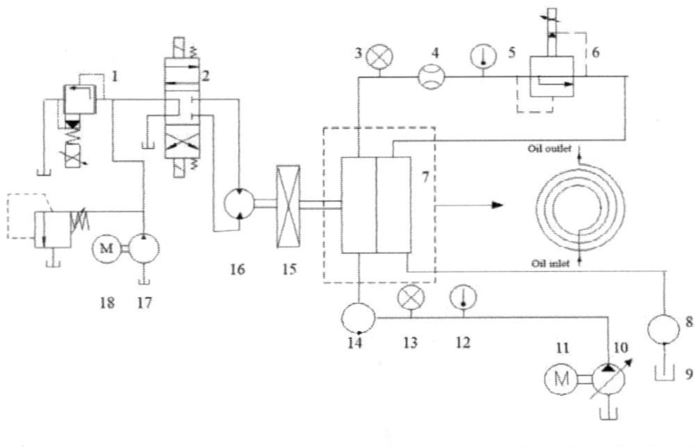

Figure 3. Scheme of the experiment: (1)Pump station proportional relief valve; (2) Pump station electromagnetic directional valve; (3)Outlet pressure sensor; (4) Flow sensor; (5) Output temperature sensor; (6)Output proportional relief valve;(7)Multi-layer winding hoses; (8)Output swivel joint; (9) Oil tank; (10) Main pump; (11) Main motor; (12) Input temperature sensor; (13) Input pressure sensor; (14) Input swivel joint; (15) Reducer; (16) Pump station hydraulic motor; (17) Pump; (18) Pump station motor

The MLWH testing system is composed of a sensor group, MLWH testing section and data collection system.

The working fluid (oil) is stored in a tank. The main motor drives the main pump which pumps the working fluid to the whole MLWH testing system. The working fluid first flows through the input sensor group at the exit of the main pump, which is used to measure the pressure and temperature of the inlet working fluid. Then, the working fluid flows through the MLWH testing section, which is composed of a hydraulic hose with an inner diameter of 13mm and is mainly divided into a reel system and pump station system. In the reel system, the MLWH testing section is set vertically on the reel. In the pump station system, the pump station is used to drive the reel to the desired layers. An output sensor group is set at the end of the MLWH testing section to measure the output pressure, temperature and flow rate of the working fluid. After the output sensor group, a proportional relief valve is set to prevent damage to the whole system due to high pressure. Finally, the working fluid flows into a symmetric MLWH before returning to the tank.

3.2. Experimental apparatus

The experimental apparatus is presented according to the experimental principles. The output sensor group is composed of a temperature sensor, flow rate sensor and pressure sensor. The proportional relief valve is utilized to ensure the safety of the whole hydraulic system. The input sensor group is composed of a pressure sensor and a temperature sensor. The reel onto which the multi-layer winding hoses is winded. The data collecting system is composed of a TTControl screen, an IFM controller and a personal computer.

4. Results and discussion

To investigate the thermal characteristic of the MLWH with different configurations, the average heat transfer coefficient values are evaluated against the Graetz number in Fig. 4. The outcomes demonstrate that with an increasing number of layers for the MLWH, the heat transfer coefficient becomes higher. In Fig. 4, the obtained simulation and experimental data are compared. It is clear that the experimentally obtained data are much higher compared to the data obtained by simulation. This is mainly because of the complexity of the experimental conditions, while the data obtained by simulation were calculated for ideal conditions. Actually, the thermal improvement in the MLWH is attributed to the geometries (i.e. curvature).

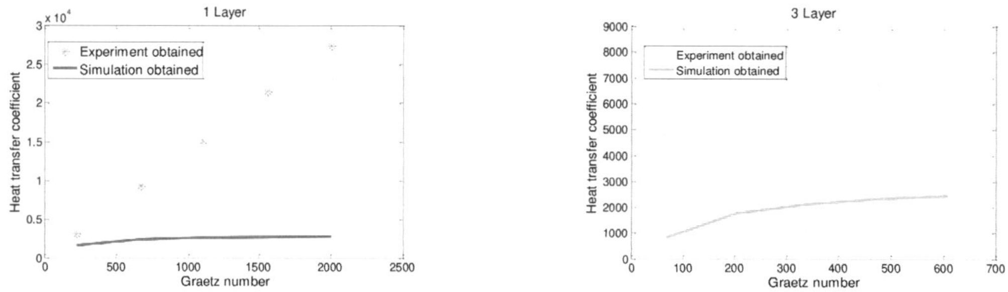

(a) 1 layer (Case 1) (b) 3 layers (Case 2)

Figure 4. Heat transfer coefficient-Graetz number

Fig. 5 shows a comparison between the data predicted by Eq.(9) and the simulation data obtained by Fluent simulations. It is clear that the data predicted by the equation agree with the simulation data while the deviation is greater than that expected through the whole range of the Graetz numbers($G_z = \frac{D_H}{L} R_e P_r$). Hence, further research should present correction equations that can be used to improve the accuracy of the predicted data.

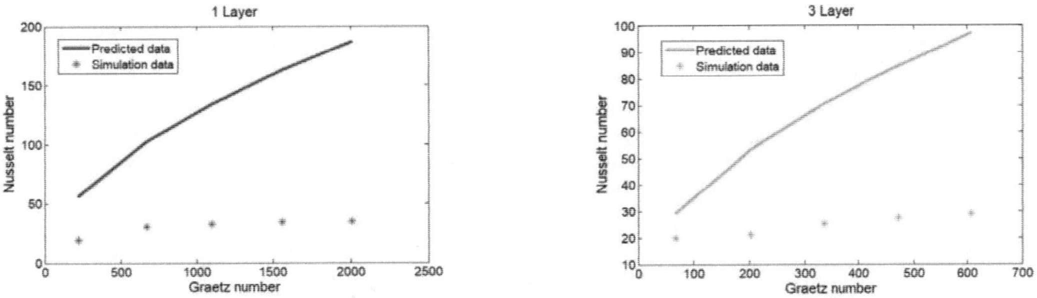

(a) 1 layer (Case 1) (b) 3 layers (Case 2)

Figure 5. Nusselt number-Graetz number

The frictional factor measures the pressure drop effect in the MLWH system. The relationship between the frictional factor and Graetz number is shown in Fig. 6. The data are obtained from the experiment. As depicted in Fig. 6, as the Graetz number increases, the frictional numbers decrease.

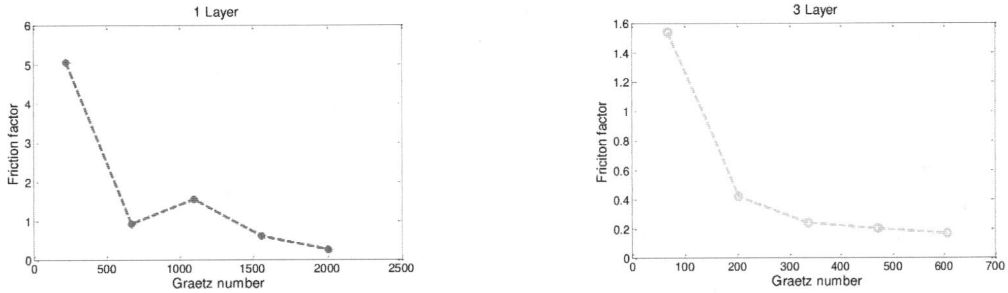

(a) 1 layer(Case 1) (b) 3 layers (Case 2)

Figure 6. Frictional factor-Graetz number

An analysis of the pressure drop for all the cases is demonstrated in Fig. 7. According to Fig. 7 (a)-(b), it is obvious that with increasing Graetz numbers, the pressure drop in the system increases at first before decreasing. For increasing number of winding layers for the MLWH, the pressure drop becomes lower. The reason for this is related to the curvature of the MLWH and the flow state of the working fluid. It is clear that the experimental data are much higher than those obtained from simulation. The situations considered in simulation are simpler than the experimental situations due to simplification of the complex flow state and external disturbances.

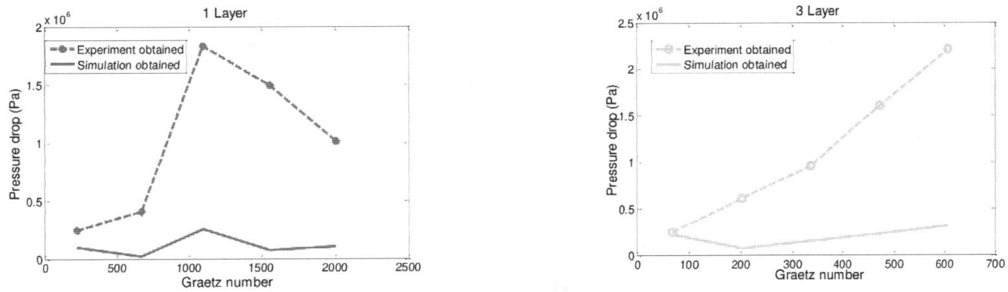

(a) 1 layer(Case 1) (b) 3 layers (Case 2)

Figure 7. Pressure drop-Graetz number

5. Conclusions

In this paper, the thermal performances and the flow characteristics of a MLWH system was studied theoretically and experimentally. Different curvatures and Reynolds numbers for the MLWH system were studied by utilizing oil as the working fluid. The conclusions are summarized as follows:

(1) The simulations and experiments all involved laminar flow. The experimental data were compared to simulations. There were many uncertainties and disturbances (not concerned) in the experiment, which led to a great deviation in the data obtained compared to simulation. More attention will be paid to future researches on this topic, with more factors taken into consideration.

(2) The simulation results showed that both the curvature and Reynolds numbers can influence the flow state of the MLWH system. The working fluid was pushed to the inner bend of the MLWH by interaction between gravity and the centrifugal force. The effect of curvature was proved to be stronger than that of the Reynolds number.

(3) The experimental results for the MLWH system showed that the curvature induced centrifugal force can influence on the heat transfer performances and pressure drop for the MLWH system.

(4) The Nusselt number utilized in this paper should be improved and future research works will focus on correction of the Nusselt number.

Our future research activities will focus on the structure-fluid interaction of the MLWH and the time delay problems in the MLWH systems.

References

[1] Kowalski K and Downarowicz D 2019 *Chemical and Process Engineering-Inzynieria Chemiczna I Procesowa* **40** pp 101-114
[2] Khodabandeh E, Safaei R, Akbari S, Akbari A and Alrashed A 2018 *Renewable Energy* **122** pp 1-16
[3] Vashisth S, Kumar V and Nigam P 2008 *Industrial & Engineering Chemistry Research* **47** pp 3291-3337
[4] Joseph X, Francisco R and Ruiquan L 2019 *International Journal of Engineering and Technology Innovation* **9**(3) pp 155-170
[5] Qiang H, Li W, Xu J and Wang Y 2019 *International Journal of Distributed Sensor Networks* **15** p 9
[6] Yoo G, Choi H and Dong W 2012 *Journal of Central South University of Technology* **19** pp 471-476
[7] Dean W 1927 *Mathematika* **4** pp 208-223
[8] Ciofalo M, Arini A and Di Liberto M 2015 *International Journal of Heat and Mass Transfer* **82** pp 123-134
[9] Akbaridoust F, Rakhsha M, Abbassi A and Saffar-Awal M 2013 *International Journal of Heat and Mass Transfer* **58** pp 480-491
[10] Sheeba A, Abhijith C M and Prakash M J 2019 *International Journal of Refrigeration-Revue Internationale Du Froid* **99** pp 490-497
[11] Khorasani S, Jafarmadar S, Pourhedayat S, Abdollahi A and Heydarpour A 2019 *Applied Thermal Engineering* **147** pp 983-990
[12] Dastmalchi M, Arefmanesh A and Sheikhzadeh G A 2017 *International Journal of Thermal Sciences* **121** pp 294-304
[13] Wen J, Huang H R, Fu Y C, Xu G Q and Zhu K 2017 *Applied Thermal Engineering* **121** pp 853-862
[14] Kurnia J C, Sasmito A P, Shamim T and Mujumdar A S 2016 *Applied Thermal Engineering* **102** pp 849-860
[15] Naghibzadeh S K, Hajmohammadi M R and Saffar-Avval M 2019 *Chemical Engineering Research & Design* **146** pp 36-47
[16] Omidi M, Farhadi M and Darzi A A R 2018 *Energy Conversion and Management* **176** pp 236-245
[17] Patil R H 2017 *Experimental Thermal and Fluid Science* **84** pp 144-155

[18] Kurnia J C, Sasmito A P and Mujumdar A S 2014 *International Communications in Heat and Mass Transfer* **57** pp 100-108
[19] Khoshvaght-Aliabadi M, Khaligh S F and Tavassoli Z 2018 *Applied Thermal Engineering* **143** pp 358-375
[20] Sreedhara B R and Surywanshi G D 2015 *International Journal of Engineering and Technology Innovation* **5**(2) pp 99-107
[21] Sun Y, Xu J, Qiang H and Lin G 2019 *Ieee Transactions on Industrial Electronics* **66** pp 8589-8599
[22] Dravid A and Smith K 1971 *AIChE* **17** pp 1114-1122

A DEMATEL approach to select green reverse logistics service providers

Shu Hua Taso[1], Lu Yang[1] and Shih Ming Ou[2]

[1]Business School, City College of Dongguan University of Technology, Guangzhou, China
[2]Department of Logistics Management, Fujian Jiangxia University, Fuzhou, China

E-mail: benao13@gmail.com

Abstract. With the development of science and technology, the problems of environmental pollution and waste of resources are becoming more and more serious. Urban logistics is one of the regions with the highest frequency of logistics activities. In the face of the destruction of the environment and its impact on human life, urban logistics need to be transformed to be more friendly to the environment. This study aims to find the key success factors for selecting green reverse logistics service providers. Based on urban logistics and reverse logistics, on the perspective of sustainability, this study collected five aspects and twenty-four factors. DEMATEL is employed to process the data of expert assessments. Referring to the causal map, "Green operations level" and "Greenness of facilities and equipment" are crucial.

1. Introduction

With the rapid development of science and technology, the continuous advance of society, and the continuous extension of economic scope, the public has developed more requirements on both the physical and spiritual levels. However, the progress of human civilization has brought a series of irreversible problems to environmental resources, such as energy shortage, serious decrease of forest resources, and sharp decline in mineral resources, etc. The contradiction between economic benefits and environmental benefits is increasingly exposed, which has seriously threatened the survival and development of human beings. Therefore, the emergence of circular economy and low-carbon economy brings the coordinated development of environmental resources and material and spiritual life of humans. Sacrificing the limited resources is "drain the pond to get all the fish" in order to pursue the rapid development of the economy, and if only protecting the environment and not pursuing the economy development is just like that you seek fish on the tree, only the organic combination of the two aspects, and the environmental improvement in development can achieve the sustainable development of resources.

In order to keep up with the trend of circular economy, China has introduced a lot of environmental resources protection policies in recent years, spent a large number of manpower, material resources and funds, proposed to build an environment-friendly and resource-saving society, and the slogan has become a household phrase of environmental protection that clear water and green mountains are like gold and silver mountains. The rise of the e-commerce industry and the logistics industry has brought a more convenient and fast and indoor lifestyle, and a series of environmental problems which brings have attracted more and more attentions. The logistics activities contrary to the traditional logistics direction, such as returns, product recalls, waste recycling and so on, are gradually increasing. This is a new module of logistics research, this form of logistics is called reverse logistics. On the premise of protecting environmental resources, one of the most wasteful activities in the logistics process, that is, reverse logistics and green activities, should be connected to reduce the influence of reverse logistics activities on the environment, such as waste disposal, recycling and reprocessing of goods, circulating processing and recycling, etc.

Similarly, China's urban logistics is facing the situation of environmental pollution and resource destruction caused by the expansion of economic scope, which proposed new challenges and requirements for urban logistics. In the past, when customers selected logistics service providers, they considered more factors, such as transportation safety, transportation speed, storage level, advanced information system, etc. However, with the development of big data and the instruction of sharing economy, actually, many service providers have a lithe difference in these important logistics characteristics. When these standards are considered by the customers as the basic standards they should reach, they will put the standards of choice on the logistics characteristics that the service providers can provide is different from other operators. Reverse logistics is in the field of emerging logistics, there are many different points from traditional logistics, now most of urban logistics service providers can not propose a more comprehensive construction form. Therefore, developing reverse logistics can be as an important means for enterprises to expand their competitive advantage, combined with green logistics, it not only creates profits and value for the enterprise itself and the city's reverse logistics suppliers, but also contributes to form a green society.

Through the basic understanding of the status of green logistics and urban logistics research, combined with the selection model of reverse logistics service providers, through the analysis of DEMATEL, this paper proposes the key factors that select green city reverse logistics service providers to explore the importance of various key factors of selecting green city reverse logistics service providers, and thus further concluded, provides the reference that the relevant companies and logistics enterprises make decision when select reverse logistics service providers.

2. Literature review

2.1. Green logistics

Green logistics refers to a behaviors, which is in a series of logistics activities about purchasing, production, transportation, warehousing, packaging, distribution, recycling and so on, from the perspective of environmental protection, its purpose is to minimize environmental pollution and the waste of resources. Through advanced logistics means, building system logistics planning and carrying out green management and control, to make the links of logistics activities more environmental and green. Green logistics covers not only green transportation, green warehousing, green packaging, green circulation and processing, etc, but also green reverse logistics.

In "Logistics Terms" (GB/T 18354-2001) published by China in year 2001, the definition of green logistics is: to prevent the logistics from harming the environment during the logistics process, meanwhile, to achieve the purification of the logistics environment, and to make the most use of logistics resource.

2.2. Urban logistics

Urban logistics is a new industry in the 21st century, domestic and international scholars defined urban logistics according to their own understanding from different perspectives, but there is still no unified expression. Cheng pointed out that urban logistics is all logistics links within a certain scope, including all physical movements, like warehouse, packaging, transportation, circulation processing and information transmission, etc [1]. Dai believes that the development of urban logistics is affected by the expansion of economic scale, and is also affected by traffic conditions and resource abuse, so as to improve the relevant logistics activities of urban logistics enterprises [2]. Zou believes that the expansion of scale economy cannot be separated from the development of urban logistics, optimizing the management of urban logistics activities, so as to make urban traffic running more smoothly, and to help to achieve a balance between urban logistics activities and urban planning operations [3]. Cao pointed out that urban logistics refers to the physical movement and distribution of goods within a certain scope [4].

Based on the research of above scholars, this paper thinks urban logistics refers to the logistics that services for the city, it promotes the economic development of the city, takes into account the systematic management of environment and transportation, and consists of this kind of activities like the spatial flow of goods in the urban area, the transportation of goods with external areas, and urban reverse logistics.

2.3. The selection model of reverse logistics operator

Lu combined with the methods of Analytic Hierarchy Process (AHP) and Data Envelopment Analysis to research the performance dynamics of service providers [5]. Xang established a neural network through modular neural network method and established an evaluation model to select third-party reverse logistics service providers [6]. Guo constructed an evaluation index system from four aspects: service quality, resource capacity, management level and cooperation index, and selected third-party reverse logistics service providers by using the comprehensive evaluation method of improved fuzzy analytic hierarchy process and grey correlation degree analysis [7]. Chen adopted a weighted fuzzy clustering method when evaluating third-party reverse logistics service providers [8]. Chen established a comprehensive evaluation index system for reverse logistics service providers through fuzzy comprehensive evaluation method, and combined the analytic hierarchy process to determine the weights of evaluation indicators, and carried out data analysis to draw conclusions [9].

3. Research method

The method used in this research is Decision Making Trial and Evaluation Laboratory (DEMATEL). DEMATEL was proposed by the Bastille National Laboratory in the United States in 1971, it is a method proposed by screening the central factors of complex systems and simplifying the analysis process of system structure. Professor Zeng Guoxiong promoted DEMATEL in the 2002 Taiwan's election, which made many scholars gradually realize the effectiveness of this method, and thus apply this method to more various studies. For example, Professor Wang established a risk analysis model for mobile e-commerce supply chain by using DEMATEL analysis method, and took Taobao as an example for empirical research [6]. Yang integrated the factors that affect the quality of logistics service by studying relevant literature, sorted the factors by DEMATEL method, found out the main factors that affect the quality of logistics service, and helped to build the index system that influences the factors of logistics service quality [10].

The calculation process of DEMATEL as shown in Figure 1.

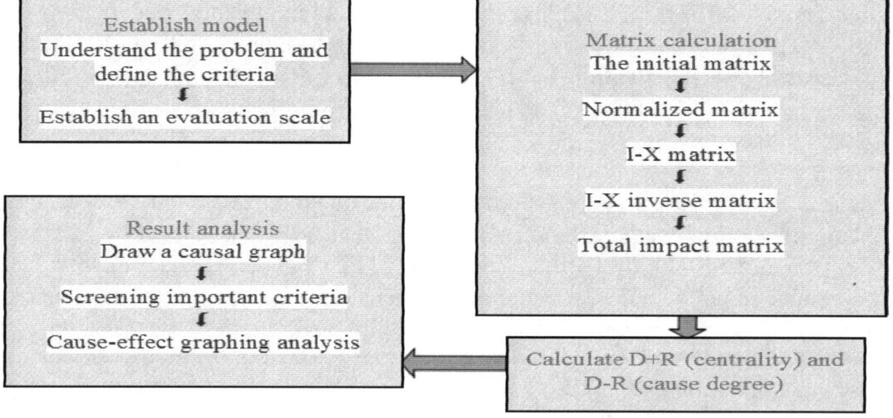

Figure 1. The calculation process of DEMATEL

4. Discussion on key success factors in selection of green city reverse logistics suppliers

4.1. Problem description

China's urban logistics started late, but it developed rapidly. Due to the influence of geographical environment and urban planning, the characteristics of urban logistics are more characterized by small batch size, high frequency, many nodes, etc. In response to the goal of building a green society, urban logistics should also develop new businesses and appropriately transform. With the continuous popularization of e-commerce and the continuous occurrence of positive logistics generated by shopping, the reverse logistics are generated when customers are not satisfied with the goods, or when recycle certain items that customer discard, both of them will cause a certain burden on the environment. In order to solve such problems, green logistics and reverse logistics should be combined, which not only helps enterprises increase economic efficiency, expand economic scale, but also contributes to building a resource-saving society.

4.2. The experiment

Through the analysis and research of green city reverse logistics service providers, the following five aspects is discussed, which are operation level, service quality, information level, cooperation attribute and green activity.

1. Operation Level

As the enterprise providing logistics services, its most basic attribute is to have the corresponding logistics characteristics, which can provide basic logistics activities. This is also the factor that the enterprise should have when customers choose a company, including these factors: transportation capacity, certain circulation processing capability, the level of network coverage, warehousing levels and the quality of employees, etc.

2. Service Quality

With the expansion of the economic scale, the needs of customers have changed from the initial simple transportation of goods and receipt of goods to hope that enterprises can provide personalized and customized services to meet the non-traditional requirements of customers and enhance customer satisfaction. Today, the third industry becomes the leading industry, and the influence of customer satisfaction on an enterprise is self-evident.

3. Information Level

The development of big data proposed by the sharing economy has put forward new requirements for the information level of logistics suppliers. Whether it has advanced information level not only affects the progress of its own logistics activities, but also affects the evaluation choices of customers.

4. Cooperation Attribute

Among the different cooperative relationships between enterprises, it must be a strategic alliance to maximize to achieve information sharing and create economic benefits, when customers want to establish strategic partnerships with suppliers, they need to have a more comprehensive and deep understanding of suppliers. In addition to the company's ability indicators, it also includes the company's popularity, corporate reputation, corporate culture, development potential, corporate strategic decision-making ability and cost control ability. Only when both sides become good strategic partners can they progress and grow together and achieve a win-win situation.

5. Green Activity

In order to build an environmental-friendly society, green logistics and reverse logistics can be combined, which requires enterprises to have a more mature level of green process operation, facilities and equipment should also be environmentally friendly, and goods processing and goods recycling should also have a high degree of green.

On the basis of reference to the literature review, according to the content of this paper, the key factors of green city reverse logistics supplier selection are divided into five aspects and twenty-four criteria to determine the research framework of this paper. This study will describe the five aspects and twenty-four criteria and connotation of these criteria of the key factors in the green city reverse logistics suppliers selection, the detailed description in the following Table 1.

Table 1. Aspects and Criteria

Aspects	Criteria	Connotation	References
A Operation level	A1 Transportation capacity	The short-distance transportation capacity of the city reverse logistics, quickly receiving the goods returned by the customer, and can also be sent out quickly	[2-4, 10]
	A2 Circulating processing level	Mainly for the goods that need to be recycled and reprocessed, which should have a certain level of circulation processing	
	A3 The level of network coverage	The area covered by the company's reverse logistics business	
	A4 Warehouse level	Including storage capacity, classification ability, special cargo storage capacity	
	A5 The quality of employees	Including professional skill level, service awareness, environmental awareness, and the experience in handling reverse logistics items	
B Service quality	B1 Response timeliness	Respond to customer needs quickly and timely, improve customer satisfaction	[2, 5, 11]
	B2 Solving ability of emergency	As for the solving ability of emergency, whether or not can react quickly and appropriately	
	B3 Service price	The reasonable price that can be accepted by customers	
	B4 After-sales service	Deal with customer complaints in time to provide certain guarantee for reprocessing goods	
	B5 Ability to provide personalized value-added needs	Provide customers with unconventional personalized needs	
C Information level	C1 Advanced level of information system	The advanced degree of enterprise information system affects the progress of reverse logistics activities	[2, 4, 7]
	C2 Logistics tracking capability	Real-time logistics tracking capabilities, query logistics information	
	C3 Information integration and sharing ability	The ability to integrate information and share information with partners	
	C4 Information security	Ensure the security of information	
Cooperative attribute	D1 Corporate popularity	Status and prestige in the industry	[7, 9, 11]
	D2 Corporate reputation	Reputation with past partners and current partners, affects partner selection	
	D3 Corporate culture	Corporate culture affects the future development of the company and is reflected in its business activities.	
	D4 Development potential	The future development potential of the company affects the choice of partners	
	D5 Corporate strategic decision-making ability	Whether the strategic decision of the enterprise is long-term and grand	
	D6 Control cost capability of enterprise	Reasonably control costs and maximize revenue with the lowest possible cost	
E Green activities	E1 Green process operation level	Green level of logistics activities	[3-4, 6, 11]
	E2 Greenness of facilities and equipment	Environmental protection level of facilities and equipment	

	E3 Greenness of handling process of goods	Whether the handling process of non-recyclable items is environmentally friendly, whether it is to minimize or even eliminate waste of resources and environmental pollution, and how to operate	
	E4 Greenness of recycling process	Whether the recycling process is environmentally friendly, whether it is to minimize or even eliminate waste of resources and environmental pollution, and how to operate	

Design the DEMATEL questionnaire based on the above five facets and the twenty-four criteria, and improve the questionnaire design through experts' and scholars' suggestion. Secondly, before completing the questionnaire, communicate with experts and scholars to fully explain the core and content of the questionnaire, so as to successfully fulfill the effective answer.

In the questionnaire survey of 12 industry and government and university experts and scholars in related fields, the collected data were drawn into a visual causality diagram based on the calculation results of DEMATEL (as the Figure 2).

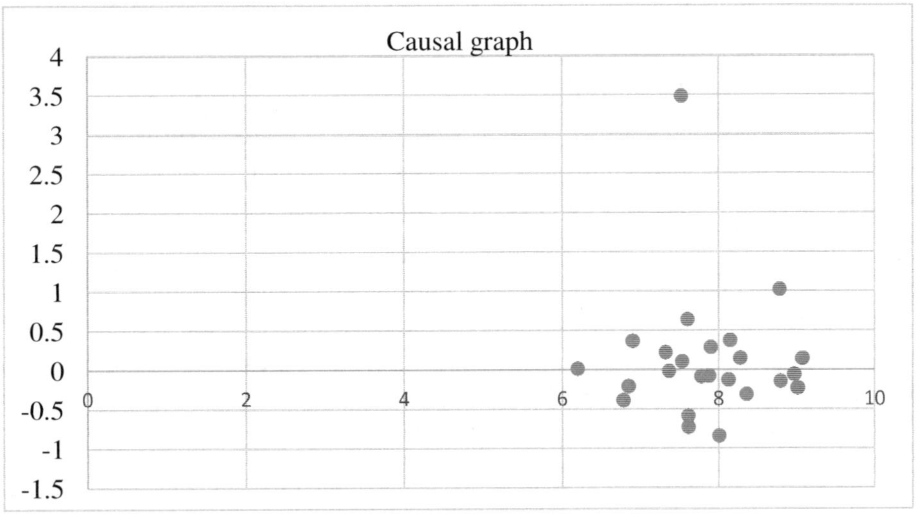

Figure 2. Causal map

When viewing the causal map, the first quadrant is the core factor area, which is characterized by high centrality, high cause degree, and has a great influence on other factors. It is a key element of the research subject and is the important part that enterprises should pay attention to. The second quadrant is the driving factor area with low centrality, high cause degree, independence, and influence on very few other factors. The third quadrant is the independent factor area, characterized by low centrality, low cause degree, and low impact on other elements. It is enough to manage these elements independently. The fourth quadrant is the affected factor area, with high centrality and low cause degree, which is easily affected by other factors. If the factors in the first, second and third quadrants are well managed, they can be linked to improve the factors in this quadrant.

The elements of the first quadrant are: network coverage level, advanced level of information system, information integration and sharing ability, and enterprise control cost capability. The elements of the second quadrant are: the quality of employees, information security, strategic decision-making ability of enterprises, the level of green process operation, the greenness of facilities and equipment, the greenness of item processing, and the greenness of recycling process. The elements of the third quadrant are: circulation processing level, warehouse level, solving ability of emergency, logistics tracking ability, corporate popularity, corporate reputation, and corporate culture. The elements of the fourth quadrant are: transportation capacity, response timeliness, service price, after-sales service, ability to provide personalized value-added needs, and development potential.

5. Conclusions

This paper makes some conclusions on the selection factors of green city reverse logistics service providers based on the current severe environmental and resource problems. By referring to the literature review of scholars and the questionnaire survey of experts and scholars, the influencing factors are summarized as five aspects and twenty-four criteria, the results are analyzed by DEMATEL, and conclusions are as follows:

1. With the rapid development of logistics enterprises and big data, the key elements considered when choosing enterprises have changed from the basic logistics characteristics such as the original transportation capacity, warehouse level and so on to pay more attention to overall information level of the enterprise, whether it is traditional enterprise cooperation or strategic cooperation that can create more benefits, and now, the role that information level plays in the logistics stage is receiving more and more attention.

2. The fifth major aspects can be clearly found in the causality diagram, that is, all the criteria contained in the green activity are in the second quadrant: the green process operation level, the greenness of the facility and equipment, the greenness of the item processing, and the greenness of recycling process. Combined with the characteristics of the second quadrant, the enterprise can independently develop this feature, which will not greatly affect the remaining existing characteristics of the enterprise, but will increase the business scope of the enterprise. Green logistics not only helps enterprises to expand economic benefits, but also responds to the circular economy, which is a win-win situation.

3. The three elements of corporate popularity, corporate reputation, and corporate culture in the cooperative attribute aspect are distributed in the independent aspect area, and the interaction with other elements is low, and the enterprise can manage them separately.

4. The elements of the affected factor area are: transportation capacity, response timeliness, service price, after-sales service, ability to provide personalized value-added needs, and development potential. It can be clearly found that these factors are not independent, for example, after-sales service requires enterprises with certain information level, high-quality staff, certain circulation processing level, more perfect feedback system, etc. Therefore, these factors cannot be "self-improved", but can be improved jointly when the enterprise improves the other three factors.

Acknowledgement

The authors would like to thank the City College of Dongguan University of Technology of the China for financially supporting this research under Contract No. 2017YZDYB04R.

References

[1] Cheng X Z 2011*Evolution research and system construction of green logistics of Zhangjiajie Tobacco Company* Central South University
[2] Dai H 2018 *Evaluation research on logistics development level of prefecture-level cities in Shandong province* Shandong University of Technology
[3] Zou Y 2011 *Research on Evaluation and Selection of Third-Party Reverse Logistics Suppliers* Nanjing University of Finance & Economics
[4] Lu J 2006 *Research on Sustainable Development of Urban Logistics* Southwest Jiaotong University
[5] Cao L 2016 *Research on Evaluation System of Third-Party Reverse Logistics Suppliers in Household Appliance Industry* HeFei University of Technology
[6] Wang G 2011 *Research on Evaluation Standard and Practice of Green Logistics* Capital University of Economics & Business
[7] Guo Q 2009 *Research on Evaluation and Selection of Third-Party Reverse Logistics Supplier* Jinan University
[8] Chen Z 2009 *Exploring the key success factors of Construction project with DEMATEL method* Chung Hua University

[9] Chen Z 2013 *Exploring the Key Factors in Improving the Swire Cold Chain Logistics Service Quality with Six Sigma and DEMATEL* Fuzhou University of International Studies and Trade
[10] Yang L 2005 *Research on the development of green logistics* Tianjin University of Finance and Economics
[11] Geng L 2010 *Research on Reverse Logistics Suppliers' Selection Based on FCM* Chongqing University

A smart model of the rural village planning

P Wu[1] and Y Lin[2]

[1]Faculty of Graduate Institute of Landscape Architecture, Tung Hai University, Taichung, Taiwan
[2]Director of Wumo Architects, Taichung, Taiwan

E-mail: 879579@gmail.com

Abstract. The smart model of rural village planning is to build a digital platform to connect agriculture and industry through big data. The process starts with the construction of a big data agricultural platform, then planning and designing the physical environment of the smart village, and the third stage is the creation of a multi-industry alliance and co-operation model. For combining agricultural demand and physical access information data, the cloud database can obtain the required information and achieve the exact goal of increasing agricultural income. In addition, agricultural tourism uses the urban planning concept to carry out village layout, and develops qualitative and quantitative regulations, ecology, landscape and tourism data in scientific calculations to control the intensity of land use that can be supported by rural areas and provide the best tourism condition and ecological environment. The project also builds a shared economic model in which enterprises contribute to assisting farmers in housing repairs and invite them to participate in operations. Thus, farmers are transformed into inheritors and educators, and they are able to participate in the operation of the village as a teacher, so that the culture of sustainable agriculture can continue.

1. Introduction

In the past history of human development, the industrial revolution has been a great leap forward in the development of human civilization, and the pace of urbanization has been accelerating and expanding. Marvin and Luque [1] pointed out that the over-expansion of population caused by urbanization, rural poverty, environmental pollution, inadequate public construction, health diseases and other issues have worsened, and urban living environment and other related issues have gradually emerged. With the rapid growth of economy, mainland China has flourished for more than 30 years since the reformed and opened up and has leapt to the core of the world economic stage. However, the hidden worry behind the huge gap between urban and rural areas has become one of the key issues that must be faced at the same time [2]. The phenomenon of domestic immigrants from urban migrant workers and rural populations to urban areas has also continued to reduce the number of rural residents [3]. In contrast to the booming development of cities and towns, rural residents have left their homes to invest in the convenience of the city in order to pursue higher economic income and self-realization [4]. The loss of rural population, slow rural growth, and local cultural faults have become the biggest worry for rural progress, so the State Council of China issued a national poverty alleviation and development work.

In the modern rural areas, due to the lack of life inheritance in the outflow of the population, the issue of how to make the farmers getting the real benefits and fair distribution on the crops, so that the farmers can surpass the urban migrant workers and even the general blue-collar class income, and attract the rural areas of the living cities is crucial. The account is reflowed, and then the layout and connection of the third industries of commerce, tourism and the secondary industries are carried out by means of sustainable management. It is hoped that this concept can achieve a balance point between development and environment, economy and ecology. Also, it has a friendly space for coexistence in rural construction and tourism development. Under the concept of complete land development model, the developed land products will be given new definitions, and the rich ecological content of the products will form a segmentation from the original market; with appropriate planning and design, the unnecessary land development process will be reduced [5]. Through effectively reducing development costs in the entire development process, a large number of cutback to the natural environment, ecological damage, shaping the attitude and responsibility of the company's environmental responsibility, and the corporate image has a significant improvement which gives feed-back to the company's innovative product value. For rural development is not just a question of farming and housing, cross-domain integration of related industries has become an inevitable result.

The empirical site of this study is Yanzi Village, Dawo Town, Gao County, Yibin City, southeast of Si-chuan Province of China, which is an agricultural county dominated by mountainous hilly terrain. Dawo Town Rural Complex Industrial Park project covers an area of about 27 hectares and contains general farmland, house site, woodland, reservoir, and road agricultural facilities. Land use category belongs to agricultural land and is a typical rural community at present and the total number of farmland households in the site is 12.35 hectares. However, there are no public facilities on the site and the gathering activities of the public are concentrated in the upper town hall of Dawo Town. The agricultural farming is carried out in full accordance with the natural season cycle [6]. One of the neighboring village is the existing tourist attraction in the town. Its cabin accommodation environment built by lake has the most beautiful scenery of the lakeshore, the life style of the agriculture in the mountains can attract some people who desire to escape from the city to rest. The circumstance is hard to compare with the complete tourism supporting industry and rural participation activities. Thus, the government is actively promoting the rural complex to raise rural tourism and recreation opportunities.

Therefore, the main objective of this study for Dawo Town Rural Complex is to improve the local crop market value on the premise of maintaining the current environment ecology and rejuvenation, including the sustainable development of agricultural management, commercial market operation, network platform, sight-seeing tourism, environmental education and cultural preservation. Establishing the sustainable countryside with recycling ecology to create wealthy villages and become a comprehensive agricultural platform for surrounding villages. Agricultural upgrading aims to introduce advanced farming technology, promote agricultural industry, build promotion hub, technology platform and supply of materials, and use this technology to drive management, technical, R&D, business planning talents to build incubators, agricultural talent bases and brands. Secondly, acquiring the optimization of land use division, and the future utilization of the land can also be operated under the most reasonable conditions. Finally, set up counselling mechanism, marketing promotion, education cooperation, job creation, modern business model and online big data operation to establish brand characteristics, integrate rural complexes, upgrade traditional rural areas into a miniature modernity place with agro-ecological value.

2. Method and material

The establishment of a sustainable agricultural integrated platform of products, tourism, education and cultural preservation for rural areas through the use of ubiquitous network technology and the Internet of Things (IoT) is necessary. The important technology such as RFID electronic tags, Artificial Intelligent (AI), 5G, provide technology which integrates users to manage big data analytics to achieve the most appropriate development and cost-effectiveness of the best management [7]. Thus, the smart model of rural village planning is a comprehensive development plan constructed with agriculture as the original structure. Combine agricultural big data in the scope of the base plan with the Internet of Things, cloud computing, and central management technology, master the production, logistics, and sales core of agricultural products, and then enhance the real income of agricultural production, and at the same time construct a rural area with smart agriculture, smart environment, and smart tourism [8]. The process for smart rural construction, each with its corresponding steps and priorities is as follows.

2.1. Agricultural value chain construction

Value chain, also known as value chain analysis, value chain model, etc. was proposed in 1985 by Michael Porter in the book Competitive Advantage [9]. Porter pointed out that companies must develop unique competitive advantages and create higher added value for their goods and services. The business strategy is the business model of the structural enterprise, and it becomes a series of value-added processes. The series of value chain of a general enterprise is mainly divided into primary activities and support activities. First, primary activities include the core production and sales program of the company: (1) Inbound logistics that is incoming materials storage and transportation belongs to the resource market. (2) Operations which is processing and manufacturing belongs to a manufacturer's market. (3) Outbound Logistics that is finished product storage and transportation belongs to a middleman market. (4) Marketing and sales belongs to the consumer market and (5) after sales service is also included. Secondly, support activities, including other activities that support core operational activities, are also known as joint operations [10].

With the development of the Internet technology and logistics industry, farmers have opened up new paths on the Internet. They operate communities and websites, join local organizations, and sell their own agricultural products directly on the Internet. From the place of origin to the consumer, there is no need to stack the price at the layer level, only the logistics cost. The advantage is that the price can be higher than the old process when the traditional large supply of wholesalers, but the price to the consumer is not more expensive than the outside. The shortcoming is that farmers must be responsible for the operating costs of goods and reduce the labor intensity. However, farmers in the study site have not used this sales model.

Therefore, setting up a statistics platform of big data, including digital management cores such as production data, market demand, price analysis, channel transportation and webpage sales platform, integrate the data required for production from production to sale, and carry out scientific assessment planning [11]. Timing to increase market acceptance and selling prices at crop harvests to avoid saturation of oversupply. The production and sales resume can be improved through the open sales platform and big data logistics to enhance the trustworthiness of agricultural products and the image of agricultural products. The market supply and product prices can stabilize the market stability of supply and demand.

2.1.1. Agriculture. Cloud big data provides production control, exquisite agriculture, mechanizing agriculture, organic farming method, coaching brand and production and sales resume certification to complete industrial upgrading.

2.1.2. Logistics. Cloud big data provides production control, exquisite agriculture, mechanizing agriculture, organic farming method, coaching brand and production and sales resume certification to complete industrial upgrading. Master the demand and supply market, combine with the analysis of the cloud market, the improved agricultural products will be exported in a friendly platform to reduce the transaction links, and the intermediate profits will be returned to the rural and rural households.

2.1.3. Cultural Innovation. After the promotion of the agricultural industry, it also grafted cultural, commercial, management and other employment opportunities, providing rural youth cultural and commercial space to prevent rural outflows.

2.2. Agricultural value chain construction

Stage two is the development of smart environment model. Based on the concept of small-scale urban planning, and the statistical analysis of the land-supportable strength, the layout of the site conditions, and the use of the time axis to estimate the past, present and future estimates of the maximum land use intensity. Control construction management configures various use zones such as homestay, commercial, public, road, industrial processing, sightseeing and tourism supporting land. Then, limit the scope of agricultural production and future tourism industry which will impact on the environment in advance, and maintain the agricultural living environment and surrounding ecologically [12]. Thus, land use management can effectively improve the control of the impact of rural and various tourism, processing, and cultural industry activities [13]. In line with big data monitoring and data acquisition, the central management core analysis can control the environmental changes in the site, which can be used as academic research materials and instantly adjust agricultural farming and tourism development models.

2.2.1. Land. Refer to the concept of urban planning, plan the rural environment to reach the upper limit of living, production, and ecological balance, and then carry out micro-urban planning to ensure the sustainable development of rural land.

2.2.2. Environment. Controlled development activities provide more ecological research data, including natural resources such as water resources utilization, wind power, forestry, etc., which can also achieve the advantages of environmental conservation and symbiosis and mutual benefit within the control scope.

2.2.3. Ecology. There are planned and controlled development activities that provide more development space for natural ecology, and indirectly provide potential and objective value for academic research, environmental parameters, tourist attractions, and better quality of life.

2.2.4. Real estate. Under the conditions of micro-city planning and environmental and ecological control, we plan to designate residential, commercial, industrial, cultural, educational, and hotel-related land to attract enterprises to invest in real estate development and provide space for living and activities needed for reversing urban immigration.

2.3. Smart management strategy

After the first stage of the smart agricultural structure to improve farmers' income, and the second stage into the overall development plan and management of the smart environment, the farmers have the economic autonomy and environmental control plan, and then carry out the third stage of smart tourism to set up rural tourism promotion. The third stage is based on the foundation of farmers' stable income and environmental and ecological sustainability. The number of tourists and the intensity of land support calculated from the big data can avoid the speculative atmosphere brought about by rural tourism or the inappropriate use of fallow, occupation, diversion, etc., fully comply with the better farmland control policies and regulations. At this time, the operating right can be returned to the village organization, the government unit or the civil construction unit that originally constructed the smart village planning model to continue operation and management.

2.3.1. Leisure and tourism. When the rural areas have reached the objective conditions of natural environment, sustainable ecology, local cultural characteristics, rural health, and etc., works can continue such as the integration of rural resources for smart tourism brand construction, controlling the number of tourists to ensure the overall tourism quality, but also provide rural residents income distribution and living environment construction funding sources [14].

Property management. Site development intensity management, environmental ecological monitoring management, rural environmental management, tourism mobile management, commercial logistics management, domestic waste and resource management are integrated into the smart village management core.

3. Result

Before the construction of the smart village, qualitative and quantitative planning of the base conditions and the site's support intensity must be carried out to determine the best operation mode and the most appropriate management mechanism [15]. According to different geographical conditions and background conditions, the planning intensity that is most suitable for the land is calculated and selected. In principle, the base development is carried out in a low-impact mode and the selected amount is the estimated load. Zoning needs to be coordinated with landscape environment analysis to avoid ecologically sensitive areas and geological sensitive areas for zoning and path planning.

3.1. Agricultural big data platform

Combine agricultural big data in the scope of the base plan with the Internet of Things, cloud computing, and central management technology, master the production, logistics, and sales core of agricultural products, and then enhance the real income of agricultural production, and at the same time construct a rural area with smart village, smart tourism, and smart environment. Moreover, according to the effective management model of the value chain, it consists of a series of business processes, each of which can be broken down into specific operations. Those activities involve personnel, finance, planning, research and development, procurement, etc. In the value activities of different enterprises, not every link creates value. In fact, only certain specific value activities truly create value. These truly value-creating business activities are the strategic links in the value chain.

If the agricultural products can complete the value chain operation by themselves, there is a way to establish the brand advantage. After the brand advantage is established, the transaction process and the transaction case of the goods can enter the database of effective control. Through the access and analysis of the database to provide the actual demand of the market and adjust the content of the goods, to control the product model to improve product yield and profit margin. Now the input field of raw materials brought into the value chain is used as the denominator of all values. The market value is increased according to its cost ratio. Each additional procedure in the value chain must increase the value once, but the market price of agricultural products is a fixed value. In the traditional market layer reselling, and then the market price to push back the cost of crops, the natural farmers' distribution is extremely limited; if the market value remains unchanged, it is inevitable to improve the income of farmers by reducing the channel program and increasing the value of raw materials.

3.2. Rural village planning and design

After the completion of the overall blueprint, the operations will be carried out step by step. In principle, site development is conducted in the concept of low-impact development. Regional division needs to cooperate with landscape environment analysis to avoid ecological sensitive interval and geological sensitive interval for zoning and path planning.

Figure 1. Site Analysis

Rural modernization, rural enterprise, rural complex, characteristic towns and other rural constructions need to deal with the problem of land use intensity. According to the Land Administration Law of the People's Republic of China, rural land is used only as a residential land or public welfare facility. The basic farmland is also subject to the total amount control and cannot be arbitrarily changed. Therefore, any rural or any land planning plan must first produce a development assessment or master plan. If the overall plan considers an open community environment rather than an internal space allocation, there must be a planning principle that can be shared by any development model. To develop a comprehensive consideration and future forecast, this plan attempts to introduce the urban planning concept and use the scientific data for land evaluation and distribution to provide the required land use intensity and location allocation for future operation and management.

The site is located between the hills of Sichuan, stretching terraces and non-commercially simple rural villages. It has a wide view to the heights, and the distant mountains set off a tranquil pastoral landscape. Compared with other magnificent scenery throughout the country, the site is more elegant and peaceful. The environmental resources include the inner landscape resources in the site and the surrounding tourist attractions. The elements of the environmental resources include the topographic and geomorphological background, the geological soil environment, the climatic resources, the hydrological environment and the river system, the flora and fauna ecology, and the current landscape (see Figure 1).

According to the legal per capita living area and cultivated land area, the maximum number of people in the planned area can be estimated, and the land use ratio of the urban planning should be set according to the population value. After calculation, the optimal construction area of the site is 3.833 hectares, the legal resident population is 1278 people, the largest tourist number is 1.517 people which is 1.5 times the resident population, the number of tourists is 511 people accounted for 40%, and the permanent population is 391 people accounted for 30.6%. The number of farmers supporting arable land is 267, accounting for 20.9%, the number of resident population including commercial and public services is 109, accounting for 9%, and when facing the largest number of tourists, the number of commuting services is 54, with the largest distribution of 27 hectares in the region. In terms of number of people, the per capita tourist area is 141 square meters. This study uses the zoning and low-impact development mode to avoid developing land with forest and slope greater than 30%, use general farmland as the basis for changing the use of zoning to reduce development behaviour and development costs, and integrate existing roads and general facilities into each zoning. The land use zoning is as table 1.

Table 1. Land Use Zoning Statistics

Item of Land use	Area	Unit	Remark
Forest	10.95	Hectare	
Water	0.72	Hectare	
Farm	11.01	Hectare	Including roads and other facilities
Housing	3.36	Hectare	
Road and others	0.96	Hectare	Including parking lots
Total	27	Hectare	

The planned area of the first phase of the Dawo Town Rural Complex Industrial Park is 27 hectares. The service facilities in each district are integrated with field management and various service facilities and are connected as a part of big data intelligent management system. The site will have a very rich landscape change during the walk on the hilly terrain. The moving path cooperates with the topography and zoning management to present different rural scenery. The planning walkway is started by the flower entry plaza. After passing through the plaza, the low-shrub and fruit trees are planted to company with the beautiful scenery. Then the smooth moving path is connected to the field observation area, and then the site is entered. The southern end begins with an uphill section of the mountainous area. You can visit the flower valley, the terraced wetland and the coffee forest. This area is planted on the mountain surface for the four seasons of flowering shrubs and fruit trees, and the back mountain surface is a vegetable grass flower plant, allowing the greatest interaction between air and the pacers, and the path moves slowly to the top of the mountain in a roundabout way with four stop areas to increase the chance of rest on the uphill section to reduce fatigue.

The division of management zones is mainly based on the existing residents' number, terrain conditions, and area of cultivated land. The rule is to define the boundaries according to the current cultivated land and the location and change of terrain as the operating principle.

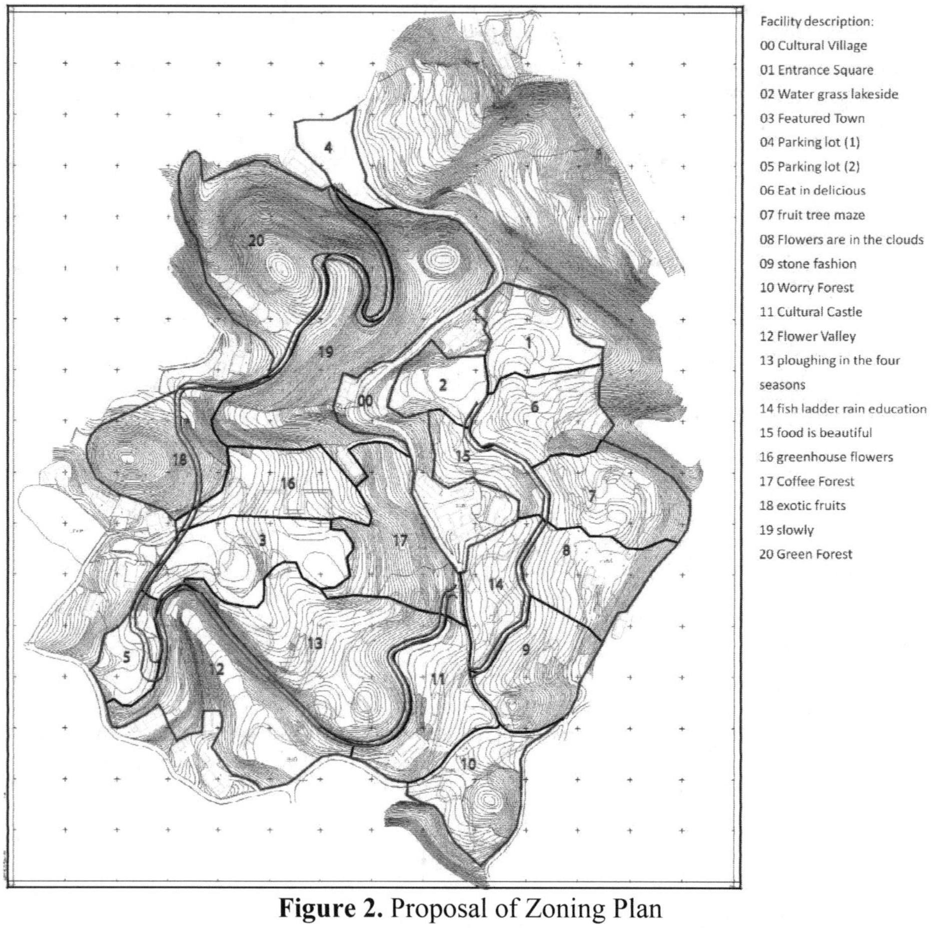

Figure 2. Proposal of Zoning Plan

Walking to the highest point of the complex park is the newly built homestay in this area. It is also the newly built town street in this district. The neighbourhood provides shopping, cultural creation, education, experience, accommodation, food and beverage snacks. The biggest feature in the activity area is the exquisite crops and seedlings planted in the greenhouse. In addition, you can go up for the ecological forestry trail. And a special restaurant for dining with afternoon tea or large events is designed there. In addition to basic diet, trading, and sanitary equipment, they are displayed in 21 areas. Other special marking items including ecological observation, explanation service, and signage stop are specifically marked (Figure 2).

When the rural areas have the objective conditions of natural environment, sustainable ecology, local cultural characteristics, health enhancement and slow life style, the integration of rural resources for the creation of smart tourism brands, to control the number of tourists to ensure the overall tourism quality, and also provide income sources for rural residents and the source of funds for residential environment construction. In addition, site development intensity management, environmental ecological monitoring management, rural environmental management, tourism line management, commercial logistics management, living waste and resource management are integrated into a smart village management system.

3.3. Sustainable management

In the middle of the operation, smart village started to promote the tourism and experience project. In the early stage of agricultural upgrading, the agricultural materials management room constructed in each district is the preparatory space for this service. The material management room can cooperate with the terrain construction principle. The upper floor is a two-story building. The first floor can be mixed used as a promotion place, service center, material room, agricultural product shop, and product story hall, while the second floor can be used as a homestay space. The homestay can provide opportunities for tourists to interact with local residents in a life experience with an educational and entertaining context and atmosphere in a way that combines local festivals and rural life experiences.

In addition, the construction of the homestay model is a concept of joint participation in the concept of shared economy. It is different from the aggressive operation method of local culture and customs after the introduction of commercial activities. The construction of the houses is funded by the management unit. The management team seeks the repair and reconstruction of the houses with owners in the village that are willing to participate and operate together. Before the construction, the two parties sign a joint operation contract, and the operator will pay for the repair and renovation. Residents provide more than one-bedroom units for management and operation of the management units, and the out-of-town villagers must retain their original living patterns and serve as the owner of the homestay and the management unit of the park. The imagery is based on the experience of the original residents.

This model of sharing keeps the living culture of rural residents and increases their non-industry income without changing the residents' farming habits. In the case of rural busy, they must also participate in the farming operations of the owner of the homestay, so that the travelers can share the inheritance from the life style experiencing the meaning of rural life and traditional culture. This model is a concept of complete rural tourism operation for rural residents, funded enterprises, travel tourists, and cultural success. It is one of the effective operational methods for reaching the goal of this project (Table 2). After the completion of the construction of the three phases, the rural operation model of the smart village is stable, and the operational management rights can be returned to the village organization, the government unit or the civil construction unit that originally constructed the smart village. The structure of smart village is based on the benefits of farmers, and is jointly operated by farmers, experts and management units. The company only serves as a consulting and platform operation, and excavates the wealth that originally belonged to farmers and is rich in agriculture, grafted tourism and cross-domain integration. It is a mutually beneficial management platform that complements each other and is the embodiment of the overall economy of agriculture and the emergence of a shared economy.

Table 2. Service Facility Estimated Turnover Calculation

Item	Unit	Quantity	Reduction	Quantity (after reduction)	Day	Unit price (USD)	Total price (USD)
Ticket	person	1278	50%	639	365	3.75	874,631.25
Story house homestay	room	10	50%	5	365	97.5	177,937.5
Featured town homestay	room	41	50%	21	365	67.5	517,387.5
Creative goods	person	639	80%	511	365	3.75	699,431.25
Food & drinks	person	639	70%	447	365	9	1,468,395
Activity	person	639	30%	192	365	3	210,240
Camping	camp	50	40%	20	365	3	21,900
Total							3,969,922.5

4. Conclusion

Rural development should make the first favor for local farmers to benefit from farming, instead of only emphasizing the derivative advantages of the surrounding industries, and encouraging agricultural tourism and integration to forget the importance of the stability of agricultural roots. The shortage of agricultural food and the misappropriation of cultivated land are commercial operations. The plan considers that the appropriate conversion of agricultural land to tourism facilities is feasible, because the tourism industry profits are indeed more profitable and easy to operate than agricultural, but any land conversion must be administered according to rules and regulations, for the basic farmland is the national life field. More attention should be paid to this part of the modification.

Furthermore, proper comment, planning result and suggestion are discussed as follows:

1. This study introduces agricultural big data in the scope of the site plan into the Internet of Things, cloud computing, and central management technologies, mastering the production, logistics, and sales core of agricultural products. In addition, according to the concept of Porter's value chain analysis, all processing or conversion programs will become external profit value, but the value of products extended by agricultural processing or tourism processing is not fully reflected in the basic raw materials. The win-win model has become a dependent industry.
2. To acquire the optimization of land use division, urban planning regulation and Lishan initiative concept which are proposed as references based on the assessment of data with master planning and considers the commons structure, while respecting the traditional community public land use rights and downstream land ecological rights for social responsibility operation planning. For the open community environment and the distribution of internal space of the enterprise, it can play a regulatory role. The future utilization and development of the complex land can also be operated under the most reasonable conditions to avoid the operational dilemma.
3. The integration of upstream and downstream industries through rural development and cross-domain integration has become an inevitable development result. The farmers' profitability is planned for all basics. The initial input cost is only the cost of technology introduction and sales channel laying and overall development. The proportion is lower than the tourism-based planning model, which can completely reduce the starting point of enterprise entry. The cost threshold and the impact of development activities on the environment are reduced, and farmers still maintain their daily routines and traditional habits. Development also allows residents and farmers to bear the brunt of the operation results and profit sharing. It can fully implement the real meaning of low-impact development, from low environmental impact to low cultural impact, enhancing the real income of agricultural production, and at the same time construct rural areas with smart countryside, smart tourism, and smart environment.

Acknowledgements

This work was supported by the Industry-University project of the ecological leisure farm in Waipu Taiwan. The authors acknowledge the financial support from Institute of Stone Park Ecological Leisure Farm.

References

[1] Marvin S and Luque A 2017 *Int. J. of Urban and Regional Research* **41**(1) pp 84-103
[2] Cheng J and Masser I 2004 *Environment and Planning B: Planning and Design* **31**(2) pp 167-194
[3] Lambin E F, Turner B I and Geist H J 2001 *Global Environmental Change* **11**(4) pp 261-269
[4] Fang H, Xu Y, Ye Z, Zhang Z, Pan S, Deng L, Luo Z and Chen S 2014 *Int. J. of Sustainable Development & World Ecology* **22**(2) pp 117-126
[5] Xüling L, Zhaoping Y and Xuegang C 2012 *Industrial Survey* **2** pp 80-84
[6] Okayama N, Tanaka N, Honda K and Matsumoto R 2017 *J. of the Japanese Forest Society* **99**(5) pp 202-209
[7] Strohbach M, Ziekow H, Gazis V and Akiva N 2015 *Modeling and Processing for Next-Generation Big-Data Technologies Springer* pp 257-282
[8] Nkomo M, Hancke G P, Abu-Mahfouz A M, Sinha S and Onumanyi A J 2018 *Sensors* **18**(10) p 3215

[9] Mudambi R and Puck J 2016 *Journal of Management Studies* **53**(6) pp 1076-1093
[10] Vilajosana I, Llosa J, Martinez B, Domingo-Prieto M, Angles A and Vilajosana X 2013 *Communications Magazine IEEE* **51**(6) pp 128-134
[11] Pretty J N and Bharucha Z 2014 *Annals of Botany* pp 1-26
[12] Harshada R, Bhede M and Arati S P 2015 *Int. Journal of Modern Trends in Engineering and Research* **2**(7) pp 956-964
[13] Nohl W 2001 *Landscape and Urban Planning* **54** pp 223-237
[14] Walker R 2001 *Ecological Economics* **37**(3) pp 357-369
[15] Li X, Ma K, Fu B and Niu S 2004 *Acta Ecologica Sinica* **24**(5) pp 1055-1062

IOP Conf. Series: Materials Science and Engineering 1113 (2021) 012013 doi:10.1088/1757-899X/1113/1/012013

Preparation of hydroxyapatite by aqueous precipitation from calcium carbonate and phosphoric acid

Byeong Woo Lee

Department of Ocean Advanced Materials Convergence Engineering, Korea Maritime and Ocean University, Busan, Korea

E-mail: bwlee@kmou.ac.kr

Abstract. Hydroxyapatite ($Ca_{10}(PO_4)_6(OH)_2$, HAp) is the representative biological mineral which is hard to attain the solution equilibrium and difficult to produce by solution synthesis. HAp finds new application fields to remove harmful elements from polluted wastewater in industrial scales. The low-cost manufacturing process should be developed using abundant available $CaCO_3$ materials. Calcium carbonate ($CaCO_3$), as an economically feasible and abundant precursor, has been used in calcium source for the HAp synthesis. The synthesis was conducted by reacting $CaCO_3$ with phosphoric acid (H_3PO_4) in aqueous solution with aid of citric acid and sodium hydroxide under moderate synthesis conditions. Compositionally pure HAp has been prepared. The process offered a relatively fast, convenient and economical route for the synthesis of HAp.

1. Introduction

Calcium phosphates have proved attractive for biological applications. Among the calcium phosphates, significant attention has been given to hydroxyapatite (HAp, $Ca_{10}(PO_4)_6(OH)_2$) due to its prominent bioactivity and biocompatibility. HAp is quite close to inorganic part of a human bone and has been widely utilized in the biomedical field, as implant and restoration materials, and in drug delivery systems [1,2]. Since HAp is a multifunctional material, HAp finds new applications including the fields of heterogeneous catalysis and toxic material fixation with use of improved adsorption or catalytic decomposition activity against heavy metals, bacteria, viruses and other biohazardous substances [3-6]. HAp or chemically modified HAp are host to many substitutions by cations, anions and anionic radicals that resemble its original constituents in size and charge. This material stabilizes metals by chemically binding them into new stable phosphate phases and other relatively insoluble phases [6]. For example, minor to major amounts of Cd, Co, Cu, Sr, Hg, Fe, Mn, Pb, and U are known to replace calcium. However, it is important to develop a low-cost synthesis procedure for implying HAp at industrial scale for the waste water treatment. Mass production of HAp with using various kinds of economical biogenic sources such as seashells and eggshells is necessary. These shells are composed of 95% calcium carbonate ($CaCO_3$), and the remainder is thermally decomposable organic matter and other compounds [7]. $CaCO_3$ is abundantly available in nature as the cheapest calcium source. Various processes have been developed for the synthesis of HAp using $CaCO_3$ from low-cost starting materials.

In the present study, $CaCO_3$ has been used as a calcium precursor by assuming that the abundant biogenic materials were used.

Various wet chemical techniques were developed for the synthesis of HAp using different calcium and phosphorus containing starting materials. Chemical precipitation from aqueous solution is the most common synthesis route to prepare HAp and the process offers advantageous features such as a simple, environmentally friendly and low-cost synthesis [8-12]. Using the precipitation process, relatively large amount of HAp can be produced without expensive and sometimes harmful organic solvents.

In order to reduce or eliminate the generation of byproducts, a wet chemical process using calcium hydroxide and phosphoric acid as starting materials was developed [8, 9].

$$10\ Ca(OH)_2 + 6\ H_3PO_4 \rightarrow Ca_{10}(PO_4)_6(OH)_2 + 8\ H_2O \quad (1)$$

According to Eq. (1), the use of $Ca(OH)_2$ and H_3PO_4 can generate HAp avoiding the generation of byproducts except water (H_2O). Since $Ca(OH)_2$ is obtained from $CaCO_3$ by heating and hydration steps, $Ca(OH)_2$ costs more than $CaCO_3$ [10]. In this view, the use of $CaCO_3$ and H_3PO_4 as starting reactants for the synthesis under moderate conditions is cost effective. The reaction byproduct is CO_2 gas, which leaves the reaction mixture in gaseous form. The expected reaction is as follows:

$$10\ CaCO_3 + 6\ H_3PO_4 \rightarrow Ca_{10}(PO_4)_6(OH)_2 + 10\ CO_2 + 8\ H_2O \quad (2)$$

In this study, the one-step synthesis of HAp from aqueous solutions without $Ca(OH)_2$ production process has been proposed based on the Eq. (2). Calcium carbonate ($CaCO_3$) has been used in HAp synthesis via simple precipitation conditions. Previously, several studies have been employed to produce HAp from calcium carbonate sources using phosphate solutions [11,12]. However, attaining the solution equilibrium is very difficult to achieve. The reaction times to form single phase HAp was too long from tens of hours up to days, which makes them unfeasible for the economical production. In the present study, the effects of additional reactants on the phase formation of HAp from $CaCO_3$-H_3PO_4 via aqueous precipitation was investigated. Unlike other precipitation processes with a long, time-consuming process, citric acid was used to resolve some unreacted calcium carbonate to increase the reaction speed in mixed solution and pH was controlled to above 6 using NaOH to form HAp phase conveniently.

2. Materials and methods

Table 1. The samples with reaction compositions and solution pH.

Sample	Reaction Composition	Reaction (Solution) pH
Hap-0	0.5 $CaCo_3$ + 0.3 H_3PO_4	6.2
Hap-1	0.5 $CaCo_3$ + 0.3 H_3PO_4 + NaOH	9.0
Hap-2	0.5 $CaCo_3$ + 0.3 H_3PO_4 + Citric acid (0.15 mole) + NaOH (→ lower pH)	6.7
Hap-3	0.5 $CaCo_3$ + 0.3 H_3PO_4 + Citric acid (0.15 mole) + NaOH (→ higher pH)	8.9

In the study, experiments were done using distilled water as solvent at room temperature. HAp (Ca/P=1.67) nanoparticles were prepared by a one-step wet precipitation route under controlled synthesis conditions. Ca/P molar ratio, 1.67, of initial reagents was adjusted in order to synthesis HAp powders. Calcium carbonate powder (98%, Junsei) and orthophosphoric acid (85 wt% in water, Junsei) were used as received. Citric acid (99%, $C_6H_8O_7 \cdot H_2O$, Junsei) was used as an additional chemical to resolve some remaining $CaCO_3$. Reaction pH was controlled by adding sodium hydroxide (97%, NaOH, Dae Jung). Undiluted H_3PO_4 (0.3 mole) was added dropwise onto a slurry containing $CaCO_3$ (0.5 mole, 50 g in 200 ml water (2.5 M)) in 20 min and, subsequently, citric acid (0.15 mole) and sodium hydroxide were added. Then the mixtures were ball milled for 6 h at 250 revolutions per minute. Table 1 shows the prepared samples from different reactants and reaction pH. The resultant precipitates were washed and filtered with distilled water and dried at 60°C. Unlike the other preparative processes for the hydroxyapatite, no long maturing process was involved. The overall precipitation process is shown in Figure 1.

The prepared powders were calcined at different temperatures ranging from 500°C to 900°C to study phase developments. The properties for the powders were characterized using X-ray diffraction (XRD, Cu-Kα radiation, λ = 1.5418 Å). The particle sizes of the synthesized powders were also calculated by the XRD peak broadening according to the Scherrer's equation (Dcal (particle size) = $0.89\ \lambda/(B \cos \theta)$: where λ is the wavelength, B is the full width at half maxima (FWHM) in radians and θ is the diffraction angle).

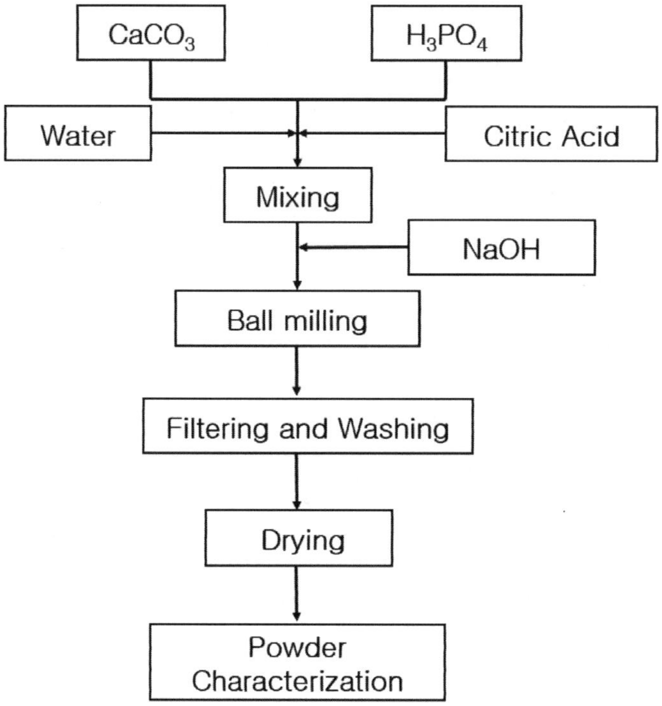

Figure 1. HAp powder synthesis procedure by precipitation from $CaCO_3$-H_3PO_4 with citric acid and sodium hydroxide.

3. Results and discussion

In the present study, HAp (Ca/P=1.67) powders were prepared by a new one-step approach of the aqueous precipitation method with use of citric acid and sodium hydroxide. Without citric acid and sodium hydroxide, the autogenerated reaction pH was as low as 6.2 and the prepared powder has been shown with various phases such as unreacted calcium carbonate, calcium hydroxide and DCPD (Dicalcium phosphate dehydrate, Ca/P=1, $CaHPO_4 \cdot 2H_2O$) as shown in Figure 2. DCPD is prone to be produced in low pH condition < pH ~6 in calcium phosphate aqueous chemistry [13]. As heat-teat temperature increase from 500°C to 900°C, desirable HAp phase was formed, but the impurities of β-TCP and free CaO were also found. Insufficient HAp synthesis condition led to formation of detectable β-TCP and free CaO according to the Eq. (3).

$$Ca_{10}(PO_4)_6(OH)_2\ (HAp) \rightarrow 3\ Ca_3(PO_4)_2\ (\beta\text{-TCP}) + CaO + H_2O \quad (3)$$

The effect of both citric acid and sodium hydroxide on the preparation of phase pure HAp is presented in Figure 3. When the powder was prepared without citric acid, a large amount of undesirable calcium carbonate which is unreacted was also found with HAp peaks (HAp-1). Single phase HAp was difficult to obtain even at relatively higher pH 9.0 without the aid of citric acid. In the case of HAp-2, 0.15 mole of citric acid was added to HAp-1 composition to resolve unreacted calcium carbonate at relatively lower pH condition of 6.7. The low pH resulting from less amount of sodium hydroxide addition led to the extra DCPD phase formation as shown in Figure 1 for HAP-0. For HAp-3, however, a sufficient amount of sodium hydroxide was added to achieve pH 8.9, leading to the single HAp phase formation. In addition, the broad and weak HAp peaks indicate poor crystallinity and/or nanometer scale crystallites of the HAp powders synthesized. The prepared powders are approximately 20 nm in particulate size calculated by the Scherrer's equation. The most distinct (002) reflections (near 2θ 26°) in the XRD patterns were taken into calculation.

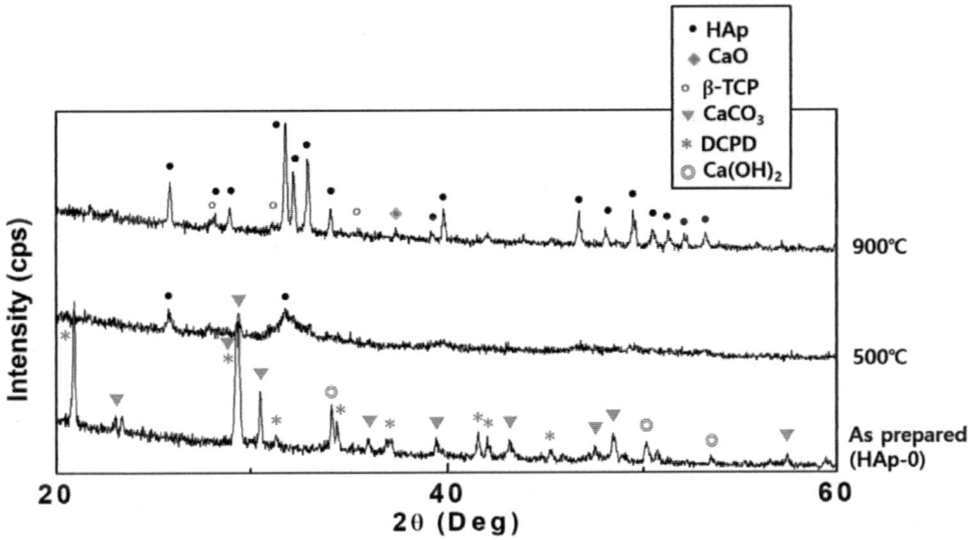

Figure 2. X-ray diffraction patterns of as-prepared and heat-treated powders from $CaCO_3$-H_3PO_4 without citric acid and sodium hydroxide (autogenerated pH 6.2, HAp-0).

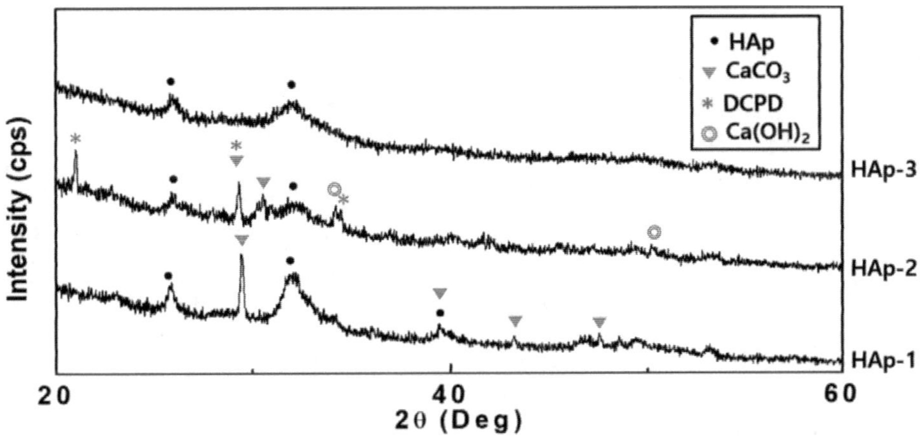

Figure 3. X-ray diffraction patterns of as-prepared powders from $CaCO_3$-H_3PO_4 under different reaction conditions with or without citric acid (HAp-1, HAp-2 and HAp-3).

Figure 4 shows the XRD patterns of the synthesized HAp powders heat-treated at 500°C and 900°C from the composition HAp-3 with both citric acid and sodium hydroxide at relatively higher pH 8.9. HAp was the only phase without any trace of impurity or secondary phases. These results confirmed that the starting reactants and pH played a significant role in obtaining pure HAp. The overall peaks of HAp become more distinct as the heat-treatment (calcination) temperature increases, and the narrow peak width at 900°C implies increased crystallinity and particulate size. The particulate size increased from 25 to 70 nm when the calcination temperature increased from 500 to 900°C, respectively. Since the crystallinity and the crystallites size of HAp can be tailored by the calcination step, the prepared HAp having different powder properties can find also various applications. It has been demonstrated that amorphous or low crystallinity HAp crystals have a better biocompatibility or an enhanced activity of removing harmful elements rather than well crystallized HAp crystallites [14,15].

In most chemical precipitation processes for calcium phosphates, the powder properties depend heavily on the processing parameters during the precipitation reaction due to the slow reaction rate and the unstable nature of the precursor solution. The morphology, phase, and crystallinity of precipitated powders were closely related to precursor concentration, precursor dropping rate, reaction pH, and maturing (aging) time in solutions. For example, to prepare single HAp phase particles, normally very slow precursor dropping rates such as a few drops per minute and a very long aging time over 24 h were essential at low aqueous precursor concentrations of less than 0.6 M Ca^{2+} [16,17]. However, in this study, nanometer scale HAp powders were conveniently prepared in a fast and simple manner: faster precursor dropping rate at higher precursor concentration of 2.5 M Ca^{2+}. Therefore, it can be proposed that the new one-step process in this study is economically feasible of HAp preparation in large quantities.

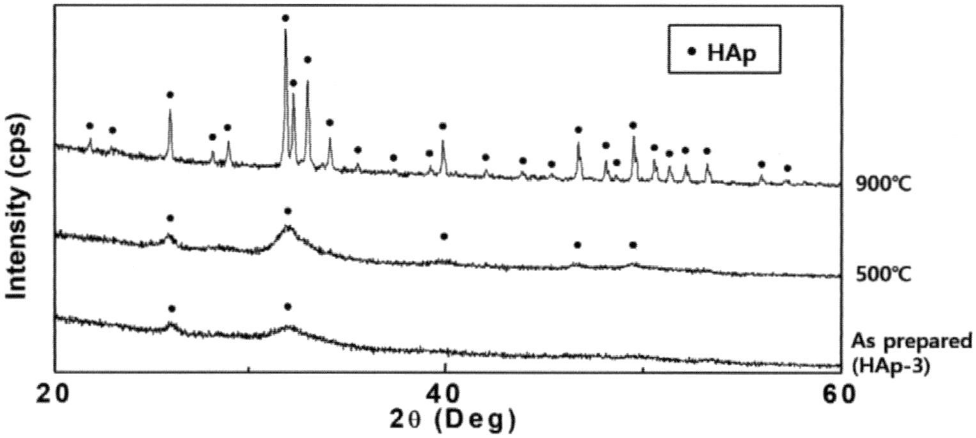

Figure 4. X-ray diffraction patterns of as-prepared and heat-treated powders from $CaCO_3$-H_3PO_4 with citric acid and sodium hydroxide under higher pH 8.9 (HAp-3).

4. Conclusions

Nanometer-sized HAp powders were conveniently prepared in a fast and simple manner from calcium carbonate and phosphoric acid with aid of additional reactants, citric acid and sodium hydroxide. The effect of both citric acid and sodium hydroxide on the preparation of phase pure HAp has been investigated. The pure HAp powders (HAp-3) were synthesized with citric acid (0.15 mole) and sodium hydroxide (pH 8.9). In the case of HAp-3, phase pure HAp without any noticeable secondary phase can be synthesized. It was also shown that, as the calcination temperature increased from 500°C to 900°C, the overall peaks of HAp become more distinct. The narrower peak width and higher peak height at 900°C suggest the increased crystallinity and particulate size. The new one-step approach of the aqueous precipitation method adopted in this study using calcium carbonate offered a relatively simple, convenient and economical route to synthesize HAp without long ripening time used in most wet precipitation studies.

Acknowledgement

This work was supported by the National Research Foundation of Korea (NRF) grant funded by the Korean government (Ministry of Science and ICT) (No. 2017M2B2A9073125).

References

[1] Zhou H and Lee J 2011 *Acta. Biomater.* **7** 2769 (https://doi.org/10.1016/j.actbio.2011.03.019)
[2] Orlovskii V, Komlev V and Barinov S 2002 *Inorg. Mater.* **38** 973 (https://doi.org/10.1023/A:1020585800572)
[3] Vukomanovic M, Zunic V, Otonicar M, Repnik U, Turk B, Skapin S and Suvorov D 2012 *J. Mater. Chem.* **22** 10571 (https://doi.org/10.1039/C2JM00136E)
[4] Fihri A, Len C, Varma R S and Solhy A 2017 *Coord. Chem. Rev.* **347** 48 (https://doi.org/10.1016/j.ccr.2017.06.009)
[5] Takagi O, Kuramoto N, Ozawa M and Suzuki S 2004 *Ceram. Int.* **30** 139 (https://doi.org/10.1016/S0272-8842(03)00061-0)
[6] Jeanjean J, Rouchaud J C and Tran L 1995 *J. Radioanal. Nucl. Chem. Lett.* **201**(6) 529 (https://doi.org/10.1007/BF02162730)
[7] Hamester M R R, Balzer P S and Becker D 2012 *Mater. Res.* **15**(2) 204 (https://doi.org/10.1590/S1516-14392012005000014)
[8] Akao M, Aoki H and Kato K 1981 *J. Mater. Sci.* **16** 809 (https://doi.org/10.1007/BF00552220)
[9] Abidi S and Murtaza Q 2014 *J. Mater. Sci. Techno.* **30** 307 (https://doi.org/10.1016/j.jmst.2013.10.011)
[10] Cahyaningrum S E, Herdyastuty N, Devina B and Supangat D 2018 *IOP Conf. Series: Mater. Sci. Engine.* **299** 012039 (https://doi.org/10.1088/1757-899X/299/1/012039)
[11] Minh D P, Rio S, Sharrock P, Sebei H, Lyczko N, Tran N D, Raii M and Nzihou A 2014 *J. Mater. Sci.* **49**(12) 4261 (https://doi.org/10.1007/s10853-014-8121-7)
[12] Verwilghen C, Chkir M, Rio S, Nzihou A and Sharrock P 2009 *Mater. Sci. Engin. C* **29**(3) 771 (https://doi.org/10.1016/j.msec.2008.07.007)
[13] Mansour S F, El-dek S I, Ahmed M A, Abd-Elwahab S M and Ahmed M K 2016 *Appl. Nanosci.* **6** 991 (https://doi.org/10.1007/s13204-015-0509-4)
[14] Müller W E G, Tolba E, Schröder H C, Muñoz-Espí R, Diehl-Seifert B and Wang X 2016 *Acta Biomater.* **31** 358 (https://doi.org/10.1016/j.actbio.2015.11.060)
[15] Vermeul V R, Szecsody J E, Fritz B G, Williams M D, Moore R C and Fruchter J S 2014 *Groundwater Monitor. Remed.* 1 (https://doi.org/10.1111/gwmr.12055)
[16] Sokolova M, Kreicbergs I, Zalite V and Berzina-Cimdinaa L 2012 *Eurasia Conf. Chem. Sci. Proc.* p 1
[17] Bang L T and Othman R 2014 *Ceramics-Silikaty* **58** 157 (http://www.ceramics-silikaty.cz/2014/pdf/2014_02_157.pdf)

IOP Conf. Series: Materials Science and Engineering 1113 (2021) 012014 doi:10.1088/1757-899X/1113/1/012014

The stabilization of state feedback for a class of systems with time delay

H P Li and Y Zhang

School of Science, Hebei University of Science and Technology, Shijiazhuang, China

E-mail: lishuxue@126.com

Abstract. In this paper we consider the stabilization of state feedback for a class of systems with time delay. At first, we investigated the stability on the system with two additive time delays. Then an appropriate LKF with double integral was constructed. Since the linear matrix inequalities, we get the sufficient condition to guarantee the system's asymptotically stability. At last, the effectiveness of the proposed method is demonstrated by a numerical example.

1. Introduction

It is well known that time delays are often encountered in various practical systems, such as engineering systems, biological, economic, neural networks and networked control systems, among others [1-6]. Since time-delay is frequently the main cause of oscillation, divergence or instability, considerable effort has been devoted to stability for systems with time delays. According to whether stability criteria include the information of the delay, they are divided into two classes: delay-independent stability criteria and delay-dependent ones. It is well known that delay-independent stability criteria tend to be more conservative especially for small size delays. More attention has been paid to delay-dependent stability.

In Recent years, networked control systems have been receiving great attention for their advantages. For example: low cost, simple maintenance, high reliability and so on. It is well known that the transmission delay and the data packet dropout are two fundamental issues in networked control systems. The transmission delay generally includes the sensor-to-control delay and the control-to-actuator delay. In most of existing papers the sensor to-control delay and the control-to-actuator delay were combined into one state delay, while the data packet dropouts were modelled as delays and absorbed by the state delay, thus formulating networked control systems as systems with one state delay. There are two fundamental issues: the transmission delay and the data packet dropout in the networked control systems. The transmission delay includes the sensor-to-control delay. τ_{sc} and the control-to-actuator delay τ_{ca}. In most of existing papers, let $\tau = \tau_{sc} + \tau_{ca}$, that is the two state delays were combined into one state delay. But it is not suitable since τ_{sc} and τ_{ca} are different delays. The stabilization for a networked control system with additive delays was investigated in the paper [7-12]. When the Lyapunov-Krasovskii functional (LKF) was constructed, each term must be positive definite, but it needn't be positive definite for each term in Lyapunov-Krasovskii functional in fact. So the results of the paper [7-12] can still be improved. An appropriate Lyapunov-Krasovskii functional is constructed in the paper [13], however, it is not necessary for all the Lyapunov matrices involved to be positive definite. Obviously requiring all the Lyapunov matrices involved to be positive definite may induce conservatism. But notice the estimation on the Lyapunov-Krasovskii functional; Jenson inequality was used, bringing the conservative result.

This paper is centered on the problem of state feedback stabilization for a class of systems with two additive input delays. First, the stability on the system with two additive delays is investigated. An appropriate Lyapunov-Krasovskii functional is constructed, and it is positive definite as a whole rather than each term. When estimating the derivative of the Lyapunov-Krasovskii functional along the system, different slack matrices are introduced according to the time delay belonging to different interval. Obtaining the stability condition of the system with less conservative. At last, illustrative example is provided to show the advantage and applicability of the developed results for network control systems.

2. Problem statement and preliminaries

Consider the following system with two additive input delays

$$\dot{x}(t) = Ax(t) + Bu(t - d_1(t) - d_2(t)) \tag{1}$$

where $x(t) \in R^n$ is the state vector, A, B are known real constant matrices, $d_1(t)$, $d_2(t)$ are two additive input delays, satisfying

$$0 \le d_1(t) \le h_1,\ 0 \le d_2(t) \le h_2, \cdots,\ d_1(t) \le \mu_1, d_2(t) \le \mu_2 \tag{2}$$

Given $u(t) = Kx(t)$ with a controller gain K, the closed-loop system is given by

$$\dot{x}(t) = Ax(t) + BKx(t - d_1(t) - d_2(t)) \tag{3}$$

Let $d(t) = d_1(t) + d_2(t)$ and system (3) can be rewritten as

$$\dot{x}(t) = Ax(t) + BKx(t - d(t)) \tag{4}$$

It is easy to get that

$$0 \le d(t) \le h,\ \dot{d}(t) \le \mu,\ h = h_1 + h_2,\ \mu = \mu_1 + \mu_2 \tag{5}$$

Two lemmas are given before the main results which are in the work of [14-15].

Lemma 2.1. $f_1, f_2, \cdots, f_N: R^m \to R^+$, if

$$\left\{ g_{ij} : R^m \to R,\ g_{ij}(t) = g_{ji}(t), \begin{bmatrix} f_i(t) & g_{ij}(t) \\ g_{ji}(t) & f_j(t) \end{bmatrix} \ge 0 \right\}$$

then

$$\min_{\left\{\alpha_i \mid \alpha_i > 0, \sum_i \alpha_i = 1\right\}} \sum_i \frac{1}{\alpha_i} f_i(t) = \sum_i f_i(t) + \max_{g_{ij}(t)} \sum_{i \ne j} g_{ij}(t)$$

Lemma 2.2. (Wirtinger inequality) For any symmetric matrix$M = M^T > 0$, then

$$-(b-d)\int_d^b \dot{x}^T(s) M \dot{x}(s) ds \le -\left(x(b) - x(d)\right)^T M \left(x(b) - x(d)\right) - 3\Omega^T M \Omega$$

where

$$\Omega = \left(x(b) + x(d) - \frac{2}{b-d} \int_d^b x(s) ds \right)$$

3. Main results

To simplify, define the following symbols:

$$\xi_0(t) = (x^T(t) \quad \dot{x}^T(t) \quad x^T(t-d(t)) \quad x^T(t-d_1(t)) \quad x^T(t-h) \quad x^T(t-h_1))^T$$

$$\xi_1(t) = (\frac{1}{d(t)}\int_{t-d(t)}^{t} x^T(s)ds \quad \frac{1}{h_1-d(t)}\int_{t-h_1}^{t-d(t)} x^T(s)ds \quad \frac{1}{h_2}\int_{t-h}^{t-h_1} x^T(s)ds)^T$$

$$\xi_2(t) = (\frac{1}{d(t)-h_1}\int_{t-d(t)}^{t-h_1} x^T(s)ds \quad \frac{1}{h-d(t)}\int_{t-h}^{t-d(t)} x^T(s)ds \quad \frac{1}{h_1}\int_{t-h_1}^{t} x^T(s)ds)^T$$

$$\eta_1(t) = \left(\xi_0^T(t) \quad \xi_1^T(t)\right)^T, \ \eta_2(t) = \left(\xi_0^T(t) \quad \xi_2^T(t)\right)^T$$

$$e_i = (0_{n\times(i-1)n} \quad I \quad 0_{n\times(9-i)n}), \ i = 1, 2, \cdots, 9$$

Theorem 3.1. The system (3) is asymptotically stable. Given h_1, h_2, μ_1, μ_2, if there exist symmetric matrices P, Q_2, Q_4, symmetric positive definite matrices Q_1, Q_3, Z_i, i=0, 1, 2, R_1, R_2, and the matrices x_i, i=1, 2, R_1, R_2, and the matrices X_i, i=1, 2, 3, M_1, M_2, G_1, G_2, with the appropriate dimension such that the following hold

$$\Omega_0 + h_1\Omega_1 < 0, \ \Omega_0 + h_1\Omega_2 < 0, \ \bar{\Omega}_0 + h_2\bar{\Omega}_1 < 0, \ \bar{\Omega}_0 + h_2\bar{\Omega}_2 < 0 \tag{6}$$

$$\begin{bmatrix} \frac{h_1}{h}P + Z_1 & -Z_1 \\ * & Z_1 + h_1 Q_4 \end{bmatrix} > 0, \ \begin{bmatrix} P + Z_2 & -Z_2 \\ * & Z_2 + hQ_2 \end{bmatrix} > 0 \tag{7}$$

$$\Xi_1 = \begin{bmatrix} \tilde{Z}_1 & M_1 \\ * & \tilde{Z}_1 \end{bmatrix} \geq 0, \ \Sigma_1 = \begin{bmatrix} Z_0 & X_1 & X_2 \\ * & Z_0 & X_3 \\ * & * & Z_0 \end{bmatrix} \geq 0 \tag{8}$$

$$\Xi_2 = \begin{bmatrix} \tilde{Z}_2 & M_2 \\ * & \tilde{Z}_2 \end{bmatrix} \geq 0, \ \Sigma_2 = \begin{bmatrix} Z_0 & Y \\ * & Z_0 \end{bmatrix} \geq 0 \tag{9}$$

where

$$\Omega_0 = \Gamma_1 + \Gamma_2 + \Phi$$
$$\bar{\Omega}_0 = \Gamma_1 + \bar{\Gamma}_2 + \Phi$$

$$\Gamma_1 = 2e_1^T P e_2 + e_1^T(Q_1 + Q_3 + Q_4)e_1 - (1-\mu_1)e_4^T Q_1 e_4 - (1-\mu)e_3^T Q_3 e_3 \\ - e_5^T Q_2 e_5 - e_6^T(Q_4 - Q_2)e_6 + e_2^T\left[h_1(Z_0 + Z_1) + h_2 Z_2\right]e_2 + e_1^T(h_1 R_1 + h_2 R_2)e_1$$

IOP Conf. Series: Materials Science and Engineering 1113 (2021) 012014 doi:10.1088/1757-899X/1113/1/012014

$$\Gamma_2 = -\frac{1}{h_1}\begin{bmatrix} e_1 - e_3 \\ e_1 + e_3 - 2e_7 \\ e_3 - e_6 \\ e_3 + e_6 - 2e_8 \end{bmatrix}^T \Xi_1 \begin{bmatrix} e_1 - e_3 \\ e_1 + e_3 - 2e_7 \\ e_3 - e_6 \\ e_3 + e_6 - 2e_8 \end{bmatrix} - \frac{1}{h_2}\begin{bmatrix} e_6 - e_5 \\ e_6 + e_5 - 2e_9 \end{bmatrix}^T \tilde{Z}_2 \begin{bmatrix} e_6 - e_5 \\ e_6 + e_5 - 2e_9 \end{bmatrix}$$

$$-\frac{1}{h_1}\begin{bmatrix} e_1 - e_4 \\ e_4 - e_3 \\ e_3 - e_6 \end{bmatrix}^T \Sigma_1 \begin{bmatrix} e_1 - e_4 \\ e_4 - e_3 \\ e_3 - e_6 \end{bmatrix} - h_2 e_9^T R_2 e_9, \quad \tilde{Z}_2 = \begin{bmatrix} Z_2 & 0 \\ 0 & 3Z_2 \end{bmatrix}$$

$$\Phi = 2(G_1^T e_1 + G_2^T e_2)^T(-e_2 + Ae_1 + BKe_3), \quad \Omega_1 = -e_7^T R_1 e_7, \quad \Omega_2 = -e_8^T R_1 e_8$$

$$\bar{\Gamma}_2 = -\frac{1}{h_1}\begin{bmatrix} e_1 - e_6 \\ e_1 + e_6 - 2e_9 \end{bmatrix}^T \tilde{Z}_1 \begin{bmatrix} e_1 - e_6 \\ e_1 + e_6 - 2e_9 \end{bmatrix} - \frac{1}{h_1}\begin{bmatrix} e_1 - e_4 \\ e_4 - e_6 \end{bmatrix}^T \Sigma_2 \begin{bmatrix} e_1 - e_4 \\ e_4 - e_6 \end{bmatrix}$$

$$-\frac{1}{h_2}\begin{bmatrix} e_6 - e_3 \\ e_6 + e_3 - 2e_7 \\ e_3 - e_5 \\ e_3 + e_5 - 2e_8 \end{bmatrix}^T \Xi_2 \begin{bmatrix} e_6 - e_3 \\ e_6 + e_3 - 2e_7 \\ e_3 - e_5 \\ e_3 + e_5 - 2e_8 \end{bmatrix} - h_1 e_9^T R_1 e_9, \quad \tilde{Z}_1 = \begin{bmatrix} Z_1 & 0 \\ 0 & 3Z_1 \end{bmatrix}$$

$$\bar{\Omega}_1 = -e_7^T R_2 e_7$$
$$\bar{\Omega}_2 = -e_8^T R_2 e_8$$

Proof: Define Lyapunov-Krasovskii functional candidate as

$$V(t) = V_1(t) + V_2(t) \tag{10}$$

where

$$\begin{aligned} V_1(t) &= x^T(t)Px(t) + \int_{t-d_1(t)}^{t} x^T(\alpha)Q_1 x(\alpha)d\alpha + \int_{t-d(t)}^{t} x^T(\alpha)Q_3 x(\alpha)d\alpha \\ &+ \int_{-h_1}^{0} x^T(t+\alpha)Q_4 x(t+\alpha)d\alpha + \int_{-h}^{-h_1} x^T(t+\alpha)Q_2 x(t+\alpha)d\alpha \\ &+ \int_{-h_1}^{0}\int_{t+s}^{t} \dot{x}^T(\alpha)Z_1\dot{x}(\alpha)d\alpha ds + \int_{-h}^{-h_1}\int_{t+s}^{t} \dot{x}^T(\alpha)Z_2\dot{x}(\alpha)d\alpha ds \end{aligned}$$

$$\begin{aligned} V_2(t) &= \int_{-h_1}^{0}\int_{t+\theta}^{t} \dot{x}^T(s)Z_0\dot{x}(s)dsd\theta + \int_{-h_1}^{0}\int_{t+\theta}^{t} x^T(s)R_1 x(s)dsd\theta \\ &+ \int_{-h}^{-h_1}\int_{t+\theta}^{t} x^T(s)R_2 x(s)dsd\theta \end{aligned}$$

We show Lyapunov-Krasovskii functional (10) is positive definite, by Jenson inequality it follows that

$$\int_{-h_1}^{0}\int_{t+s}^{t} \dot{x}^T(\alpha)Z_1\dot{x}(\alpha)d\alpha ds \geq \frac{1}{h_1}\int_{-h_1}^{0} [x(t) - x(t+s)]^T Z_1 [x(t) - x(t+s)] ds$$

$$\int_{-h}^{-h_1}\int_{t+s}^{t}\dot{x}^T(\alpha)Z_2\dot{x}(\alpha)d\alpha ds \geq \frac{1}{h}\int_{-h}^{-h_1}\left[x(t)-x(t+s)\right]^T Z_2\left[x(t)-x(t+s)\right]ds$$

On the other hand,

$$x^T(t)Px(t)=\frac{1}{h}\int_{-h_1}^{0}x^T(t)Px(t)ds+\frac{1}{h}\int_{-h}^{-h_1}x^T(t)Px(t)ds$$

$$V(t)\geq \frac{1}{h}\int_{-h_1}^{o}x^T(t)Px(t)ds+\frac{1}{h}\int_{-h}^{-h_1}x^T(t)Px(t)ds+\int_{-h_1}^{0}x^T(t+s)Q_4x(t+s)ds+\int_{-h}^{-h_1}x(t+s)^TQ_2x(t+s)ds$$
$$+\frac{1}{h_1}\int_{-h_1}^{0}\left[x(t)-x(t+s)\right]^T Z_1\left[x(t)-x(t+s)\right]ds+\frac{1}{h}\int_{-h}^{-h_1}\left[x(t)-x(t+s)\right]^T Z_2\left[x(t)-x(t+s)\right]ds$$
$$=\int_{-h_1}^{0}\begin{bmatrix}x(t)\\x(t+s)\end{bmatrix}^T\begin{bmatrix}\frac{1}{h}P+\frac{1}{h_1}Z_1 & -\frac{1}{h_1}Z_1\\ * & \frac{1}{h_1}Z_1+Q_4\end{bmatrix}\begin{bmatrix}x(t)\\x(t+s)\end{bmatrix}ds$$
$$+\int_{-h}^{h_1}\begin{bmatrix}x(t)\\x(t+s)\end{bmatrix}^T\begin{bmatrix}\frac{1}{h}P+\frac{1}{h}Z_2 & -\frac{1}{h}Z_2\\ * & \frac{1}{h}Z_2+Q_2\end{bmatrix}\begin{bmatrix}x(t)\\x(t+s)\end{bmatrix}ds$$

$V(t)$ is positive definite by (7). Derivative $V_1(t), V_2(t)$ along the trajectory of (3) yields

$$\begin{aligned}\dot{V}_1(t)&\leq 2x^T(t)P(Ax(t)+BKx(t-d(t)))+x^T(t)(Q_1+Q_3+Q_4)x(t)\\&+x^T(t-h_1)(Q_2-Q_4)x(t-h_1)-x^T(t-h)Q_2x(t-h)\\&-(1-\mu)x^T(t-d(t))Q_3x(t-d(t))-(1-\mu_1)x^T(t-d_1(t))Q_1x(t-d_1(t))\\&+\dot{x}^T(t)(h_1Z_1+h_2Z_2)\dot{x}(t)-\int_{t-h_1}^{t}\dot{x}^T(s)Z_1\dot{x}(s)ds-\int_{t-h}^{t-h_1}\dot{x}^T(s)Z_2\dot{x}(s)ds\end{aligned}$$

$$\begin{aligned}\dot{V}_2(t)&=h_1\dot{x}^T(t)Z_0\dot{x}(t)+x^T(t)(h_1R_1+h_2R_2)x(t)\\&-\int_{t-h_1}^{t}x^T(s)R_1x(s)ds-\int_{t-h}^{t-h_1}x^T(s)R_2x(s)ds\\&-\int_{t-h_1}^{t}\dot{x}^T(s)Z_0\dot{x}(s)ds\end{aligned}$$

The following are discussed based on different values of $d(t)$.
Case1: $d(t)\in[0,\ h_1]$ by Lemma 2.1 and Lemma 2.2, we have

$$-\int_{t-h_1}^{t}\dot{x}^T(s)Z_1\dot{x}(s)ds=-(\int_{t-d(t)}^{t}\dot{x}^T(s)Z_1\dot{x}(s)ds+\int_{t-h_1}^{t-d(t)}\dot{x}^T(s)Z_1\dot{x}(s)ds)\leq-\frac{1}{h_1}\alpha_1^T(t)\Xi_1\alpha_1(t) \quad (11)$$

where

$$\alpha_1(t) = \left(\alpha_{11}^T(t) \quad \alpha_{12}^T(t) \quad \alpha_{13}^T(t) \quad \alpha_{14}^T(t)\right)^T$$
$$\alpha_{11}(t) = x(t) - x(t-d(t))$$
$$\alpha_{12}(t) = x(t) + x(t-d(t)) - \frac{2}{d(t)}\int_{t-d(t)}^{t} x(s)ds$$
$$\alpha_{13}(t) = x(t-d(t)) - x(t-h_1)$$
$$\alpha_{14}(t) = x(t-d(t)) + x(t-h_1) - \frac{2}{h_1 - d(t)}\int_{t-h_1}^{t-d(t)} x(s)ds$$

Similarly,

$$-\int_{t-h}^{t-h_1} \dot{x}^T(s)Z_2\dot{x}(s)ds \le -\frac{1}{h_2}\beta_1^T(t)\tilde{Z}_2\beta_1(t)$$
$$\beta_1(t) = \left(\beta_{11}^T(t) \quad \beta_{12}^T(t)\right)^T \tag{12}$$

where

$$\beta_{11}(t) = x(t-h_1) - x(t-h)$$
$$\beta_{12}(t) = x(t-h_1) + x(t-h) - \frac{2}{h_2}\int_{t-h}^{t-h_1} x(s)ds$$

Similarly,

$$-\int_{t-h_1}^{t} \dot{x}^T(s)Z_0\dot{x}(s)ds = -(\int_{t-d_1(t)}^{t} \dot{x}^T(s)Z_0\dot{x}(s)ds + \int_{t-d(t)}^{t-d_1(t)} \dot{x}^T(s)Z_0\dot{x}(s)ds + \int_{t-h_1}^{t-d(t)} \dot{x}^T(s)Z_0\dot{x}(s)ds)$$
$$\le -\frac{1}{h_1}\varphi_1^T(t)\Sigma_1\varphi_1(t) \tag{13}$$

where

$$\varphi_1(t) = \left(\varphi_{11}^T(t) \quad \varphi_{12}^T(t) \quad \varphi_{13}^T(t)\right)^T$$
$$\varphi_{11}(t) = x(t) - x(t-d_1(t))$$
$$\varphi_{12}(t) = x(t-d_1(t)) - x(t-d(t))$$
$$\varphi_{13}(t) = x(t-d(t)) - x(t-h_1)$$

Applying Jenson inequality obtains

$$-\int_{t-h_1}^{t} x^T(s)R_1x(s)ds = -\int_{t-d(t)}^{t} x^T(s)R_1x(s)ds - \int_{t-h_1}^{t-d(t)} x^T(s)R_1x(s)ds$$
$$\le -d(t)v_{11}^T(t)R_1v_{11}(t) - (h_1 - d(t))v_{12}^T R_1 v_{12}(t) - \int_{t-h}^{t-h_1} x^T(s)R_2x(s)ds \le -h_2 v_{13}^T(t)R_2v_{13}(t)$$

For any constant matrices G_1, G_2 with the appropriate dimension, it follows that

$$(G_1^T(t)x(t) + G_2^T(t)\dot{x}(t))^T(-\dot{x}(t) + Ax(t) + BKx(t-d(t))) = 0$$

We have

$$\dot{V}(t) \le \eta_1^T(t)(\Omega_0 + d(t)\Omega_1 + (h_1 - d(t))\Omega_2)\eta_1(t) \tag{14}$$

Case2.

$$d(t) \in [h_1, h]$$

Then

$$\dot{\mathrm{V}}(\mathrm{t}) \le \eta_2^{\mathrm{T}}(\mathrm{t})(\bar{\Omega}_0 + (\mathrm{d(t)} + \mathrm{h}_1)\bar{\Omega}_1 + (\mathrm{h} - \mathrm{d(t)})\bar{\Omega}_2)\eta_2(\mathrm{t})$$

where

$$\beta_2(t) = \begin{pmatrix}\beta_{21}^T(t) & \beta_{22}^T(t)\end{pmatrix}^T$$
$$\beta_{21}(t) = x(t) - x(t-h_1)$$
$$\beta_{22}(t) = x(t) - x(t-h_1) - \frac{2}{h_1}\int_{t-h_1}^{t} x(s)ds$$

Similarly,

$$-\int_{t-h}^{t-h_1} \dot{x}^T(s)Z_2\dot{x}(s)ds \le -\frac{1}{h_2}\alpha_2^T(t)\Xi_2\alpha_2(t)$$

where

$$\alpha_2(t) = \begin{pmatrix}\alpha_{21}^T(t) & \alpha_{22}^T(t) & \alpha_{23}^T(t) & \alpha_{24}^T(t)\end{pmatrix}^T$$
$$\alpha_{21}(t) = x(t-h_1) - x(t-d(t))$$
$$\alpha_{22}(t) = x(t-h_1) + x(t-d(t)) - \frac{2}{d(t)-h_1}\int_{t-d(t)}^{t-h_1} x(s)ds$$
$$\alpha_{23}(t) = x(t-d(t)) - x(t-h)$$
$$\alpha_{24}(t) = x(t-d(t)) + x(t-h) - \frac{2}{h-d(t)}\int_{t-h}^{t-d(t)} x(s)ds$$

Similarly,

$$-\int_{t-h_1}^{t} \dot{x}^T(s)Z_0\dot{x}(s)ds \le -\frac{1}{h_1}\varphi_2^T(t)\textstyle\sum_2 \varphi_2(t)$$

$$\varphi_2(t) = \begin{pmatrix}\varphi_{21}^T(t) & \varphi_{22}^T(t)\end{pmatrix}^T$$
$$\varphi_{21}(t) = x(t) - x(t-d_1(t))$$
$$\varphi_{22}(t) = x(t-d_1(t)) - x(t-h_1)$$

From Lemma 2.1, it follows that

$$-\int_{t-h_1}^{t} x^T(s)R_1x(s)ds \le -h_1(\frac{1}{h_1}\int_{t-h_1}^{t} x(s)ds)^T R_1(\frac{1}{h_1}\int_{t-h_1}^{t} x(s)ds)$$

$$-\int_{t-h}^{t-h_1} x^T(s)R_1x(s)ds = -\int_{t-d(t)}^{t-h_1} x^T(s)R_2x(s)ds - \int_{t-h}^{t-d(t)} x^T(s)R_2x(s)ds$$
$$\le -(d(t)-h_1)v_{21}^T(t)R_2v_{21}(t) - (h-d(t))v_{22}^T(t)R_2v_{22}(t)$$

$$v_{21}(t) = \frac{1}{d(t)-h_1}\int_{t-d(t)}^{t-h_1} x(s)ds$$
$$v_{22}(t) = \frac{1}{h-d(t)}\int_{t-h}^{t-d(t)} x(s)ds$$

then

$$\dot{V}(t) \le \eta_2^T(t)(\bar{\Omega}_0 + (d(t)+h_1)\bar{\Omega}_1 + (h-d(t))\bar{\Omega}_2)\eta_2(t) \tag{15}$$

We have $V(t) < 0$ whatever $d(t) \in [0,\ h_1]$ or $d(t) \in [h_1,\ h]$ by convex combination, if (6) holds, then system (3) is asymptotically stable.

4. Numerical examples

Example 4.1. Consider the system (3) with the following parameters

$$A = \begin{bmatrix} -2 & 0 \\ 0 & -0.9 \end{bmatrix}$$
$$B = \begin{bmatrix} 1 & 0 \\ 0 & 1 \end{bmatrix}$$
$$K = \begin{bmatrix} -1 & 0 \\ -1 & -1 \end{bmatrix}$$
$$\mu_1 = 0.1$$
$$\mu_2 = 0.8$$

Admissible upper bound h_2 are listed in Table 1 with given h_1 by theorem 3.1.

Table 1. Lipase search from the data mining of putative open reading frame with over 250 amino acids in length based on the NCBI BLAST result.

h_1 / h_2 / Method	1	1.2	1.5
[7]	0.415	0.379	0.248
[8]	0.512	0.406	0.283
[9]	0.583	0.519	0.421
[10]	0.595	0.463	0.312
[11]	0.873	0.673	0.373
[12]	0.988	0.836	0.563
Theorem 3.1	1.076	0.891	0.574

It can be seen from the table1 that Theorem 3.1 in this paper provides a much larger upper delay bound than other approaches, showing the effectiveness of our approach.

5. Conclusions

This paper is centered on the problem of state feedback stabilization for a networked control system with two additive input delays. The sufficient condition for the asymptotic stability is given through constructing an appropriate Lyapunov-Krasovskii functional, comprehensive utilization of convex combination, inverted convex combination, Wirtinger inequality and Jenson inequality. Then based on the stability results state feedback stabilization is studied. One numerical examples are given to show the effectiveness and advantage of the proposed results.

Moreover we can further develop results if drawing in triple integrals, constructing augmented Lyapunov-Krasovskii functional or using the delay decomposition Lyapunov-Krasovskii functional.

Acknowledgments
The project supported by the Natural Science Foundation of Hebei Province of China under Grant No.A2019208336.

References

[1] Wang G 2011 *J. Math. Anal. Appl.* **383** pp 573-584
[2] Gu K, Kharitonov V L and Chen J 2003 *Stability of time-delay systems* (Boston: Birkhäuser)
[3] Peng C and Yang T 2013 *Automatica* **49** pp 1326-1332
[4] Zhang X M and Han Q L 2013 *IET Control Theory Appl.* **8** pp 226-234
[5] Li T, Guo L, Sun, C and Lin C 2008 *IEEE Trans on Neural* **19** pp 726-730
[6] Jiang X, Han Q L, Liu S and Xue A 2008 *IEEE Tran. Automat. Contr.* **53** pp 1025-1032
[7] Lam J, Gao H and Wang C 2007 *Syst. Control Lett.* **56** pp 16-24
[8] Gao H, Chen T and Lam J 2008 *Automatica* **44** pp 39-52
[9] Li P 2014 *ISA Trans.* **53** pp 258-266
[10] Shao H, Zhang Z, Zhu X and Miao G 2014 *Abstr Appl Anal.* **2014** 923436
[11] Ge X 2014 *Applied. Math. Comput.* **241** pp 42-46
[12] Shao H 2009 *Automatica.* **45** pp 744-749
[13] Shao H and Zhang Z 2016 *Appl Math Comput.* **265** pp 748-758
[14] Park P, Ko J and Jeong C 2011 *Automatica.* **47** pp 235-238
[15] Seuret A and Gouaisbaut F 2013 *Automatica.* **49** pp 2860-2866

Building an asynchronous HTML5-related competency-based guided e-learning system

Cheng Hsiu Li

Department of Information Management, National Taitung Junior College, Taitung, Taiwan

E-mail: lcs@ntc.edu.tw

Abstract. Improving learner behavior and attitude is one method for improving learning quality. However, every student has unique characteristics, which makes the prediction of how learning behavior alters learning effects challenging. Humans obtain knowledge through cognitive learning, which can be effectively taught using asynchronous e-learning systems. However, these systems require self-study and include tedious processes that can affect learner motivation and the overall learning effect. This study implemented an HTML5-related competency-based guided e-learning (CBGE) system with an embedded competency-based guided learning (CBGL) mechanism to implement a personalized learning environment. Moreover, learner behaviors were recorded in a log dataset for data mining. The results revealed that the participants were satisfied with the CBGL mechanism and e-course design. The cognitive ability of the participants significantly improved after the experiment. Moreover, the growth in cognitive ability of the participants who completed the guiding process was significantly higher than that of the participants who did not finish the guiding process. This indicates that cognitive ability was improved through the CBGE system and the completion of the CBGL process provided a significantly more pronounced learning effect. A decision tree technique was also employed to construct a predictive model of the learning effect that could help learners achieve the learning objective. The predictive model revealed that learners who used the CBGE system could not achieve the learning objective without passing the stage tests a minimum of 2.5 times. Thus, the number of times that the stage tests were passed was the critical factor in achieving the learning objective.

1. Introduction

Improving learners' behavior and attitude is one method for improving learning quality [10]. However, every student has unique characteristics, which makes the prediction of how learning behavior alters learning effects challenging [9].

Humans obtain knowledge through cognitive learning. Currently, numerous online test systems exist for cognitive learning. Most of these systems involve asynchronous drills and include practice e-learning systems that employ cognitive question banks. These systems provide immediate feedback by using sequential training processes to determine the cognitive knowledge of the learner. Such systems are often used in education to assist in teaching. However, asynchronous e-learning without teacher assistance coupled with tedious learning processes negatively affects learner motivation and outcomes. This is a critical concern for many teachers; thus, studies that seek to overcome these problems are of clear importance.

In the 1970s, academicians determined that humans obtain cognitive abilities, such as knowledge and memory, according to the principles of the theory of behaviorism [2]. Cognitive development is the first stage of humans recognizing the world around them. Humans continuously gain knowledge through cognitive stimuli, cognitive responses, and experience feedbacks (S–R–F). A computer-assisted testing system is a digital drill and practice learning tool. Digital drill and practice learning is a type of e-learning method that integrates behaviorism and is usually used in cognitive ability learning. The cognitive stimulus and cognitive response (S–R) association model is formed from continuous stimuli and responses and is strengthened by feedback. The behaviorism model is illustrated in Figure 1. Appreciation and reward provide positive reinforcement and punishment restrains wrong behavior. The principle of contiguity emphasizes immediate feedback and states that events that occur together are likely to become associated. The principle of exercise emphasizes continuous practice as a route to attaining mastery. Therefore, applying the correct response in the shortest time through continuous exercise can be emphasized through a curriculum designed to strengthen the S–R association. In the drill and practice e-learning system, questions act as a stimulus and answers act as a response. The system adds points when the answer is correct and deducts points (negative feedback) when the answer is incorrect. The reward and punishment come from the immediate feedback (as in the principle of contiguity). Moreover, learners using this system must practice continuously (as in the principle of exercise). The drill and practice e-learning system can be a useful tool for accelerating the S–R–F process in cognitive knowledge learning. Li employed this process to propose an algorithm known as the proficiency-based learning algorithm. This algorithm is quite suitable for knowledge-oriented courses, which are designed to employ several items with standard answers [17]. However, because the system emphasizes continuous practice to increase efficacy, it can cause boredom among students and does not motivate learners or encourage self-study [7].

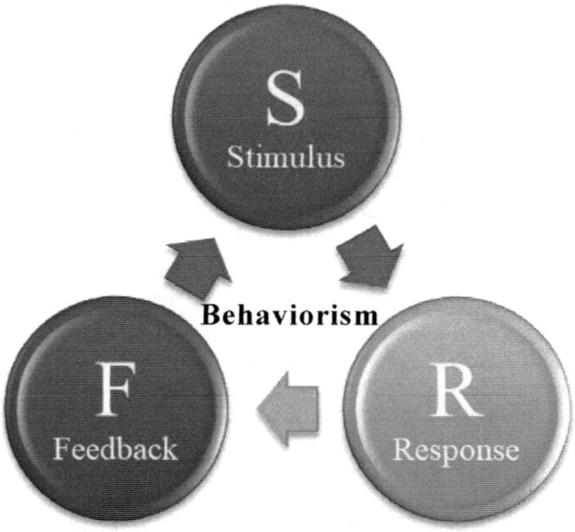

Figure 1. Behaviorism model.

Computer-assisted instruction (CAI) is generally used in interactive functions to introduce learning materials and provide an individual or individualized teaching environment. The benefits of CAI include increased interaction opportunities, individualized instructional requirements, immediate feedback, study motivation, and easy-to-monitor learning. Personalized learning is widely used in education [5]. In the out-of-classroom learning process of online learning systems, students can select personalized activities. These systems progressively increase the difficulty of activities and provide a personalized learning schedule. As students progress through their schedule, they experience a feeling of satisfaction and increased motivation. This is a form of personalized purpose that enables learners to have a more favorable learning experience and outcome [8]. The inclusion of stages allows a personalized design that is similar to a computer game, which gradually promotes users through different levels [3, 6]. Competency-based guided learning (CBGL) theory emphasizes that students be the masters of the learning process and also emphasize individual differences among students. CBGL theory involves stages. The cognitive learning material is usually always presented in a fixed question bank. When the learning material is in a fixed range and can be separated into several units, e-learning systems using CBGL can be based on the different levels of every unit to schedule a personalized learning path and effectively guide learners through their study [4]. The stimuli and building methods in e-learning environments directly affect students' learning, and students' attitudes to learn clearly affect the learning effects in such environments [1]. When e-learning environments record users' behaviors, a large, logged dataset is available to researchers for analysis for exploring users' behavior models [11, 16].

Therefore, this study used Internet technologies to develop a web-based asynchronous drill and practice e-learning system employing an HTML5-related cognitive course. The system embedded a CBGL mechanism to develop a personalized learning environment that guided learners to study in stages to promote positive learning effects. The logged dataset recorded learners' learning behaviors. Then, a decision tree technique was employed to analyze the logged dataset to explore why the boring and lengthy learning processes of asynchronous drill and practice e-learning systems result in poor learning effects. Finally, based on the findings, a predictive model was developed to determine successful learning among learners using such systems.

2. Related work

To tutor technicians, teachers use the test to check the performance of students' study. Testing methods have changed from traditional paper tests to computer-aided testing. By using the system, teachers can provide results instantly and be unrestricted vis-à-vis time and place. In this study, the discipline of level C technician for computer software application was used to the experimental material. This study proposed a proficiency-based learning algorithm by observing the training processes of technicians. The proposed algorithm was embedded in the computer-aided testing system. The newly developed system contained a proficient learning mechanism. This system is quite suitable for question-based knowledge-oriented courses. However, the system only provides a fixed sequence of activities in the learning path for every learner and repeats this sequence until learners are proficient. The learning material is not adaptive. Moreover, because the system emphasizes continuous practice to increase learning efficacy, boredom can be caused among learners and their motivation can be affected [17].

Hsu and Li revealed that when the learning material is in a fixed range and can be separated into several units, e-learning systems using the CBGL algorithm can be based on different levels of every unit to schedule a personalized learning path and effectively guide learners through their study. Before study, a learner is guided to a pretest, and the learner's cognitive ability for each unit is analyzed. The algorithm then generates an adaptive learning path according to the cognitive abilities of the learner. The system guides the learner to learn proficiently with the learning path. The question bank interface is also redesigned for open and closed answers so that learners do not memorize answers when they study and they can internalize and absorb. If the cognitive abilities of some units tested in the pretest already attain the expected goal, then the system does not guide learners to study. By using this method, the study period can be reduced without affecting the learning outcomes [4].

Predictive studies have been widely used in education to provide strategic information related to teaching, learning, and researching which variables predict certain educational outcomes [23, 24, 25]. Data mining allows the discovery and abstraction of knowledge from one or many large-scale databases, data warehouses, and other information sources [11, 12, 13, 21]. A typical procedure of knowledge data discovery includes five stages: data preprocessing, data transformation, data mining, data visualization, and data interpretation [15]. Data mining has been applied to web-based educational systems to discover unique data patterns [11, 14, 19, 22]. The gradual increase in the availability and quality of educational datasets and data mining algorithms indicates that educational data mining has become a crucial research domain that combines the fields of education and information or computer sciences [10, 14, 18, 20, 21]. The decision tree method is a well-known and standard approach in data mining and has been widely used in data science since the 1980s. The decision tree method has been widely used in many studies pertaining to education [26, 27, 28, 29, 30].

3. Proposed methods

This section introduces the system model, cognitive e-course design, the design of the adaptive, personalized learning cognitive e-course, and the method used to record user behavior logs. This study employed PHP6, HTML5, CSS3, JavaScript, and MySQL to implement an HTML5-related competency-based guided e-learning (CBGE) system. The system website can be found at http://lcs.dk.ntc.edu.tw/~lcs/mtahtml5etask_ng/.

3.1. System model

The system model is illustrated in Figure 2. The system contains four major elements: a user, asynchronous HTML5-related CBGE system, webpage server, and database server. The system was based on an asynchronous drill and practice e-learning system. The e-learning course was focused on HTML5-related cognitive knowledge by using a question bank. The CBGL algorithm was integrated into the cognitive e-course so that the system could arrange the learning stages according to learners' cognitive abilities. This method guided the learners in a self-study mode without any assistance from a teacher.

For studying, users were permitted to use any device, such as notebook personal computers, tablets, and smartphones, that was capable of accessing the webpage. Because the system was accessed through the Internet, users were able to log into the system to study at any location with Internet access.

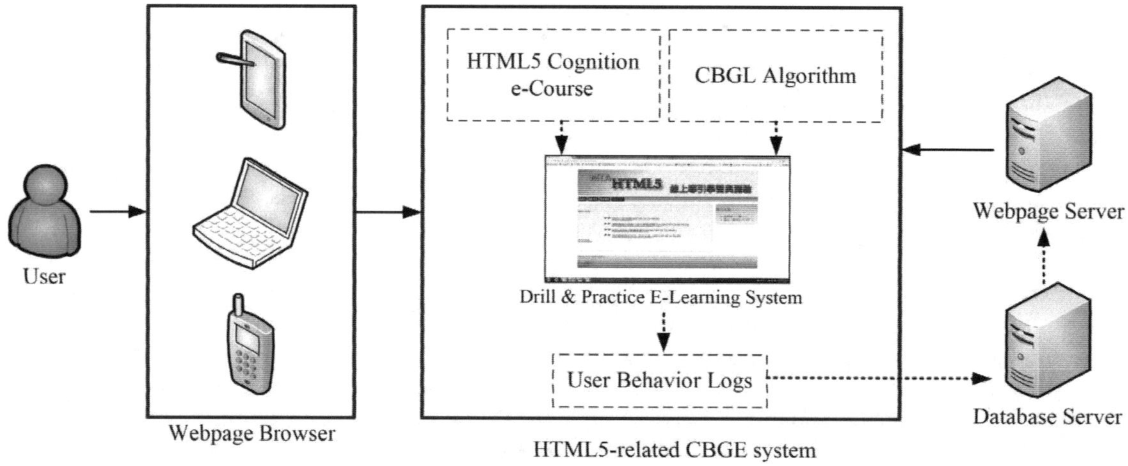

Figure 2. System model.

3.2. Design of the cognitive e-course and personalized learning

Online test systems are often used to verify learning effects by using question banks as the learning material.

Programming is a standard course in the departments of information engineering and information management and is also becoming a common subject in many other areas. Numerous programming languages exist. HTML5 is the newest front-end programming language for website development. This study employed a cognitive question bank based on HTML5 as the learning material and the CBGL algorithm to categorize the content into five learning units that included 257 questions. The system then developed the e-course interface based on the question bank. Training courses for

HTML5 programming include two parts—cognitive knowledge and practical technique. These parts cover HTML5, CSS3, JavaScript, and jQuery. Cognitive knowledge can be learned using an e-learning system to increase teaching efficiency; this allows teachers to focus on teaching practical techniques.

Students and teachers rarely access e-learning systems simultaneously during asynchronous e-learning. Therefore, the system must explain the meaning of its answers and assist the students as a teacher might do. Constructivism theory was employed to design a system interface that clearly indicated the meaning of the answers. Two types of interfaces were designed—closed and open solutions (Figure 3). The closed solution interface sought to directly explain the reason behind the answer. By contrast, the open solution interface used hyperlinks connected to learning resources on the Internet to provide students with their own space to think about and comprehend the related information. Through this method, students could come to understand the answer rather than simply memorizing the answer. This learning method enabled them to realize a cognitive link through their personal experience.

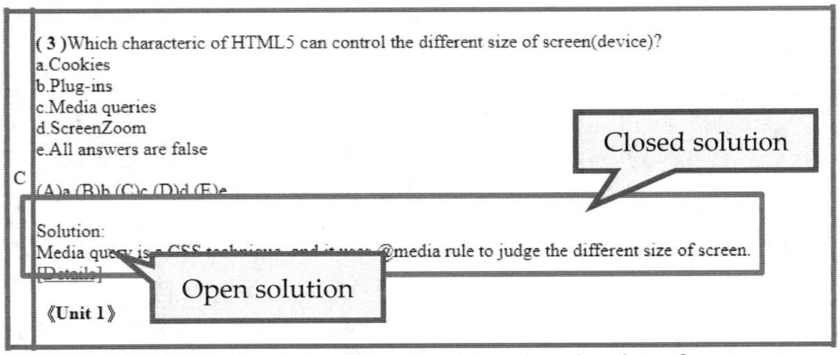

Figure 3. Design of the cognitive learning interface.

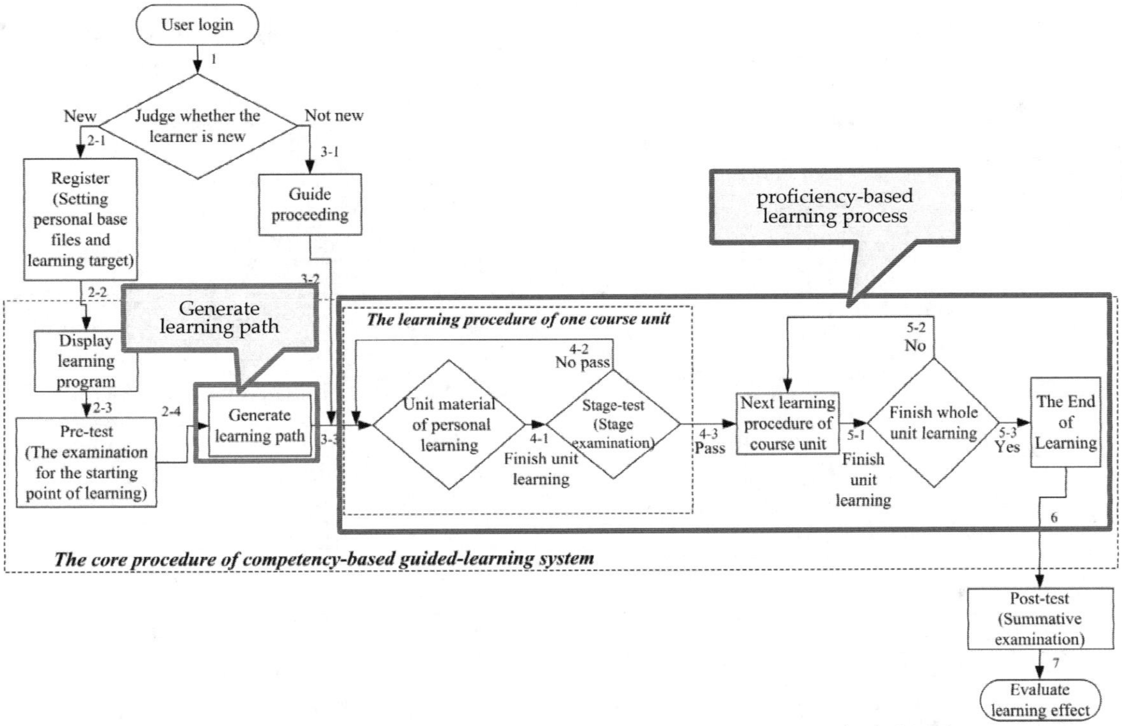

Figure 4. Algorithmic process of the CBGL system (Hsu and Li, 2015).

If the cognitive learning material is contained within a question bank, which has a fixed range and is separated into five units, then the CBGL algorithm can be used to generate personalized learning paths according to the different cognitive abilities of its users, with the system guiding users to study in stages [4]. The CBGL algorithm was the basis of the proposed system design. Every learner was assigned a pretest to gauge their HTML5 cognitive abilities before learning, and the system generated a personalized learning path based on the pretest results.

Hsu and Li proposed an algorithmic process for a CBGL system (Figure 4) [4]. Upon completing registration, their system guides learners directly to the pretest stage. The system randomly displays questions from the question bank, and learners proceed through the test. The result of every unit is recorded in the system database. The obtained test result is then computed, and the learning path is generated using Equation (1), which then proceeds to the proficiency-based learning process. The learners proceed to the next unit after they attain a required level of cognitive ability. Otherwise, learners have to repeatedly practice the unit until they attain the necessary standard.

$$\textit{Personal learning path} ::= \textit{Bubble sort rank}(Score(i)\,|_{1\le i\le N}) \qquad (1)$$

3.3. Design of recording user behavior logs

In this study, the learners' learning behaviors were recorded into a log dataset. Subsequently, educational data mining technology was used to analyze the reasons why the learners may have found the learning process to be lengthy or boring. From this analysis, a predictive model was constructed for successful learning.

The algorithmic process of the CBGE system (Figure 4) was analyzed, and the users' behaviors were coded to rules for exploring the learning effect through data mining. Sixteen learning behavior styles were summarized, and the coded rules of users' behaviors are listed in Table 1.

The system used the MySQL database to record users' behaviors while studying. The database first constructed a log table with ID, user, time, and catalog columns. To record the users' behaviors, a function was designed (Figure 5) and embedded into every webpage in the system. Users' behaviors could be written into the columns of a table for analysis through data mining by using this function.

Table 1. Rules of users' learning behaviors.

Behavior Style	Coded Rule
Proceeding to login the guided learning process	Guided-learning Login
Proceeding to do the pre-test 1	Pre-test1 testing
Proceeding to do the pre-test 2	Pre-test2 testing
Proceeding to do the pre-test 3	Pre-test3 testing
Proceeding to do the pre-test 4	Pre-test4 testing
Proceeding to do the pre-test 5	Pre-test5 testing
Proceeding to browser the guided learning progress	Browser the Guided-learning Progress
To be confident for question N	Reading Guided-learning QuestionN
To read again for question N	Re-reading Guided-learning QuestionN
Proceeding to the stage test of current unit	Stage-test Testing
Finishing the stage test	Stage Success
Failing in the stage test and guide to learn the unit again	Fail the Stage-test and Go Back to Guided-learning
Passing through the stage test	Pass the Post-test1 Test
Proceeding to the final test	Post-ending-test testing
Finishing the final test and calculate the result	Finishing Post-ending-test and Calculating the Result
Browsing the result of final test	Post-ending-test Result Browsering

```
$insertLOGs = sprintf("INSERT INTO logs (`user`, `time`, catelog) VALUES (%s, %s, %s)",
                GetSQLValueString($_POST['email'], "text"),
                GetSQLValueString(date("Y-m-d H:i:s"), "text"),
                GetSQLValueString("Reading Guided-learning Question".$_POST['item'], "text"));
mysql_select_db($database_hdexam, $hdexam);
$Result = mysql_query($insertLOGs, $hdexam) or die(mysql_error());
```

Figure 5. Function of recording users' learning behaviors.

4. Experimental results and analysis

SPSS was used to analyze the results of system satisfaction and the teaching experiment. Moreover, the decision tree of Rattle GUI for R was used to analyze the learners' learning behaviors and construct a predictive reference model of the learning effect.

4.1. Conceptual framework of teaching experiment

This study employed Internet technologies to implement a web-based HTML5-related CBGE system. This system embedded a CBGL mechanism to adaptively guide learners to study more efficiently. However, because the learning process of asynchronous drill and practice e-learning systems can be lengthy and tedious, the overall learning effect is often negatively affected. Therefore, the present study focused on whether the completeness of the guiding procedure is related to the learning effect.

This study conducted a teaching experiment to evaluate the difference between the learning effects of the participants who completed the guiding process (finished guided learning group) and the participants who did not complete the guiding process (unfinished guided learning group). The teaching experiment analyzed system satisfaction and learning effects. The learning effect analysis included two dimensions: the whole learning effect and the learning effect related to the completeness of the CBGL guiding procedure. Thus, a null hypothesis was created. The conceptual framework of the teaching experiment is displayed in Figure 6.

H1: In the HTML5-related CBGE system, there is no difference between the learning effects of the finished and unfinished guided learning groups.

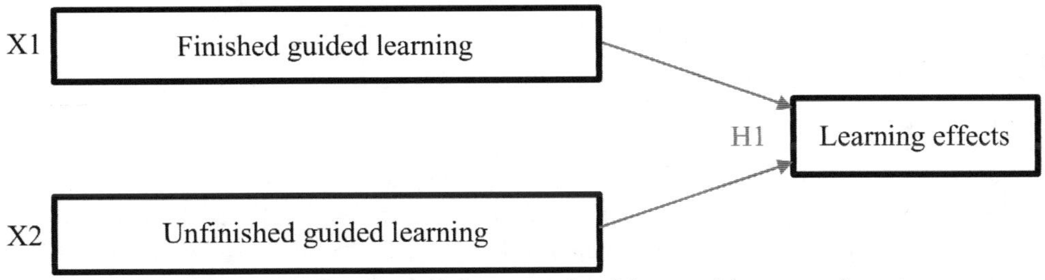

Figure 6. Conceptual framework of the teaching experiment

4.2. Scheme of the teaching experiment

The overview of the study is as follows:

A. Experimental period

The teaching experiment was conducted from September 20, 2017 to November 15, 2017.

B. Experimental participants

The participants were 39 first-year students studying computer programming at National Taitung Junior College.

C. Experimental course

The learning material in the experiment was an HTML5-related cognitive knowledge e-course.

D. Experimental design

(A) All the participants used the asynchronous HTML5-related CBGE system to study online.

(B) All the participants proceeded through the learning process without any teacher assistance and continued their study according to the system's adaptive guidance.

(C) After finishing the experiment, the participants completed a system satisfaction survey. Then, the learning effect was analyzed, and educational data mining was conducted to examine the learners' behaviors.

4.3. Teaching experiment analysis

1. System satisfaction analysis

A system satisfaction survey was conducted to determine the learners' perception after using the system and whether the system fulfilled the research objectives. The adaptively guided learning system satisfaction questionnaire designed by Hsu and Li was selected [4], and a 5-point Likert scale was used to examine the learners' satisfaction. The system design focused on the CBGL algorithm and the interface of the HTML5 cognitive e-course. Therefore, the questionnaire focused on these two dimensions. The total number of valid responses to the questionnaire was 37 for an effective response rate of 94.9%. The content of the questionnaire is presented in Table 2.

Table 2. Analysis of the system satisfaction survey results.

Dimension	Question	Mean	Standard deviation	Average dimension score
Design of CBGL guided learning	I can use the system to study positively.	4.03	.645	3.76
	The system enables me to learn.	3.81	.776	
	I feel the learning process is fun.	3.59	.762	
	I feel the learning process is too lengthy.	3.30	.878	
	I feel the system can guide me to learn effectively.	4.08	.640	
Design of e-Course	When I do not understand the meaning of a question, I check the explanation of the solution.	4.11	.699	4.02
	The explanation of the solution completely solves my confusion pertaining to the question.	3.92	.722	

Note: $n = 37$.

According to the result of the system satisfaction survey, the total average score was higher than 3.89. This value indicates that the participants were satisfied with the system design. The average score of the CBGL guiding learning design was 3.76. This value indicates that the participants were satisfied with the CBGL guided learning mechanism. They used the system to study positively and felt that the guidance of the system enabled them to learn effectively. However, they also felt that the learning process was slightly fun and lengthy. These problems occurred very often in the drill and practice learning process, and they affected learners' learning performance. These problems will be considered in a future study.

The average score of the e-course design was 4.02. This value indicates that the participants were very satisfied with this type of e-course interface. The participants checked the explanations of the solutions when they did not understand the meaning of the questions, and the solutions could completely solve the confusion of the questions.

2. Learning effect analysis

The study focused on verifying whether the completeness of the guiding procedure is related to the learning effect and whether finding the guided learning process made any difference to the learning effect. Thus, two dimensions of the learning effects were examined: the whole learning effect and the learning effect related to the completeness of the CBGL guiding procedure. Moreover, the learning effect of unfinished guided learning, the learning effect of finished guided learning, and the difference between the learning effect of finished and unfinished guided learning were analyzed for the learning effect related to the completeness of the CBGL guiding procedure.

A. Whole learning effect

Repeated measures were conducted on the results obtained from the 39 participants who completed the learning process, and the significance of mean posttest and pretest scores was examined using the *t* test results of paired samples. Then, an evaluation was conducted to determine whether the learning effect was improved using the CBGE system.

The results presented in Table 3 reveal that the mean posttest and pretest scores were 58.64 and 25.54, respectively. Thus, the average score increased by 33.1 after using the e-course system. This reflects an increased score distribution. Paired sample *t* test yielded a *t* value of 8.442 and a significance of .000. Differential t test analysis results indicated that the experimental group attained a significantly higher posttest score ($p < .001$). Therefore, the cognitive abilities of the experimental group were significantly improved after the experiment.

Table 3. Analysis of participants' posttest and pretest scores.

Variable	N	Mean	SD	*t*
Experimental group				
Post-test scores	39	58.64	25.71	8.442***
Pre-test scores	39	25.54	6.20	

Note: ***$p < .001$

B. Analyzing the learning effect related to the completeness of the CBGL guiding procedure

(A) Learning effect of unfinished guided learning

Repeated measures were conducted on the results of the 24 participants who did not finish the guided learning process, and the mean posttest and pretest scores were also tested to determine their significance by using the paired sample *t* test. Then, an evaluation was conducted to determine whether the learning effect was improved in the CBGE system when the guiding process was not completed.

The results listed in Table 4 reveal that the mean posttest and pretest scores were 44.46 and 25.33, respectively. Thus, the average score increased by 19.13 after using the e-course system. This reflects an increased score distribution. The paired sample *t* test yielded a *t* value of 4.748 and a significance of .000. Differential *t* test analysis results revealed that the unfinished guided learning group attained a significantly higher posttest score ($p < .001$). Therefore, the cognitive abilities of the unfinished guided learning group significantly improved after the experiment.

Table 4. Analysis of the posttest and pretest scores of the unfinished guided learning group.

Variable	N	Mean	SD	*t*
Unfinished guided learning group				
Post-test scores	24	44.46	21.66	4.748***
Pre-test scores	24	25.33	6.66	

Note: ***$p < .001$

(B) Learning effect of finished guided learning

Repeated measures were conducted on the results of the 15 participants who finished the guided learning process, and the mean posttest and pretest scores were also tested to determine significance by using the paired sample *t* test. Then, an evaluation was conducted to determine whether the learning effect was improved in the CBGE system when the guiding process was completed.

The results listed in Table 5 reveal that the mean posttest and pretest scores were 81.33 and 25.87, respectively. Thus, the average score increased by 55.46 after using the e-course system. This reflects an increased score distribution. The paired sample *t* test yielded a *t* value of 19.850 and a significance of .000. Differential *t* test analysis results revealed that the finished guided learning experimental group attained a significantly higher posttest score ($p < .001$). Therefore, the cognitive abilities of the finished guided learning group significantly improved after the experiment.

Table 5. Analysis of the the posttest and pretest scores of the finished guided learning group.

Variable	N	Mean	SD	*t*
Finished guided learning group				
Post-test scores	15	81.33	11.28	19.850***
Pre-test scores	15	25.87	5.58	

Note: ***$p < .001$

(C) Learning effect differences between finished and unfinished guided learning groups

The study used grade growths to verify the learning effect differences. An independent sample t test was conducted to test the significance in grade growth (posttest scores minus pretest scores) among the 15 participants who completed the learning process and 24 participants who did not complete the learning process. Then, an evaluation was conducted to determine whether the learning effect was significant between the two groups.

The results listed in Table 6 reveal that the mean grade growths of the finished and unfinished guided learning groups were 55.47 and 19.13, respectively. An independent sample t test yielded a t value of 7.414 and a significance of .000. Differential t test analysis results revealed that the grade increase between the finished and unfinished guided learning groups attained a significantly higher score ($p < .001$). Therefore, the increase in the cognitive ability between the finished and unfinished guided learning groups was significantly different after the experiment.

This indicates that participants' cognitive ability was improved using the CBGE system and that completing the CBGL learning process yielded a more significantly pronounced learning effect.

Table 6. Analysis of the grade growth in finished and unfinished guided learning groups.

Variable	N	Mean	SD	*t*
Grade growths				
Finished guided learning group	15	55.47	10.82	7.414***
Unfinished guided learning group	24	19.13	19.73	

Note: ***$p < .001$

4.4. Predictive reference model of learning effect

According to the experimental design, the learning behaviors of the participants in the CBGL learning process were recorded in the system's log tables. Then, the dataset was analyzed to develop a predictive reference model of the learning effect. The model can be used as a reference for other learners to help them achieve the learning objectives in the CBGE system.

The learners' learning behaviors are defined in Table 1. Therefore, the test conditions were explicit. A decision tree was employed using the data mining tool. This technique used an ID3 algorithm to choose the selected attribute nodes and obtain the classified information of the nodes by using information gain.

After the experiment, the log table contained 13,443 records of learning behaviors. The table of recorded log data of behaviors is presented in Figure 7. According to the CBGL guided process, the study selected seven coded rules for the test conditions: browse the guided learning progress, stage success, fail the stage test and go back to the guided learning, pass posttest 1, read the guided learning questions, reread the guided learning questions, and finish the posttest and calculate the result.

The Rattle GUI for R was used to analyze the learners' learning behaviors and construct a predictive reference model of the learning effect. The predictive reference model of the learning effect was built by using the decision tree and is illustrated in Figure 8. The analysis results revealed that the information gained from "stage success" was the highest. The optimized decision tree was constructed by only selecting this condition for the test node—whether learners can achieve the learning objective (a mark of 60) can be predicted by monitoring only one condition.

id	user	time	catelog
1	lcs@ntc.edu.tw	2017-08-30 15:55:49	Browser the Guided-learning Progress
2	lcs@ntc.edu.tw	2017-08-30 16:25:51	Browser the Guided-learning Progress
3	lcs@ntc.edu.tw	2017-09-20 10:50:21	Pre-test1 testing
4	lcs@ntc.edu.tw	2017-09-20 10:51:51	Browser the Guided-learning Progress
5	lcs@ntc.edu.tw	2017-09-20 10:52:14	Reading Guided-learning Question224
6	lcs@ntc.edu.tw	2017-09-20 10:52:27	Re-reading Guided-learning Question209
7	lcs@ntc.edu.tw	2017-09-20 10:54:10	Browser the Guided-learning Progress
8	denny0913323640@gmail.com	2017-09-20 10:55:05	Pre-test1 testing
9	a0972730966@gmail.com	2017-09-20 10:55:05	Pre-test1 testing
10	assd65333@yahoo.con.tw	2017-09-20 10:55:10	Pre-test1 testing
11	q20875669@gamil.com	2017-09-20 10:55:15	Pre-test1 testing
12	peter1001peter@gmail.com	2017-09-20 10:55:26	Pre-test1 testing
13	tinatsai8940@gmail.com	2017-09-20 10:55:05	Guided-learning Login
14	b0983281919@gmail.com	2017-09-20 10:55:27	Pre-test1 testing
15	gn510221a@gmail.com	2017-09-20 10:55:32	Pre-test1 testing
16	d93xjp6520@gmail.com	2017-09-20 10:55:37	Pre-test1 testing
17	okok951753852@yahoo.com.tw	2017-09-20 10:55:42	Pre-test1 testing
18	s835566@gmail.com	2017-09-20 10:55:44	Pre-test1 testing
19	love05040810@gmail.com	2017-09-20 10:55:44	Pre-test1 testing
20	j0989920076@gmail.com	2017-09-20 10:55:46	Pre-test1 testing

Figure 7. Recording the users' behaviors in the database.

The predictive reference model revealed that when the learners studied in the CBGE system, they could not achieve the learning objective without passing the stage tests 2.5 times.

Thus, the number of times the stage tests were passed was the crucial factor in determining whether the learning objective was achieved. According to these findings, learners using such a system attempt five learning stages and should pass stage tests at least 2.5 times to achieve their learning objective.

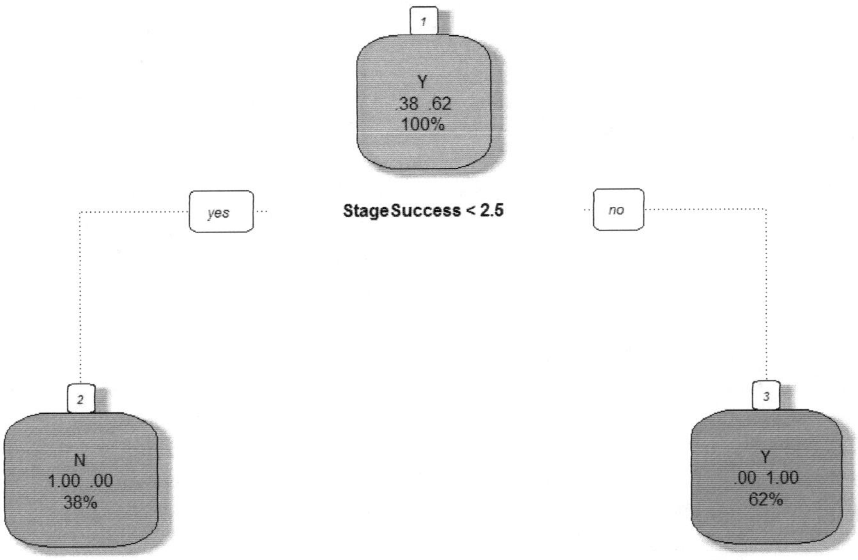

Figure 8. Predictive reference model of the learning effect.

5. Conclusions

This study used the following Internet technologies to implement a CBGE system with an HTML5-related cognitive e-course: PHP6, HTML5, CSS3, JavaScript, and MySQL. The system embedded a CBGL mechanism to implement a personalized learning environment that guided learners to study in stages while recording their learning behaviors in a log dataset for data mining.

The results of the system satisfaction survey revealed that the participants were satisfied with the designed system. The participants actively used the system to study, and they felt that the system provided effective guidance. Moreover, although they found the learning process to be lengthy, they felt the guiding procedure was interesting and could engage learners. They were also very satisfied with the e-course design, which provided guidance through closed and open solutions whenever the participants could not understand the questions.

According to the learning effect analysis, the cognitive abilities of the participants significantly improved after the experiment. Moreover, the increase in the cognitive ability of those who finished the guided learning process was significantly higher than that of those who did not finish the guided learning process. This indicated that the participants' cognitive abilities improved after the guided learning process. The learning effect was significantly more pronounced if the process was completed entirely.

A decision tree technique was used to construct a predictive model of the learning effect to help the learners achieve the learning objective. The predictive model revealed that the learners who used the CBGL system could not achieve the learning objective without passing the stage tests a minimum of 2.5 times. Thus, the number of times that stage tests were passed was the critical factor in achieving the learning objective.

Because the completion of the CBGL learning process was the key factor that affected the learning effect, future studies can focus on improving the tedious and lengthy learning process that is typical of the asynchronous drill and practice e-learning systems. A gamified design may be successful for this purpose.

References

[1] Bergman E M, Sieben J M, Smailbegovic I, de Bruin A B, Scherpbier A J and van der Vleuten C P 2013 *Anatomical sciences education* **6** pp 114-124
[2] Burke R L 1982 *CAI sourcebook* (NJ: Pretice-Hall, Englewood Cliffs Press)
[3] Deterding S 2011 Meaningful play: Getting gamification right *Google Tech Talk*
[4] Hsu W C and Li C H 2015 *Interactive Learning Environments* **23** pp 106-125
[5] Mejova Y, Borge-Holthoefer J and Weber I 2015 Bridges into the Unknown: Personalizing Connections to Little-known Countries *In Proceedings of the 33rd Annual ACM Conference on Human Factors in Computing Systems* pp 2633-42
[6] Ryan R M and Deci E L 2000 *American psychologist* **55** pp 68
[7] Skinner B F 1968 *The technology of teach* (NY: Appleton Century Crofts Press)
[8] Xu H, Song D, Yu T and Tavares A 2017 *EURASIA Journal of Mathematics Science and Technology Education* **13** pp 3001-3018
[9] Sorour S E, Goda K and Mine T 2017 *Educational Technology & Society* **20** pp 73-86
[10] Xu B and Recker M 2012 *Educational Technology & Society* **15** pp 103-115
[11] Romero C and Ventura S 2007 *Expert Systems with Applications* **33** pp 135-146
[12] Han J and Kamber M 2006 *Data mining: Concepts and techniques (2nd ed.)* (CA: Kaufmann, San Francisco Press)
[13] Pahl C and Donnellan D 2002 Data mining technology for the evaluation of web-based teaching and learning systems *In Proceedings of the E-Learn 2002 World Conference on E-Learning in Corporate, Government, Healthcare, and Higher Education* pp 747-752
[14] Baker R S J D and Yacef K 2009 *Journal of Educational Data Mining* **1** pp 3-17
[15] Witten I H and Eibe F 2000 *Data mining: Practical machine learning tools and techniques with Java implementations (2nd ed.)* (CA: Morgan Kaufmann, San Francisco Press)

[16] Romero C, Gutierez S, Freire M and Ventura S 2008 Mining and visualizing visited trails in web-based educational systems *In Proceedings of the 1st International Conference on Educational Data Mining* pp 182-186
[17] Li C H 2018 *MATEC Web of Conferences* **169** 01002
[18] Chiou S F, Wang H Y and Lo J W 2017 *Proceedings of Engineering and Technology Innovation* 7 pp 31-36
[19] Chen S Y and Liu X 2011 *Interactive Learning Environments* **19** pp 179-192
[20] Mouri K, Suzuki F, Shimada A, Uosaki N, Yin C, Kaneko K and Ogata H 2019 I*nteractive Learning Environments* DOI: 10.1080/10494820.2019.1619594
[21] Asif R, Merceron A, Ali S A and Haider N G 2017 Computers & Education **113** pp 177-194
[22] Sathick J and Venkat J 2015 *The International Review of Research in Open and Distributed Learning* **16** https://doi.org/10.19173/irrodl.v16i2.2093
[23] Osborne J W 2000 *Research& Evaluation* **7** Available online: http://pareonline.net/getvn.asp?v=7&n=2
[24] Knowles J E 2015 *Journal of Educational Data Mining* 7 pp 18-67
[25] Miller L D, Soh Leen-Kiat and Samal A 2015 *Journal of Educational Data Mining* **7** pp 117-150
[26] Gomes Cristiano M A and Almeida Leandro S 2017 *Practical Assessment, Research & Evaluation* **22** Available online: http://pareonline.net/getvn.asp?v=22&n=10
[27] Kolo D K, Adepoju S A and Alhassan J K 2015 *International Journal of Education and Management Engineering* **5** pp 12-19
[28] Hamoud A K, Hashim A S and Awadh W A 2018 *International Journal of Interactive Multimedia and Artificial Intelligence* **5** pp 26-31
[29] Al-Barrak M A and Al-Razgan M 2016 *International Journal of Information and Education Technology* **6** pp 528-533
[30] Rizvia S, Rientiesa B and Khojab S A 2019 *Computers & Education* **137** pp 32-47

Research on zero waste garment engineering design in sustainable environment

Jhan Rong Shih, Chien Jen Hung and Feng Tzu Chiu

Fuzhou College of Foreign Languages and Trade, Fuzhou, China

E-mail: 2753006833@qq.com

Abstract. With life becoming more convenient and rich, the pollution caused by human beings is far beyond our imagination. This study focuses on the clothing industry, which is closely related to human life, and its ecological pollution and waste of resources. It attempts to combine environmental issues to create clothing that can convey the concept of a sustainable environment and a more friendly Earth 2050: The Future of Energy. Different from the previous green environmental protection design, the current fashion trend is not only to promote environmental protection, but also to emphasize blue economy. In environmental protection, find out the industrial and economic values that meet various costs to achieve sustainable operation. In this study, the concept of environmental sustainability is brought into the clothing project. The goal is that the manufacturing process will not produce cloth waste. All cloth will go through the life course of clothing. The clothing project of "zero waste"will be built. The consumed resources will be fully utilized. In this research design, water resources, which is the most threatening issue of human sustainable environment, are used as the main design axis with blue as the image. Pollution and optimization issues faced by water resources are applied to clothing design thinking, presenting a series of clothing with zero waste structure, and implementing good behaviors of "sustainable" environment in life.

1. Introduction

This research aims at the progress of science and technology, which not only accelerates the economic growth, but also changes the pattern of human life.

While people enjoy the comfort and convenience brought by economic development, our natural environment has been destroyed bit by bit, such as the hole in the ozone layer, the disappearance of rainforest, the greenhouse effect etc. ecosystem imbalances [1]. In the face of environmental crisis, as a fashion designer, how to reduce and slow down the impact of clothing on the environment, and hope to remind consumers to pay attention to the related issues of sustainable and friendly environment. Timo Rissanen, an assistant professor of fashion design and sustainability at parson School of design, pointed out that about 15% of the fabrics were left behind in the process of clothing cutting and were discarded before entering the market, causing environmental problems and wasting resources.

Therefore, this study hopes to explore the relevant importance of clothing design and production in a sustainable environment, to understand the environmental impact factors caused by the production and consumption patterns of the clothing industry up to now. And more importantly, from the thinking of the design before clothing forming, bring the concept of environmental protection and sustainability into the actual act of fashion engineering, through the use of materials and fabrics, style design, and layout structure.

It is not only necessary to reduce the waste of resources and harm to the environment, but also to design clothing with the function of fashion and the concept of environmental sustainability.

2. Literature review

Understand the development of clothing industry, the general situation of environmental changes and the environmental problems faced by today, and summarize the causes of environmental problems caused by clothing industry.

2.1. Development of clothing industry

Clothing consumes a lot of natural resources in the process of production, and discharges many pollutants into the natural environment.

By understanding how designers lead the trend after the industrial revolution, the development of synthetic fiber cloth, the rise of fashion democratization in the clothing industry, and the development process of clothing industry in the era of transnational cooperation, especially in the middle of the 20th century, the invention of material science Human engineering research, mechanical automation etc. The results of the scientific and technological revolution have been widely used in the clothing industry, which makes the development of ready-made clothing leap forward rapidly. In a short period of 200 years, it has overthrown the previous all manual production mode [2]. With the popularization of information, multi-media make fashion information spread more rapidly, through newspapers, the Internet, press conferences and so on. For the promotion of fashion to the extreme, the popularity of ready-made clothing has become faster and faster, and its rise and rapid development almost dominated the clothing industry since the late 20th century, so it is called “the era of ready-made clothing”. It became the beginning of rapid fashion in the 21st century [3].

2.2. Environmental situation

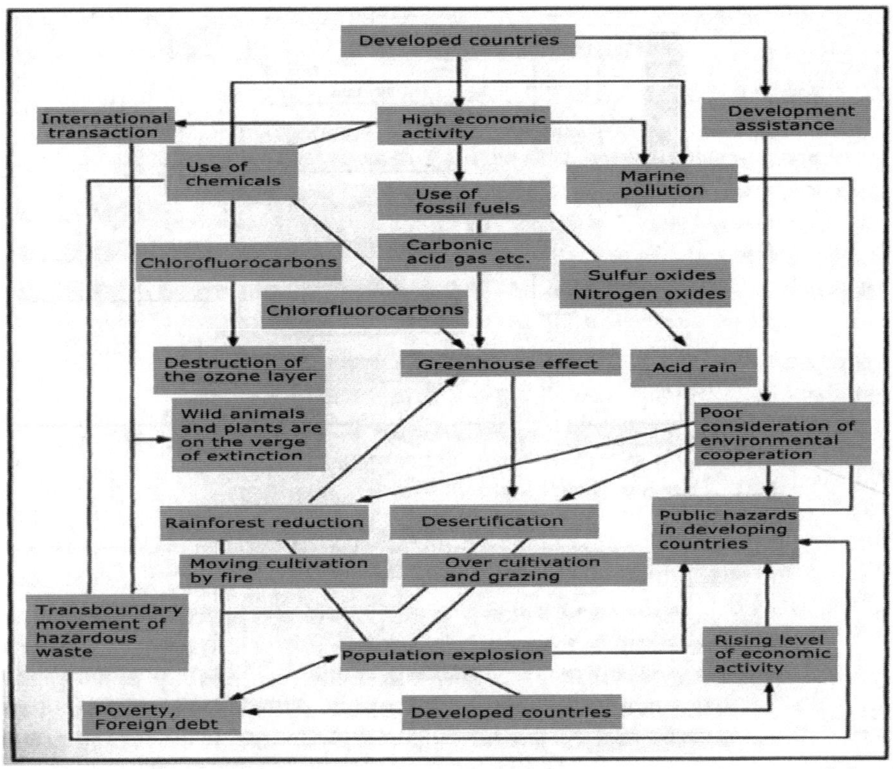

Figure 1. Diagram of the relationship between human behavior and environmental deterioration
Photo source: editing group of environmental education course, 2008

After the industrial revolution, due to the high development of machinery and the large-scale development of energy, it provides a huge power for production and life. Then many countries, in order to increase the output value, are desperate to tap natural resources. Due to the population growth, human demand promotes the continuous improvement of science and technology. After the industrial revolution, human damage to the environment is far more than any previous era. At present, the main environmental problems come from a large number of resource consumption, artificial waste and accumulation of chemical compounds, which exceed the capacity of natural decomposition and regeneration. Under the condition that nature cannot bear, human pollution accelerates the impact on the environment, and the threat to human living space gradually changes from a single region to a regional, or even expands to a global environmental problem, such as global warming. The pollution and depletion of water resources, and the relationship between human behavior and environmental degradation is shown in Figure 1.

2.3. Sustainable development of environment

The United Nations General Assembly first proposed the concept of “sustainable development” in 1980. It called on the world to study the interrelationship among nature, society, ecology, economy and natural resources, so as to ensure the sustainable development of the earth. Nowadays' sustainable development covers four aspects [4]

2.3.1. The concept of natural ecology

Ustainable development was first put forward by ecologists. The sustainable development of ecology is to protect and strengthen the production and rebirth of environmental ecosystem.

2.3.2. Economic development

According to the World Resources Research Institute in 1992, the sustainability of economy should be based on not damaging natural resources and without compromising the quality of the environment.

2.3.3. Social Development

The Sustainable Survival Strategy of protecting the Earth in 1991 states that the sustainability of society is to survive without exceeding the capacity of the system for survival, and to improve the quality of life, emphasizing that “life form and production scale must balance the earth's carrying capacity”.

2.3.4. Science and technology

In 1992, the World Resources Institute proposed that sustainable science and technology is based on the production of minimal waste and pollution.

2.4. Fast Fashion and Environment

The word (Fast Fashion) originates from the rapid development of ready-to-wear in Europe in the 20th century, while the United States calls it “Speed to Market”. The British Guardian has also created a new word “McFashion”. “Mc” is taken from MacDonald's, to express the rapid sale of fashion like McDonald's. Traditional clothing brands usually take several months from design release to circulation and sale, while fast fashion brands can go through the production cycle in about 50 days at first. With the development of Internet information and transportation, they can even be quickly circulated in a week; they are close to the fashion; they are more affordable and less expensive, so they can complete the product development, production to circulation and sale.

The life cycle of clothing, from raw material planting, adoption, fiber production, manufacturing to the use of final consumers, uses a lot of natural resources: the global textile industry uses about 37.8 billion liters of water every year. The World Bank estimates that textile dyeing, finishing and treatment account for 17% - 20% of the total industrial water pollution, and there are nearly 8000 kinds of synthetic chemicals used in the process of converting raw materials into textiles. Most of them will be discharged to the fresh water system after use. Fashion and textile industry is one of the most energy consuming industries in the world. Fast fashion design is closely related to fashion, and the price is low, so that the public can enjoy fashion in a timely and easy manner. However, due to the low price, manufacturers reduce the production process, and make fabrics with unqualified formaldehyde content, color fastness and fiber content. The cost control of fast fashion causes product quality problems. The main reason for this problem is that poor quality has almost become a feature of fast fashion brands, which makes the life cycle of clothing extremely short. Because fast fashion brands launch new clothing almost every week, huge amount of textile waste will be generated under the condition of mass production, consumption and discarding. In 2005, Timo Rissanen, an assistant professor of fashion design and sustainability at parson School of design, pointed out that about 15% of the fabrics were left behind in the process of clothing cutting and were discarded before entering the market, causing environmental problems and wasting resources.

2.5. Green action of clothing industry

More and more people realize that environmental protection requires everyone to take practical actions. Fashion has always been in the controversy of environmental protection, which has promoted the development of green fashion. The concept of environmental protection is really popular with consumers. With the trend of environmental protection, many brands have launched green fashion products. Because of the promotion of many folk forces, green fashion is brought to the world, which makes consumption bring not only the beauty of appearance, but also a brand-new and healthy outlook on life, and environmental sustainability will become the future trend [5]). In the fashion industry, sustainable development is a mere discussion, but there are few actions to be taken, said Christina Dean, a non-profit charity founded in Hong Kong in 2007. There is not enough action to reduce waste. Therefore, Redress is committed to education to reduce textile waste and save energy, and cooperates with many organizations to promote the sustainable development of the fashion industry in Asia, and hopes to inspire fashion designers and students through its annual "Eco Chic Regeneration Fashion Design Competition" held every year from 2011, so as to change the fashion landscape by creating a market that is in line with the public and change the fashion pattern with the least waste [5] .

3. Research methods

Starting from the treatment of raw materials, a piece of clothing is bleached and dyed to become yarn and fabric.

After garment production and logistics, it is recycled and even burned to decompose. There is carbon emission in every link. What can a garment designer do in the face of environmental crisis? In the tailoring of ready to wear fabric, about 15% of the fabric is wasted. If the designer can pay attention to the wasted part as well as the material in the design, it will not only increase the value but also reduce the waste. Through the literature to find out the direction for discussion, to reduce the waste of cloth, or even to create fashion with zero waste as the goal, using the literature analysis method, to analyze the contemporary popular trend, to explore the clothing case with "zero waste cutting" structure, and the design method that can be applied to no waste, and to use its characteristics as a reference for subsequent clothing design.

3.1. Contemporary trends

WGSN, the international popular trend prediction website, has put forward "ecological awakening" in its dominant trend in recent years (as shown in Figure 2). Through the awakening of human awareness of the environment, it explores how the whole society can solve the adverse impact on the ecology. Then put forward the "The Great Reset reset" in autumn and winter (as shown in Figure 3) - we will press the restart button to reverse our life, design and work. Consumers will reduce their pursuit of brand and turn to environmental protection life. The future values will focus on the product life cycle and human characteristics rather than only economic benefits. It is mentioned in the seasonal theme "mentality" released by spring and summer trend (as shown in Figure 4) [6]. Science and technology make many industries progress, expand new business opportunities, and the importance of connectivity is beyond doubt, but at the same time, it promotes reverse movement - hand in hand with nature again to create a healthy body, mind and environment.

From the analysis of ongoing and future trends in recent quarters, environmental sustainability has become an important and necessary trend. Through various technologies, we must enable us to treat the environment in a new way, regard nature as a partner, integrate environmental care into the design process, and promote us to protect the world we live in and create a sustainable industry [7].

Figure 2. Ecological Awakening
Picture source: WGSN

Figure 3. large reset
Picture source: WGSN

Figure 4. mentality
Picture source: WGSN

3.2. Sustainable clothing design method

The basic elements of clothing design are the application and combination changes of body structure, color matching and materials and fabrics. According to the two techniques of "reconstruction" and "upgrade and reconstruction" put forward by Redress and the design of "zero waste structure clothing", by prolonging the life cycle of textile fabric, the accumulation of textile waste can be reduced.

"Restructure" is to decompose the existing and no longer worn clothes, reconstruct them and combine them into new clothes; "upgrade" is to recycle the unused fabrics and transform them into products of higher value. "Zero waste structure clothing" is through the design of clothing molding structure, that is, the design of the so-called "clothing version", in order to avoid 15% to 21% of the leftovers cut by the garment industry in the production of cut clothing fabrics, so as to achieve the goal of zero waste (as shown in Figure 5 and Figure 6). Timo Rissanen, an assistant professor in the Department of fashion design and sustainability at Parsons School of design, and one of the members of Redress, studied a group of designers between the 19th and 20th century. For example, from the bias cut (oblique cutting method) invented by Madeleine Vionnet, it was found that the drape of fabric twill was used to fit the human body, and the characteristics of different weave directions between fabrics were used to increase zero waste clothing The feasibility of its creation mainly uses the "jigsaw puzzle" and "cutting method" [8].

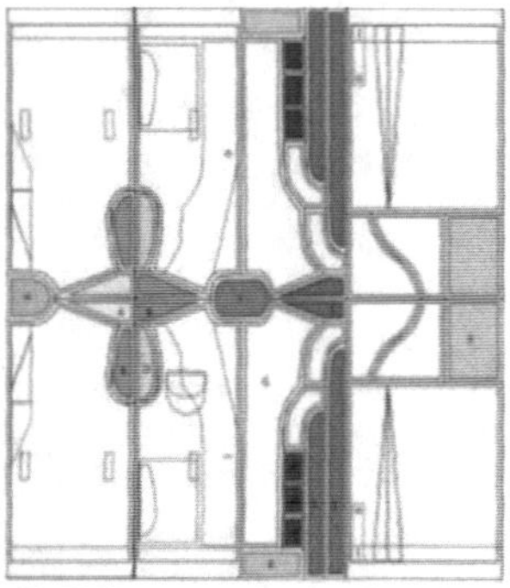

Figure 5. Winsome Lok clothing pattern
Picture source: Eco Chic Design Award official website

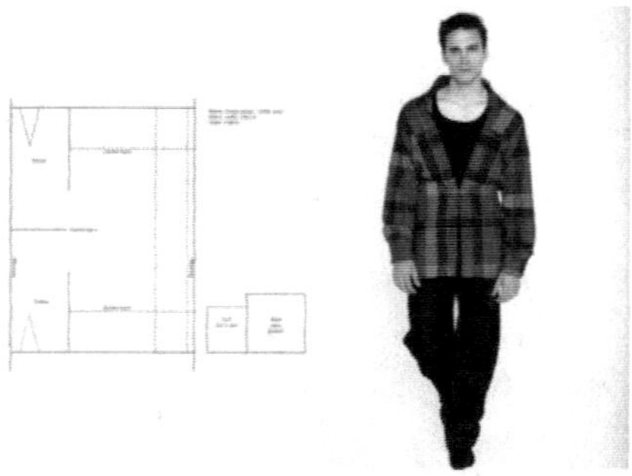

Figure 6. Timo Rissanen zero-waste structural clothing
Picture source: Alabama Chanin official website

IOP Conf. Series: Materials Science and Engineering 1113 (2021) 012016 doi:10.1088/1757-899X/1113/1/012016

4. Results and Discussion

Through the literature discussion, it is understood that the environmental problems caused by the clothing industry are mainly three major crises. "global warming", "water shortage" and "textile waste". From these three crises, it is necessary to think about the sustainable environmental direction that the theme design should aim at, analyze the fashion style and issues of the contemporary popular trend, and the basic elements of the clothing design, production and shaping, which will be friendly to the environment, positive and not a waste of resources. And the designer has a high autonomy of "zero waste tailoring" design method, develop this research design theme, and specifically create fashionable clothing with the concept of sustainable environment.

4.1. Creative theme

Different from the green used to convey environmental protection in the past, in addition to environmental protection, blue pays more attention to the regeneration and non-waste of the environment, and according to the literature, the water resources crisis is a major element of contemporary environmental sustainability. Therefore, the creation of this study echoes the sustainability of blue with the theme of water resources - "optimization" (Figure 7), hoping to highlight the anaerobic asphyxiation of organisms due to the destruction of water resources. Communicating sustainable environmental behavior requires immediate vigilance.

The application of zero waste cutting clothing structure is full of uncertainty in the design process, so the design must be completed in the process of production, rather than the traditional pattern of clothing drawing construction. This feature and challenge is that it can't be completely conceived in the first draft of design. It is necessary to design the contour of clothing first, and roughly cut distribution, and then think about how unused fabric can not be wasted, except for configuration In detail design, it can be used as the design change of clothing function adjustment to make the fabric used up without waste. The preliminary design of the clothing style (as shown in Figure 8) is based on the loose straight version and umbrella shaped silhouette, responding to the static and flowing water, which can be worn with different upper and lower clothing items to present a variety of faces.

Figure 7. Creative Design theme: NO2
Picture source: Drawing in this study

Figure 8. draft clothing style
Picture source: Drawing in this study

4.2. Zero waste theme and Fashion Design

After conceiving the costume's outline,the design must be completed in a zero waste configuration. This study divides various methods into three major modes. The first one is "After conceiving the outline, the subsequent overall style design is completed in the way of Stereo Cutting" which is shown in Figure 9.The second one is "After positioning the main large section of the fabric, then arrange the details setting of the redundant cloth"shown in Figure 10.

The third is "according to the size of the fabric and the tailoring line adjust the pattern of the existing clothing".As shown in Figure 11. The above three modes are used to modulate the structure of the clothing version.

Subsequently, according to the design theme, arrange the matchability within the fabrics, details and the single product (Figure 12), there with to meet the needs of wearing in various life situations and provide practical convenience for achieving sustainable wearing.

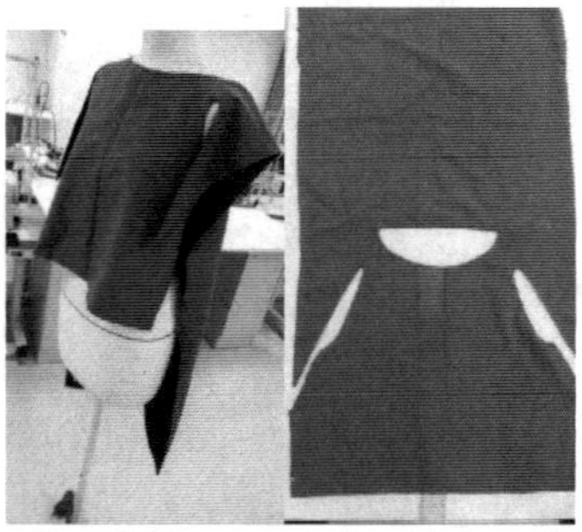

Figure 9. After conceiving the outline, the subsequent overall style design is completed in the way of Stereo Cutting
Picture source: Shooting in this study

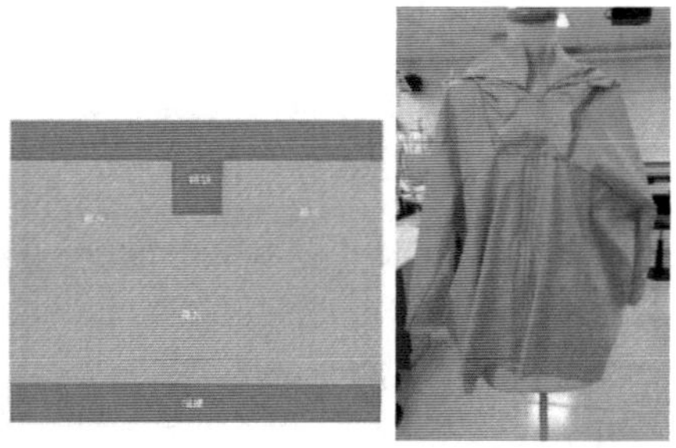

Figure 10. After positioning the main large section of the fabric, then arrange the details setting of the redundant cloth
Picture source: Shooting in this study

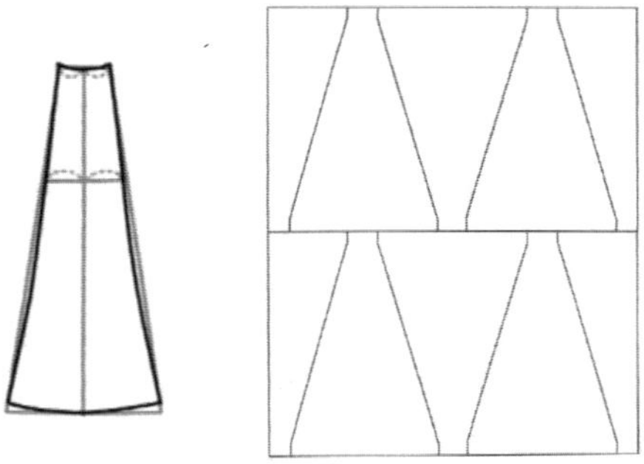

Figure 11. According to the size of the fabric and the tailoring line adjust the pattern of the existing clothing
Picture source: Drawing in this study

Figure 12. Design drawing of " NO2" series
Picture source: Drawing in this study

4.3. Series clothing design finished products

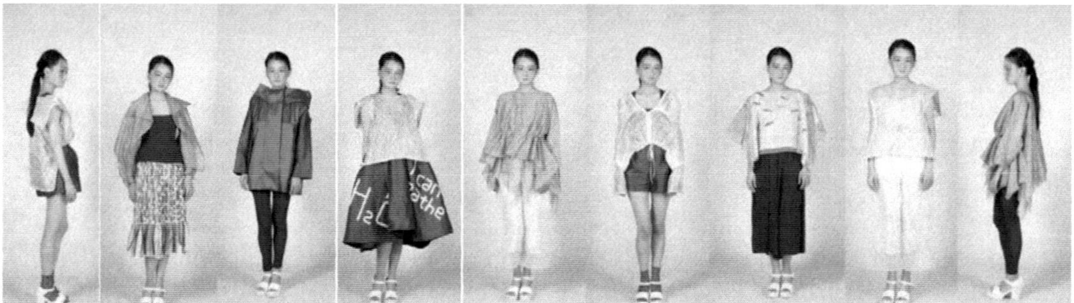

Figure 13. Finished clothing designed with zero waste version in"NO2" series
Picture source: Shooting in this study

In this study, a series of clothing styles (Figure 13) are all designed with zero waste version to echo the sustainable behavior of the environment, not only actually reduce textile waste, but also no waste of resources after many samples, finalization, production and other procedures, after the completion of the work, models are invited to wear, with appropriate makeup and hair style, to create an atmosphere consistent with the concept of creation. With the combination of the pieces, it will present a variety of wearing styles.

5. Another section of your paper

The environmental damage caused by fast fashion has been overloaded. Many clothing brands will incorporate sustainable development into their business strategies, and put forward many recycling programs to remanufacture waste products into new products.

But among these waste fabrics, about 15% to 21% of the rags are cut in the manufacturing process, consuming a lot of resources, but they are scrapped without entering the life process of clothing. This study implements the goal of environmental protection and fashion with a zero-waste layout structure. The complete application of cloth in the production of clothing not only reduces the production of textile waste, but also does not waste the resources consumed in the fabric production process.

This series of creative design will be able to form a zero waste fabric clothing structure method, combined with the water source crisis image of water quality optimization, and finally present 13 complete clothing items. The combination of the items can create many ways to wear, and successfully enable the wearer to wear a variety of clothes according to different occasions of life. In the implementation of this research and design, it is found that because of the zero waste structure, it is not conducive to the use of different fabric splicing design in the material of a single garment, so the use of fabric length and short, transparent and opaque to show the sense of hierarchy of the garment, and also to show the clarity and pollution of the water quality. As this series of clothing is made of existing fabrics in the market, the effect of the cloth table is difficult to directly associate with the picture of the improvement of water quality. Although the overall effect can show the flow and stillness of water, it lacks the heavy sense of warning after water quality optimization. The theme emphasizes that the “anaerobic suffocation” of water resources is expressed by using color and text symbols, and the image of endangering water resources is also weak. in the follow-up, if digital printed fabrics and fabric surface texture processing designed for the theme can be added, it will be better able to show the optimal image of human destruction of water quality. The outline and structure of each style completed in this series are mostly the whole loose or the whole fit. If able to show a comparison of fit in the clothing style, it will be more able to strengthen the sense of tightness that can not be breathed due to lack of oxygen, and even the tense atmosphere that will suffocate.

The creation of this study uses zero-waste version of clothing design to echo the sustainable behavior of the environment, which not only actually reduces textile waste, but also does not waste resources. It is suggested that if the follow-up development can use functional fabric as the main fabric, the use of clothing to adjust human body temperature, reduce the use of electric energy, with recycled material fabrics or natural organic fabrics, will be more actively in line with the main idea of environmental sustainability.

In this study, the pattern of garment single products recorded in the creation can provide a basis for subsequent design to modify, and self expect the sustainable design and development of fashionable clothing more in line with the sustainable environment. It also provide a reference for garment designers who want to carry out environmental protection issues in the future to have zero waste pattern structure of garment forming, which not only expect more designers to join in creating popular products with the concept of sustainability, but also hope that human beings will implement the good behavior of the “sustainable”environment in their lives.

References

[1] Ye X 2010 Key report on Anti-warming, New Naturalism Co., Ltd., Taipei pp 64-169
[2] Zhu Q 2007 The british industrial revolution- the project to change the world, Shakespeare culture, Inc., Taichung pp 11-14
[3] Lin S 2005 The development and evolution of modern ready-to-wear industry, Decoration **11** p 50

[4] Editing group of environmental education courses 2008 Environmental Protection and Life, Princeton International Co., Ltd., Taipei pp 4-330
[5] EcoChic DesignAwared 2016 from: http://www.ecochicdesignaward.com/
[6] WGSNProspective vision 2016 from: https://www.wgsn.com/en/
[7] Zheng T 2010 Environmental protection leads global fashion, green China Version A **1** pp 54-56
[8] Timo Rissanen Zero waste structural clothing 2016 from: https://www.redressdesignaward.com/learn/zero-waste

Fuzzy-DEMATEL based competitive analysis of green suppliers in a supply chain

Shu Hua Taso[1], Wenru Chen[2] and Shih Ming Ou[3]

[1]Business School, City College of Dongguan University of Technology, Guangzhou, China
[2]Department of Finance, W Guangzhou, Guanzhou, China
[3]Department of Logistics Management, Fujian Jiangxia University, Fuzhou, China

E-mail: benao13@gmail.com

Abstract. With the increasingly fierce competition among enterprises, the role of suppliers and core companies has shifted from "competitors" to "partners". Under such a co-opetition relationship, how to be sustained success in a supply chain without protection or subsidies is a key issue for the survival and development of suppliers. This study aims to explore the factors that influence the competitiveness of suppliers. Based on Baosteel, firstly, this study constructed a hierarchical structure including 3 aspects and 13 factors. After acquiring 12 experts' assessments, Fuzzy-DEMATEL was employed to process the data. Finally, causal map was plotted according to the outputs. Referring to the causal map, we found that "Green Manufacturing Process" and "Use of Environmentally Friendly Materials" are the most influential factors in term of total influence and net influence. Therefore, these two factors can be regarded as critical factors.

1. Introduction

With the continuous development of the Internet, China has ushered in the best era since the reform and opening up, it is possible to buy needed products without leaving home, which also has driven the rapid development of the manufacturing industry. At the same time, people's consumption has been upgraded and living standards have been improved. Therefore, more attention has been paid to healthy and green living environment, thus the issues that environmental protection and sustainable development are advocated. Due to the rapid development of the economy, the pollution problems brought by the manufacturing industry to the environment have seriously affected people's living environment.

The 21st century is a green era, and protecting the environment is the responsibility of each of us. Countries are also introducing stricter environmental protection policies and strengthening environmental protection and management, for example, the EU officially launched the EU greenhouse gas emission trading scheme (EUETS) in 2005 to control the emission of carbon dioxide (CO_2). China joined the World Trade Organization in 2001, and green barriers are an important factor for the survival and development of China's manufacturing industry in the international trading market. In addition, China's environmental protection tax law, which was enacted in January 2018, shows that enterprises is facing stricter environmental requirements, great environmental pressure and the development trend of future green economy. Manufacturing companies are striving to improve the efficiency of green supply chains. As an important part of the supply chain, upstream suppliers are one of the important factors whether enterprises can effectively achieve the performance of green supply chain. Therefore, what green competitive advantages the upstream green suppliers should have is the main reference indicator for the core enterprises to select suppliers. This paper, through understanding and analyzing the competition factors of green suppliers, aims to find out the key factors that have the greatest impact, and provide some reference when suppliers are developing competitive strategy and core enterprises are fulfilling their green commitments.

2. Literature review

2.1. Green suppliers

In 1966, Dickson GW sent questionnaires to 273 procurement managers to conduct a survey, and sorted out the evaluation factors of suppliers and analyzed the importance of evaluation factors, the results showed that quality is a "very important" factor, and the 9 factors such as price, delivery on time, turnover and so on are "relative important" factor, and the 12 factors such as packaging capacity and corporate attitude and so on are "important" factors, and only one is "less important", that is reciprocal arrangement. In 1991, Weber performed secondary sort of Dickson's 23 factors by reading 74 supplier selection literatures from 1967 to 1990, the top 10 factors were: price, delivery on time, quality, equipment capability, geographic location, technical capability, management and organization, industry status, reputation and historical performance. In 1997, Noci believed that the supplier's environmental management strategy should be given priority, and gave quantitative evaluation indexes such as enterprise green capability, environmental management efficiency, green image and life cycle, etc. After that, the research on suppliers was added the environmental indexes. In 1998, Walton discussed the criteria for selecting suppliers when considering environmental factors, mainly including these green norms such as disclosure and records of environmental incidents, toxic substances and waste management, compliance with legal requirements, product friendliness and so on. From the above literatures, it can be seen that early scholars' researches on the selection of green suppliers are relatively one-sided, it only considered the impact of enterprises on the environment. And then, through the continuous extension of environmental impact, scholars sorted out more comprehensive evaluation factors of green supplier selection. For example, in 2013, Tseng and Chiu proposed to take environmentally friendly design, green production planning, and green R&D technology and so on as evaluation factors.

2.2. The application of DEMATEL

DEMATEL was first proposed by Gabus and Fontela in the 1970s to analyze the relationship between various factors in complex systems. DEMATEL can more directly identify the importance of various factors in the system and the influence of various factors. Therefore, in recent years, it has attracted the attention of scholars at home and abroad and has been applied in different fields.

Wu and Lee (2007) combined the fuzzy theory and empirical research with DEMATEL to figure out how to decompose certain abilities into parts to enrich the abilities of global managers and further promote the developmental method of global managers' abilities. Wu Hongyi, Zhang Youming and Zhou Jiarong (2014) analyzed professional consumers' purchasing factors and general consumers' purchasing factors by using DEMATEL, obtained the key factors affecting professional consumers to buy sports and casual shoes, and proposed corresponding commercial suggestions for related sports and casual shoe manufacturers. Li Xiaoying and Zhu Qinghua (2015) combined DEMATEL and ANP when establishing the evaluation model of green suppliers. After DEMATEL was used to determine the impact and causal relationship of evaluation, ANP was further used to determine the weight relationship, and they provided a scientific evaluation model for enterprises to choose green suppliers.

3. Research method

Decision Making Trial and Evaluation Laboratory (DEMATEL) was proposed by Fontela and Gabus at the Battelle Association of the Geneva Research Center in the early 1970s. The method combines graph theory and matrix to calculate the degree of influence between complex factors, and then uses numbers to concretely indicate the degree of causal influence between factors, and then we can put forward suggestions through in-depth analysis. Early DEMATEL was mainly used to study difficult and complex world-level problems, such as environmental pollution, resource issues, ethnic issues, etc. Because the DEMATEL method has excellent functions in factor research analysis and specific identification, and now, it is widely used in many fields such as enterprise internal management, market operation, and supply chain, etc.

The using steps as below: establish the system of evaluation factors, use the data of expert scoring, construct the initial matrix, and then normalize the initial matrix to obtain the normalization matrix, and calculate the total impact matrix. Finally, indexes such as centrality and cause degree and so on are obtained by analyzing the total impact matrix. From a mathematical point of view, DEMATEL can be expressed in the following way.

Step 1: Define criteria (impact factor) and establish facets and hierarchical structure of criteria

Step 2: To obtain evaluation the extent to which experts have an impact on pairwise comparisons between criteria

Questionnaire survey was conducted among h experts, and the evaluation data of expert k was obtained and sorted out as: $F^k = \left[f_{ij}^k\right]_{n\times n}$, among them, $1 \le i,j \le n$, n is the number of criteria, $f_{ij} = 0$, if i=j, that is, the diagonal element of F is set to 0.

Step 3: Construct an initial impact matrix

Use $f_{ij} = \frac{1}{h}\sum_{k=1}^{h} f_{ij}^k$ to calculate the initial impact matrix $F = \left[f_{ij}\right]_{n\times n}$.

Step 4: Normalize the initial impact matrix

Calculate $\lambda = Min\left[\frac{1}{Max_{1\le i\le n}\sum_{j=1}^{n}|f_{ij}|}, \frac{1}{Max_{1\le j\le n}\sum_{i=1}^{n}|f_{ij}|}\right]$, among them, $1 \le i,j \le n$.

Normalize the initial impact matrix $Z = \lambda F$

Step 5: Derive total impact matrix

Total impact matrix $T = \lim_{l\to\infty}\left(Z + Z^2 + Z^3 + \cdots + Z^l\right) = Z(I - Z)^{-1}$

Step 6: Draw a causal diagram and analyze the results accordingly

$(D+R)$ is the horizontal axis, $(D-R)$ is the vertical axis, and then draw a causal diagram and analyze

4. Empirical research

4.1. The background introduction of case company

Haier Company, as the top 10 home appliance enterprise in China's green enterprise management, far exceeds the basic level of the same industry in the indicators of green supplier management, and has strict selection standards for green supplier. As a green supplier of Haier, Baosteel Group must have the green competition factors what the other competitive enterprises lack.

Baosteel group is the world's leading modern steel joint enterprise, strictly abides by the national environmental protection laws, regulations and standards, fulfills international environmental conventions, adheres to green manufacturing, and takes the road of sustainable development. In terms of environmental protection and resource utilization, Baosteel has independently studied the international advanced technology of high-efficiency heat transfer of coke oven riser tube (CATHE), and reduces the energy consumption of coke oven process by 6.9kgce/t. At the same time, it has independently developed the dioxin source treatment technology, which has the rate of dioxin emission reduction up to 92.2%. In addition, in cooperation with industrial scientific research institutions, it developed the flue integrated micro-dust electro coagulation technology and complete equipment technology, and PM2.5 decreased by 51.9%. And Baosteel uses artificial wetland system to treat the wastewater, and the wastewater can be reused after treatment, thus achieves zero exclusion of this part.

In terms of green products, Baosteel group uses full lifecycle management to track products in an all-round way, and produces environmentally friendly products to reduce the consumption of downstream users and extend the service life. At the same time, the company produces products based on the principle of environmental protection, and controls the emission of toxic and hazardous substances, ensures the effective utilization of energy and resources, and adopts third-party enterprises for green certification in terms of use time, appearance and secondary recycling of product. At present, eight products of Baosteel group have been certified by Intertek Green Leaf Label.

4.2. The establishment of competitive factors facets and criterion

First, we should refer to the relevant literature on the competition factors of green suppliers at home and abroad, and formulate factors and facets. These documents include Zhu Qinghua and Geng Yong (2004) proposed 12 specific indicators in view of green procurement, such as the use and purchase of hazardous materials, the consideration of finished product recycling rate, and ISO14000 certification, etc. Tseng and Chiu (2013) added green performance indicators when selecting green suppliers, including 12 environmental standards such as environmentally friendly design, total quality management, clean production, and internal green production design, etc. Based on the research results of green suppliers' selection in recent years, Wang Xingxing (2007) summarized a comprehensive system of environmental performance measurement standards, which includes the following eight aspects: green image, pollution control, environmental management system, green design, green capacity, green products, green recycling, and pollution treatment costs.

Based on the above literature and combined with the research and analysis of green competitiveness factors of Baosteel group, the following are analyzed and discussed from three aspects of green quality, green products and green technology.

- Green Quality

The green environment of the upstream supplier refers to the specific measures and policies implemented by the enterprise to reduce the environmental damage when the production and operation of the enterprises destroy the environment. Such as waste treatment, environmental management system, commitment to the protection of the social environment, production products in line with market and national environmental laws and regulations, etc.

- Green Products

Green products refer to energy saving, pollution reduction and environmental protection and recycling of products in the production process. When choosing green products, manufacturers generally pay attention to the following aspects: product recycling and recycling rate, product use and maintenance cost, energy consumption of product production, green process flow, and product stability.

- Green Technology

Green technology refers to the technology which can reduce environmental pollution and energy consumption. The commonly used technologies include: the use of environmentally friendly materials by enterprises, green process flow, non-toxic manufacturing process, and clean technologies.

4.3. Data acquisition, analysis and results

On the basis of referring to the relevant literature of this research, on November 1st, 2018, after communicating with the managers who had five years' experience in supply chain in Baosteel Company, the doctoral scholars in logistics management and the staff of Environmental Protection Bureau, we initially got the competitive factors related to green suppliers. On November 15th, 2018, we deeply communicated and discussed with the experts again, and finally determined the three major facets and 13 factors of the green supplier competitiveness factors (as shown in the following Table 1), and then determined the research framework of this research.

Table 1. The green supplier competitiveness factors

Aspects	Factors	Factor description
Green environment	Waste treatment	The way in which supplier enterprises deal with waste discharge and the treatment effect
	Environmental management system	Environmental protection work plan and management system formulated by supplier companies for protecting the environment
	Enterprise social responsibility	Measuring the extent to which supplier enterprises meet environmental targets and the achievement of their commitment to protecting the social environment
	Environmental protection law	Whether the raw materials and functions used by the supplier enterprises' products comply with the market and relevant national environmental laws and regulations
Green products	Material recycle and recycling rate	Recycling and recycling rate of materials from supplier enterprises
	Product use and maintenance cost	Products use and maintenance costs after purchasing products from supplier enterprises
	Energy consumption	Various energy costs actually consumed by supplier enterprises in production activities
	Research and Development of green technology	The costs incurred by supplier enterprises in developing green technologies such as green packaging and green processes, etc.
	Product reliability	The stability of availability and quality from supplier's products over time
Green technology	Environmentally friendly materials	The use technology of environmental protection material from supplier enterprises
	Green manufacturing process	The use of green manufacturing processes of enterprises is mainly to study and adopt process schemes with less material and energy consumption, less waste, and less environmental pollution.
	Clear technology	The costs of the treatment of environmental sanitation and using new energy equipment by supplier enterprises in treatment
	Non-toxic manufacturing process	Supplier companies use non-toxic processes and non-toxic materials in the manufacturing process

The DEMATEL questionnaire was designed based on the above three facets and 13 factors. On November 20, 2018, experts were invited to fill out the questionnaire, and the questions and contents of the questionnaire were fully explained to answer the questions in the questionnaire, so as to ensure that the experts fully understood the questionnaire, and we obtained effective questionnaires.

After obtaining the expert evaluation results, the data is processed by the DEMATEL algorithm, and draw a causal diagram accordingly.

According to the above figure, the factors in the first quadrant include green manufacturing process, environment-friendly materials, clean technology, research and development of green technology, non-toxic manufacturing process, environmental protection law, and enterprise social responsibility. Among them, green manufacturing processes and environmentally friendly materials are factors that are highly important and have a large impact on other factors, so they can be defined as key factors. The factor in the second quadrant is only energy consumption, which is the least important factor, and has a supplementary influence on the selection of green suppliers. The factors in the third quadrant include product reliability and product use and maintenance cost, which are low important but not easy to be affected, and are called independent factors. The factors in the fourth quadrant include material recycling and recycling rate, environmental management system, and waste disposal, which are low important and easy to be affected.

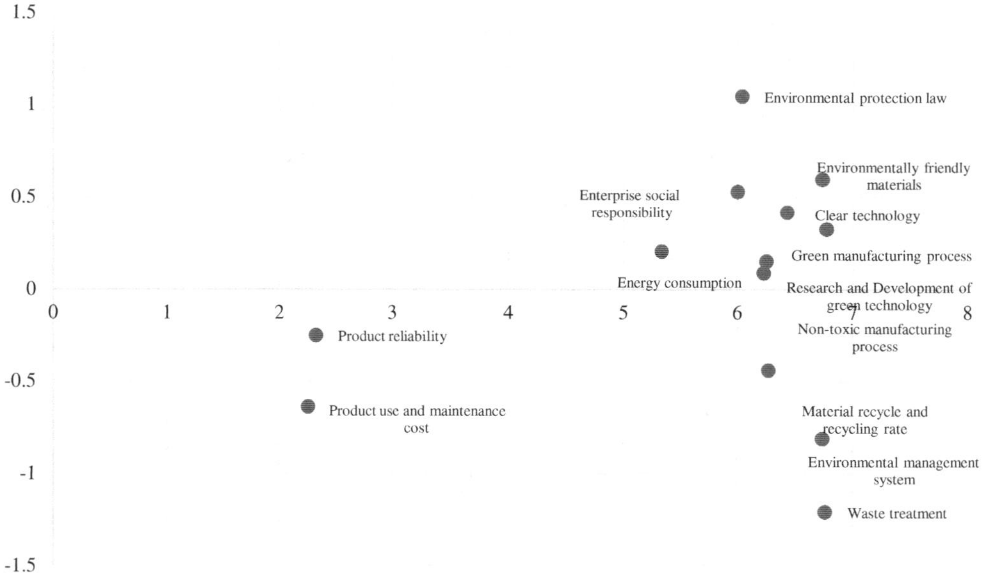

Figure 1. Causal map

4.4. Practical implications

Observing Figure 1, it can be seen that green manufacturing process and environmentally-friendly materials are the key factors for core enterprises in the supply chain to choose green suppliers. In the process of production and operation, supplier enterprises not only need non-toxic and harmless raw materials, but pay attention to the auxiliary manufacturing process in the green manufacturing process also should meet the relevant regulations on environmental protection. For example, in the second quarter of the Children's Toys Products Quality Report released by the China Baby and Toys Association in 2018, it was mentioned that 81.63% of China's exports to the EU were recalled, the reasons for the failure are as follows: benzene content exceeding the standard, small parts unqualified, heavy metal content exceeding the standard, etc. Therefore, it can be seen that whether the parts and semi-finished products provided by the upstream suppliers meet the market environmental protection regulations greatly affects whether the products can be accepted by the market. As a representative of green home appliance enterprises, Haier actively responds to the national enterprise environmental protection policy and pays high attention to the market environmental qualification rate of parts and components provided by upstream suppliers. Therefore, for upstream suppliers, the enterprise's green manufacturing process capacity and the use of environment-friendly materials are important ways for supplier enterprises to improve their "green competitiveness".

5. Conclusion

In this study, through in-depth interviews with experts and reference to a large number of literatures, the green supplier competition factors are summarized into three major facets and thirteen criteria, and we study the causal relationship analysis of green suppliers' competition factors accordingly. According to the analysis of the DEMATEL's empirical results, the following conclusions are drawn:

(1) Green manufacturing processes and environmentally-friendly materials are the most important factors for the competitiveness of green suppliers, for upstream manufacturers, green manufacturing processes and environmentally friendly materials are the first considerations for companies.

(2) Product reliability, product use and maintenance cost, material recycle and recycling rate are the most easily affected by other factors.

(3) Among the 13 factors, special attention should be paid to the environmental protection law, because its cause degree is much higher than the others, indicating that it has the greatest impact on other factors.

From the above conclusions, this paper proposes the following suggestions for the reference of relevant enterprises.

(1) Green suppliers focus on the understanding of "green" and its application in production. When improving green competitiveness, green suppliers should add "green" standards based on traditional supplier selection criteria (quality, price, on-time delivery, etc.). Based on the research in this paper, it is suggested that green supplier enterprises can greatly improve product stability, increase product durability, reduce product recycling times, and improve product "green" quality by improving green manufacturing process capacity and utilization rate of environment-friendly materials, thus optimize their green competitiveness.

(2) Under the background of market globalization, enterprises must abide by the environmental protection laws of different countries and regions if they want to have a bigger market. According to the research in this paper, the environmental protection law has far more influence on other factors than other factors. Therefore, it is suggested that enterprises actively comply with environmental protection laws and regulations, which is not only beneficial to the development of enterprises green culture and establishing the concepts of green management and green production, but also beneficial to developing the market and enhance the green competitiveness of enterprises.

Acknowledgement
The authors would like to thank the City College of Dongguan University of Technology of the China for financially supporting this research under Contract No. 2017YZDYB04R.

References

[1] Dickson G W 1996 *J. Purchasing* **2** 1
[2] Weber C A 1991 *Euro. J. Ope.Res.* **50** pp 2-18
[3] Noci G 1997 *Euro J. Purchasing Supply Manage.* **4** pp 6-13
[4] Walton V S 1998 *Inter. J. Purchasing Material Manage* **4** pp 2-10
[5] Tseng M L and Chiu A S F 2013 *J. Cleaner Production* **40** pp 22-31
[6] Gabus A and Fontela E 1973 *Collective Responsibility* **10** pp 11-18
[7] Wu W W and Lee Y T 2007 *Exp. Syst. Appl.* **32** pp 499-507
[8] Wu H Zhang Y and Zhou J 2006 *J. Sports, Health and Leisure* **5** pp 74-86
[9] Li X and Zhu Q 2015 *Technol Manage.* **5** pp 43-49
[10] Zhu Q and Geng Y 2004 *Chinese Manage.* Sci. 6 pp 238-246
[11] Wang X 2017 *J. Econ. Res.* **4** pp 194-196

Fuzzy cognitive map based vulnerability analysis of agricultural products cold chain logistics: a perspective of carbon footprint

Rongxia Zhou and Shih Ming Ou

School of Business Administration, Fujian Jiangxia University, Fuzhou, China

E-mail: zhourongxia@fjjxu.edu.cn

Abstract. As an important part of the development of a low-carbon economy, developing low-carbon logistics is an important measure to implement transform the economic development mode and cope with global warming. In order to effectively realize the low-carbon operation of agricultural cold chain logistics, a fuzzy cognitive map is employed to analyze the vulnerability of agricultural cold chain logistics on the perspective of carbon footprint. With the analysis above, the influencing factors of carbon emissions in cold chain logistics operations can be identified. Therefore, the carbon reduction strategy of the cold chain logistics of agricultural products can be obtained. Our findings can be as decision-making reference for relevant departments and enterprises to better formulate cold chain low carbon emission strategies and promote green logistics to achieve high-quality development.

1. Introduction

In recent years, global warming has constituted a research threat to the sustainable development of natural ecosystems and human society, and in this context, low-carbon economy has emerged. The Chinese government has also actively responded to international carbon reduction needs, and has adopted legal, administrative, economic and other government measures to ensure the implementation of energy conservation and emission reduction tasks for enterprises in various industries. Low-carbon logistics is an important part of the low-carbon economy, and should actively promote the low-carbon green development of the logistics industry.

With the rapid development of China's economy and the improvement of people's living standards, consumers' demand for fresh agricultural products continues to grow, and the quality requirements are increasing, which in turn drives the development of cold chain logistics. As a large agricultural country, China has about 400 million tons of fresh agricultural products entering the circulation field every year. Cold-chain logistics is required to escort agricultural products from procurement, processing, transportation, storage, to distribution. As a result, more fossil fuels are consumed, resulting in an increase in carbon emissions. The cold chain is the business that consumes the most energy and carbon emissions in the logistics industry. How to introduce the concept of emission reduction into the entire logistics management strategy, how to make the cold chain comply with the low carbon requirements of economic development, and adapt to the policy orientation of the country's cold chain logistics to help the poor get rid of poverty, which is a subject worthy of serious exploration.

2. Materials and methods

At present, cold chain logistics has developed relatively mature abroad, but it is still in its infancy in China. From the perspective of literature search, the research fields at home and abroad mainly focus on the following aspects:

(1) The development status, existing problems and countermeasures of agricultural cold chain logistics, Such as Neil V Ass. M. D [1] (2006), Kun Liang (2015) [2], etc.

(2) Agricultural product cold chain logistics operation management. Specifically, it includes the control of agricultural product cold chain logistics quality, cost control, optimization of cold chain logistics system, cold chain distribution location and route optimization, and so on. For instance, A. Osvald, Stirn [3], H. Nazif, L. S. Lee (2010) [4], Ran Ban (2018) [5], Yaming Zhang and Yanming Li(2012) [6].

(3) Measurement of carbon emissions from cold chain logistics of agricultural products and low carbon development strategies, such as Tristram O West and Gregg Marland [7], Kai-Kang, Jie-Han (2019)[8], Xing-Xu, Renwang Li(2018) [9], etc.

In summary, domestic and foreign experts and scholars have used different methods to study the cold chain logistics of agricultural products based on different perspectives, and accumulated rich results. However, there are not many studies on the low carbon development of cold chain logistics, and the only research is too general and abstract, and the reference is not very strong. Taking the vulnerability of cold chain logistics as the entry point and based on the carbon footprint perspective, combining low-carbon environmental protection with cold chain logistics, and discussing the results of cold chain low-carbon operation is almost blank. Especially in the new situation of China's active development of low-carbon economy and the help of cold-chain logistics to help the poor, there is an urgent need for practical research in this area. In view of this, this paper takes the carbon footprint as the research perspective, aims at reducing the carbon emissions of cold chain logistics of agricultural products, and carries out the analysis of the fragility of cold chain logistics of agricultural products, and then discusses the carbon reduction strategy of the cold chain logistics operation of agricultural products. It has theoretical and practical significance.

3. Research method

3.1. Introduction to fuzzy cognitive map

Fuzzy Cognitive Map (FCM) was proposed by Kosko and other scholars in 1986, which fused Axelord cognitive graph theory and Zadeh's fuzzy set theory. Cognitive Map (CM) is a graph model used to express and cause causal relationships between concepts. Since the cognitive graph model cannot quantify the degree of causality between conceptual nodes, in order to overcome the defects of cognitive graphs, fuzzy numbers are included, and fuzzy cognitive maps are generated.

The fuzzy cognitive graph is an extension of the cognitive graph. It quantifies the degree of causality between concepts by introducing fuzzy measures into the traditional cognitive graph model. Therefore, fuzzy cognitive graph is a soft computing method, which is the product of the combination of fuzzy logic and network. Because of its dual advantages of fuzzy logic and neural network, and strong mathematical reasoning ability, good interpret ability and intuitive knowledge representation, FCM has attracted extensive attention and research from scholars and has been applied in multiple Different fields, including teaching research [10], engineering control [11], business modeling and decision support [12], risk assessment [13], medical diagnosis [14-15].

3.2. Construction of fuzzy cognitive map

A fuzzy cognitive map is a graphical representation of knowledge, a fuzzy, signed, and feedback graph. Its graphical architecture consists of nodes, symbols, directions, and weighted arcs. Nodes represent events, and directed arcs between nodes represent causal relationships between events. The weight of each directed arc varies, representing the extent to which the previous node affects the next node. The basic structure of the fuzzy cognitive map is shown in Figure 1.

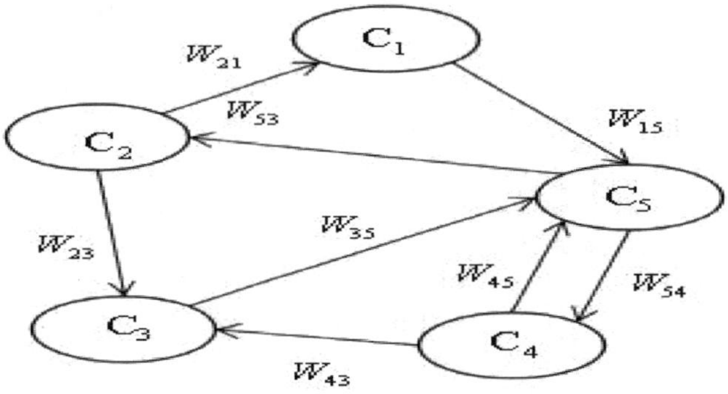

Figure 1. Schematic diagram of fuzzy cognitive map

3.2.1. Determination of concept nodes

A node in a fuzzy cognitive graph is called a concept or concept node, and is used to represent the actions, causes, results, and trends of the system, which denoted by C. In the fuzzy cognition, there is a set of finite concept nodes, that is $C = \{C_1, C_2, \cdots, C_n\}$, if the state of node C_i changes, the state of node $C_j$changes, thenC_i is the reason node of C_j, C_j is the result node of C_i.

3.2.2. Determination of weight

The degree of influence relationship between the conceptual nodes is represented by the connection weights marked $W_{ij} = \{i, j = 1, 2, \cdots, n\}$ on the directed edges, and $W_{ij} \in [-1, 1]$. Among them, $W_{ij} \in (0, 1]$ indicates that there is a positive causality between C_i and C_j, $W_{ij} \in [-1, 0)$ indicates that there is a negative causality betweenC_i and C_j,and $W_{ij} = 0$ indicates that there is no causality between the two.The fuzzy causal relationship between conceptual nodes in FCM is represented by a weighted adjacency matrix.

$$E = \begin{bmatrix} w_{11} & \cdots & w_{1n} \\ \vdots & w_{ij} & \vdots \\ w_{n1} & \cdots & w_{nn} \end{bmatrix} \quad (1)$$

3.2.3. Weighted aggregation

A represents the factor weight matrix given by the kth expert, integrates the opinions of multiple experts and finds the arithmetic average, and obtains the associated weight matrix Ec of the whole system FCM.

$$E_C = \frac{1}{m}\sum_{k=1}^{m} E_k \quad (2)$$

3.2.4. Inference of concept change

(1) Select the threshold function

The function of the threshold function is to prevent the variable value C_i on the node from exceeding the domain during the operation, and its shape represents the operation mode of the entire system association mode.There are three types of threshold functions, including Bivalent, Trivalent and Logistic Signal. The threshold functions of the first two forms are relatively easy to obtain stable or oscillating results. The more commonly used threshold function B is the Bivalent form, which has the following form:

$$B: \ C_i(t+1) = \begin{cases} 1, if \ C_i(t) \times E_{C_i} > 0 \\ 0, if \ C_i(t) \times E_{C_i} \le 0 \end{cases} \tag{3}$$

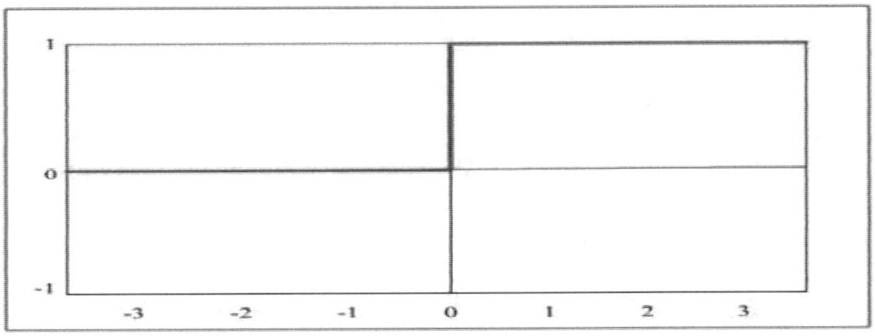

Figure 2. Threshold function of FCM bivalent morphological

In the above formula, $C_i(t+1)$ represents the matrix of the variable state values of the node at the (t+1) th operation, $C_i(t)$ represents the matrix of the variable state values of the node at the t th operation, Ec_i represents the factor weight matrix of the node.Here, the threshold function is set to zero. The significance of the threshold function is that after calculating the matrix result, the state values of all variables greater than 1 are adjusted to 1, and the state values of all variables less than 0 are adjusted to 0, as shown in Figure 2.

(2) Establish learning rules

The learning rule is used to judge whether the causal weight value between concepts in the operation process needs to be changed, so as to promptly correct the causal relationship of the node due to time change. In a simple fuzzy cognitive graph, the weight value of the causal edge can be set to a fixed value, and the variation rule of the concept node can utilize the basic concept of matrix operation, that is, multiply the initial value matrix by the factor weight matrix and then pass the threshold function. To filter out the obvious factors, and finally make the relationship between the concepts in the system clear and clear. Its calculation formula is:

$$C_i(t+1) = B(C_i(t) * E_{C_i}) \tag{4}$$

(3) Setting the initial value

In order to facilitate the calculation and analysis in the operation process, generally only one target variable is selected, and its initial value is set to 1, and the operation of the change inference is started until convergence. The nodes of the fuzzy cognitive graph may produce three kinds of results after the change inference: stability, oscillation or divergence. The oscillation state is a finite cycle, and the node value will oscillate and change with the cycle of the number of operations.

4. Empirical research

4.1. Case background introduction

F Agricultural Products Group Co., Ltd. (hereinafter referred to as F Group) is a group trading company in the southeastern region of China. It takes the development, construction, management and wholesale market as its core business, then extends its business by base planting and breeding, logistics and distribution, and fresh terminal operation. Now it has 8 holding companies including food processing, fresh food distribution, grain trade and cold storage business, and has opened a number of professional agricultural and sideline products wholesale markets. The group has won many honorary titles such as the National "Cooking Basket" reassuring engineering excellent enterprise, the national 100 large agricultural product circulation enterprises, and the national food safety demonstration unit.

4.2. Correlation analysis between cold chain logistics and carbon footprint

From the production field to the consumption field, agricultural products need to be harvested, pre-cooled, processed and packaged, warehousing, transportation, wholesale and retail, and so on. It is necessary to provide full-length cold chain monitoring and management. As shown in Figure 3, the upper part represents the entire supply chain process of agricultural products, the lower part represents the cold chain logistics operation process of agricultural products, the solid arrow indicates the operation flow, and the hollow arrow indicates the carbon footprint correlation. In order to help the whole process of cold chain energy conservation and emission reduction, reduce consumption and increase efficiency, it is necessary to integrate the carbon footprint theory into the logistics operation process of F Group, combined with the measurement method of carbon footprint, to measure the carbon footprint in all links of cold chain operation, and realize the cold chain. The statistics of carbon emissions throughout the operation provide reference for enterprises to formulate energy-saving and emission reduction strategies.

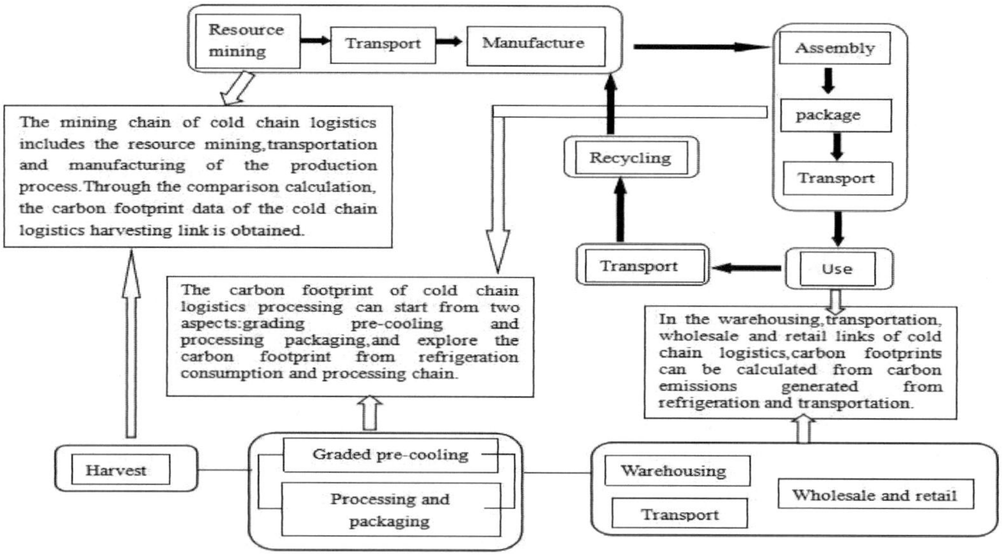

Figure 3. Association model of cold chain logistics operation process and carbon footprint

4.3. Determination of concept nodes

Determining the concept node is critical, which is directly related to the construction of the fuzzy cognitive graph model. Combining the characteristics and structure of chain logistics operation and carbon emission problems, the relevant factors are initially screened out as the conceptual nodes needed to construct the FCM model. In order to ensure the accuracy and effectiveness of the concept nodes, the author invited a number of industry experts and scholars and industry insiders to form a group of experts to discuss, and finally the ten concept nodes for constructing the FCM model are determined: carbon emissions, service levels, clean fuel usage, transportation batches, transportation road conditions, refrigerated truck types, storage temperature control, handling, transportation and packaging (as shown in Table 1).

Table 1. Description of the concept node

Concept node	Description
A.carbon emission	Average greenhouse gas emissions from the production, transportation, use and recycling of agricultural products.
B.Service Level	The customer has a psychological expectation and expectation of the logistics service elements obtained and the formation of such elements.
C.Clean fuel usage	The ratio of the amount of clean energy consumed in the transportation and

	handling of cold chain logistics to the total energy use.
D.Transport batch	Total amount of agricultural products transported
E.Traffic condition	Congested or unblocked urban traffic roads.
F.Refrigerated car type	Cold chain logistics uses different types of refrigerated vehicles to consume different energy and carbon emissions when transporting agricultural products.
G.Storage temperature control	Extend the shelf life of the item by adjusting and controlling the temperature of the agricultural product storage environment.
H.Loading and unloading	The agricultural products are transported at both ends to load and unload the products,and carry them to a certain location
I.Transport	Agricultural products are repackaged or reloaded during the shipping process
J.Package	The overall name of containers,materials and auxiliary materials used in accordance with certain techniques to protect agricultural products,facilitate storage and transportation, and promote sales during circulation.

4.4. Weight determination of concept nodes

4.4.1. Weight training

In order to determine the connection relationship between the concept nodes, the author obtains the influence degree $W_{ij} = (i,\ j = 1,\ 2, \cdots, n)$ between each node by issuing a questionnaire to the expert group. The description of the form is as follows: if the two concepts are positively correlated, the value entered is between 0 and 1, that is $W_{ij} \in (0,\ 1]$; if the two concepts are negatively correlated, the value filled in is between Negative 1 and 0, that is $W_{ij} \in [-1,\ 0)$; if the two concepts have no influence, they are directly filled with 0.The initial weight of the model is given by the experts according to the specific situation of F Group cold chain logistics operation.

4.4.2. Obfuscation of raw data

The data collected by the questionnaire is simply processed using the formulas (1) and (2) to obtain the associated weight matrix of the FCM system, and specific steps are as follows:

(1) For the numerical vector of each concept node$V_{ij} = (i,\ j = 1,\ 2, \cdots, n)$, the numerical vector is the pointer to the i-th concept node, the i-th expert scores, finds the maximum value in the vector element, that is, the highest score of the expert for all concept nodes, defined as $MAX(\mathrm{v})$, assigning a value to 1,that is, order $MAX(\mathrm{v})V_{ij} = 1$.

(2) For the value vector of each concept node $V_{ij} = (i,\ j = 1,\ 2, \cdots, n)$, look for the minimum value in the vector element, that is, the lowest score that the expert hits all concept nodes, defined as $MIN(\mathrm{v})$, assigning a value to 0, that is, order $MIN(\mathrm{v})V_{ij} = 0$.

(3) For each data of the concept node, the lowest score and the highest score are substituted into the fuzzy formula (5) perform the fuzzification on [0, 1], and the fuzzy result of all the data can be obtained.

$$XV(v) = \frac{V_i - \mathrm{MIN}(v)}{MAN(v) - \mathrm{MIN(v)}} \tag{5}$$

(4) In the statistical questionnaire, the expert gives the score value of each concept node and finds the simple arithmetic average, which is used as the initial value of the fuzzy cognitive graph model That is, the state evaluation value). The processed data is summarized in Table 2.

Table 2. Weight values between conceptual nodes

	A	B	C	D	E	F	G	H	I	J
A	0.00	0.18	0.44	0.60	0.45	0.56	0.40	0.20	0.30	0.32

B	0.12	0.00	0.36	0.32	0.29	0.44	0.38	0.40	0.36	0.44
C	0.48	0.18	0.00	0.13	0.23	0.36	0.16	0.08	0.16	0.10
D	0.64	0.12	0.16	0.00	0.46	0.44	0.32	0.40	0.50	0.46
E	0.56	0.26	0.20	0.56	0.00	0.10	0.34	0.30	0.52	0.34
F	0.70	0.41	0.34	0.46	-0.04	0.00	0.64	0.34	0.46	0.46
G	0.52	0.54	0.19	-0.06	0.14	0.44	0.00	0.08	0.08	0.12
H	0.22	0.42	0.08	0.54	0.12	0.20	0.36	0.00	0.74	0.44
I	0.26	0.42	0.10	0.38	0.46	0.52	0.48	0.58	0.00	0.44
J	0.25	0.40	0.12	0.25	0.12	0.38	0.36	0.48	0.48	0.00
Rating	0.82	0.36	0.40	0.54	0.46	0.52	0.50	0.40	0.38	0.40

4.5. The establishment of fuzzy cognitive maps

Bringing the relationship strength between the concepts after fuzzification into the basic structure of the fuzzy cognitive map, we can construct a fuzzy cognitive map of the cold chain logistics operation vulnerability analysis based on the carbon footprint perspective for the case, as shown in Figure 4.

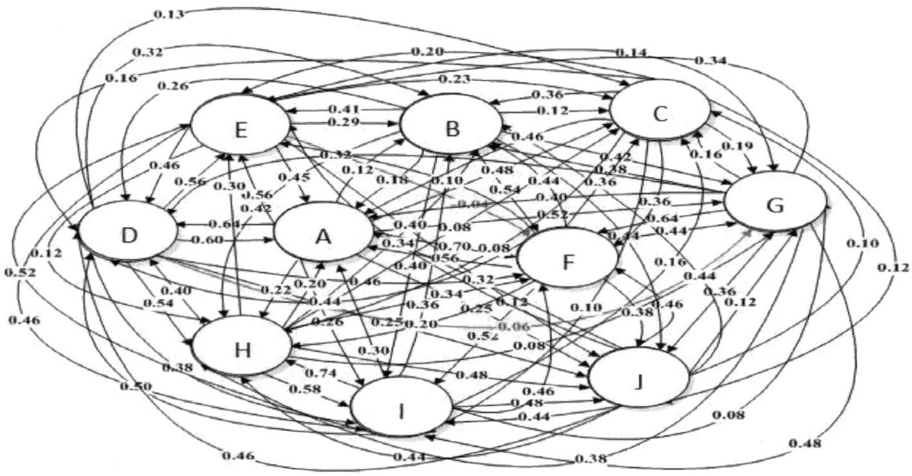

Figure 4. Fuzzy cognitive map of the case company

4.6. Causal Analysis of Fuzzy Cognitive Map

In order to further identify the vulnerability of agricultural product cold chain logistics from the perspective of carbon footprint, it is necessary to expound the causal relationship between the conceptual nodes and carbon emissions in the fuzzy cognitive map:

As shown in Figure 4, there is a positive causal relationship between the nine concept nodes and the carbon emissions of agricultural cold chain logistics. Among them, D (transportation batch), F (refrigerated vehicle type) and carbon emissions have a strong causal relationship, and the agricultural product transportation volume and The type of refrigerated truck directly affects the amount of carbon emissions; E (transportation road conditions), C (clean fuel usage rate), G (storage temperature control), I (transport), J (packaging), etc. Moderate causality, rational planning of transportation routes, improvement of traffic conditions, improvement of clean fuel utilization rate, and reasonable regulation of storage temperature have a strong incentive to reduce carbon emissions; B (service level), H (loading and handling) and Carbon emissions show a weak causal relationship, indicating a lower impact on carbon emissions.

In summary, the refrigerated transport links have the most carbon emissions. The carbon emissions caused by refrigerated transport during transportation are also related to the transportation distance and transportation quality. The reduction of carbon emissions can be started from the formulation of transportation routes and the optimization of technology in the manufacturing process; the staged pre-cooling and processing and packaging links in the cold chain logistics correspond to the "transport, assembly and packaging in the production process", in this regard, optimized balance between machines can effectively reduce carbon emissions; warehousing, transportation and wholesale and retail links of cold chain logistics correspond to the "use" in the production process of the product, proper planning of the space and time of cold chain storage and the establishment of a recycling system are one of the effective ways to reduce carbon emissions.

5. Conclusions

Based on the analysis of the correlation model between agricultural product cold chain logistics operation process and carbon footprint, this paper constructs a fuzzy cognitive map of agricultural product cold chain logistics operation vulnerability based on carbon footprint perspective with FCM theory and expert questionnaire method. The FCM architecture is based on carbon emissions and service levels in agricultural cold chain logistics operations, focusing on the relationship between key concept nodes and carbon emissions and the associated weights to properly handle the impact of agricultural cold chain logistics operations. The causal relationship between the various influencing factors of carbon emissions, which makes the construction process and results more in line with the actual situation, and provides effective decision-making reference for relevant departments and enterprises to develop low-carbon operation policies or measures for cold chain logistics. In the process of constructing FCM, the threshold impact analysis is used to understand the impact of threshold on FCM formation, so as to construct the required impact weight more objectively.

From the results of the case analysis, it is suggested that the operation of the low-carbon chain of cold chain can consider efforts to improve energy efficiency, optimize transportation routes, strengthen cold chain logistics technology innovation, and build a complete cold chain logistics transportation system.

References

[1] Neil V 2006 U. California Press.
[2] Liang K 2015 *Logistics Techn.* **3** 79-82
[3] Osvald L and Stirn Z 2008 *Food J. Food Eng.* **85** 285-95
[4] Nazif H and Lee L S 2010 *American J. App. Sci.* **1** 95-101
[5] Ban R 2018 *Techn. Econ. Manage. Res.* **12** 28-32
[6] Zhang Y Li Y and Liu H 2019 *Stat. Decision* **3** 176-81
[7] Tristram O W and Marland G 2002 *Agri. Ecosystems Environ.* **1** 123-30
[8] Kang K and Han J 2019 *Com. Eng. App.* **2** 259-65
[9] Xu X Li R and Wu X 2018 *Com. Int. Manu. Sys.* **2** 533-8
[10] Cole R J and Persichitte K A 2000 *Int. J. Intelligent Sys.* **15** 1-25
[11] Groumpos P P and Stylios C D 2000 *Chaos Solitons & Fractals* **11** 329-36
[12] Min H Q Hui J and Lu Y S 2006 *Int. Con. Intelligent Agent Techn.* 323-5
[13] Zhai D and Zhang J 2008 *Stat. Decision* **2** 161-3
[14] Papageorgiou E I and Froelich W 2012 *IEEE Eng. Med. Bio. Soc.* **16** 143-9
[15] Salmeron J L Rahimi S A and Navali A M 2016 *Neurocomputing* **232** 104-12

Applying fuzzy statistics to evaluate and allocate resources in seven star grotto national park area

Chaojun Li[1], Kuo Kuang Chang[1] and Shih Ming Ou[2]

[1]Tourism and Historical Culture College, Zhaoqing University, Zhaoqing, China
[2]Department of Logistics Management, Fujian Jiangxia University, Fuzhou, China

E-mail: benao13@gmail.com

Abstract. The purpose of this study is to use an objective function algorithm to establish the fuzzy number parameters of the fuzzy semantic scale, and then find the way to improve and the development and management of tourist attractions. This paper divides the development of the Seven Star Grotto National Park into three periods: 2001-2005, 2006-2010 and 2011-2015. Using expert scoring and field interviews, combined with the Seven Star Grotto National Park resource profile and Zhaoqing development profile, we quantitatively measured the overall resource level, resource value, scenic conditions and regional conditions of different time period.

1. Introduction

Seven Star Grotto National Park is located in Zhaoqing City, Guangdong Province, a national historical and cultural city. Seven Star Grotto National Park has a history of thousands of years and is famous both at home and abroad for its beautiful landscapes with southern characteristics. It was listed as the first batch of national key scenic spots by the State Council in 1982 [1], which was also the only key scenic spot in Guangdong at that time. Over the past 30 years, the scenic spots in various regions have become hot, and Guangzhou Zhuhai Changlong, Shenzhen World Window, Jinxiu Zhonghua, Kaiping Diaolou, Shaoguan Danxia Mountain and other scenic spots in the surrounding areas have achieved great success, while the Seven Star Grotto National Park is tepid. The dynamic evaluation of the value of its tourism resources is an important aspect of exploring the reasons why the Seven Star Grotto National Park is tepid.

Tourism resource evaluation refers to the process of comprehensive evaluation and appraisal of the value of tourism resources in a certain area and its external development conditions from the perspective of rational use and protection of tourism resources and maximum social and economic benefits [2]. The key difficulty in tourism research and development is to effectively evaluate the resources of the scenic spot. The dynamic evaluation is an evaluation model, where in order to accurately analyze the investment and make decisions, it is necessary to convert the cash flow generated at different moments into equivalent values, and then compare and evaluate the relationship between funds and time on the same basis. Since the 1970s, there have been scholars in China to evaluate the development of tourism resources [3]. The quantitative research on tourism resources evaluation in China is rich in content but the comprehensive dynamic research is lacking.

On the basis of comprehensive understanding of the types, quantity, quality, characteristics, scale, formation reasons and development potential of the resources in the Seven Star Grotto National Park, systematically and dynamically grasping the progress status, existing problems and development trends of the protection and utilization of tourism resources can provide a reliable basis for determining the direction, timing and focus of tourism resources development [4].

2. Seven star grotto national park tourism resources overview

Tourism resources refer to the part of the geographical environment that has tourism value, that is, the environmental factors that tourists are interested in and the material conditions that can be utilized during the tourism process [5]. According to the reasons for its formation, it can be divided into two parts: natural and artificial. Natural, that is, natural tourism resources, refer to the part of natural scenery that is beneficial to human health, growth, and appreciation, and exists naturally. Artificial, that is, humanistic tourism resources, which are the cultural achievements or artistic shapes created by human beings in various historical periods, including historical sites, contemporary architecture, modern culture and art, customs and folk cultural activities, shopping tasting events, sports competitions, etc. and have obvious historical, national and local characteristics [6]. Tourism resources can be visible to the naked eye, or can be invisible to the naked eye and only be felt, the former such as daylight, beach, the latter such as the city's personality characteristics and personalized atmosphere [7].

2.1. Unique natural tourism resources

Seven Star Grotto National Park is unique in geography. It leans on the mountains in the north and surrounded by water on three sides in the east, south and west. The water area is about 4 million square meters. The lake water is divided into five parts by the lake banks. The embankment is more than 20 kilometers long, and the roads are covered by the trees and the silvery lake and green hills are complemented by each other. The scenery is excellent. Because it consists of seven rocks, and their arrangement is like the Big Dipper, hence it was named as such. The seven rocks are named in order from east to west as Langfeng, Yuping, Shishi, Tianzhu, Chanchu, Xianzhang, and Apo. Since the Tang Dynasty, the reputation of "the first wonder of Lingnan" has been fascinating. The natural tourism resources of Seven Star Grotto National Park are mainly divided into three sub-categories: geography landscape, water scenery, and biological landscape.

2.2. Rich in humanities and tourism resources

There are many cultural relics in the Seven Star Grotto National Park. The main buildings include Shidong Ancient Temple, Yuhuang Temple, Sanxian Temple, Xuankong Pavilion, Shuangqing Pavilion, Nanhua Pavilion, Shiyou Pavilion, Qixing Bridge, Juxing Bridge, and Lansheng Memorial Archway, etc. After 1949, a large number of pavilions and hotels were built to make the Seven Star Grotto National Park more beautiful. The scenic spots of the Seven Star Grotto National Park are famous, and tourists are constantly visiting.

The inscriptions on precipices in the Seven Star Grotto National Park are worth mentioning here as part of the human resources. The inscriptions on precipices of the Seven Star Grotto National Park are a cluster of a large number of inscriptions on precipices preserved in the south of China. There are more than 500 walls of poems of the poets in Tang, Song, Yuan, Ming, Qing Dynasty and contemporary poets, such as Li Shen, Li Yong, Yu Dayou, Zhu De, Chen Gongyin and Ye Jianying, etc. The celebrity poems add a lot of humanistic atmosphere to the scenic spot and have unique effects and pioneering value in academic research and tourists' viewing.

3. Evaluation methods and data sources

3.1. Indicator system

The selection and construction of indicators is the core of dynamic evaluation. Due to the characteristics of tourism resources, such as stage and variability, the evaluation of tourism resources should also be adapted. This paper follows the principle of combining systematicness and scientificity, combining dynamic and relative stability, combining comparability and operability, and refers to the relevant works of literature, and divides the study of the Seven Star Grotto National Park into three stages: 2001-2005, 2006-2010, 2011-2015, and the dynamic evaluation indicator system of tourism resources was constructed from the target layer, the comprehensive layer, and the element layer. The evaluation of tourism resources value is the evaluation of the quality and the excellence of tourism resources, including seven main indicators: 1. Aesthetics, 2. History and culture, 3. Science, 4. Peculiarity, 5. Scale and composition status, 6. Tourism function, 7. Tourism environment capacity [8]. This paper divides the resource value evaluation into four values: culture, recreation, science and observation according to the actual situation. The scenic condition evaluation includes three aspects: scenic spot combination, scenic tourist facilities, and scenic spot management. The regional conditions consist of four parts: regional economic development level, central city size, traffic accessibility and inter-regional combination of tourism. Each factor was subdivided into a total of 19 indicator factors to evaluate the development of tourism resources in the Seven Star Grotto National Park. The value of appreciation needs to be measured in terms of both peculiarity and enjoyment. Cultural value is measured from three aspects: historical and cultural value, religious sacrificial culture value and folk custom value. Scientific value is measured from two aspects: scientific research value and science popularization education value. Recreation value is measured from three aspects: tourism resource portfolio integrity, travel period and environmental capacity. The combination of attractions mainly refers to the combination of resources. The tourism facilities are measured from the two aspects: infrastructure and project facilities. The scenic spot management includes the period of travel and environmental capacity, etc. [9]. As shown in Figure 1.

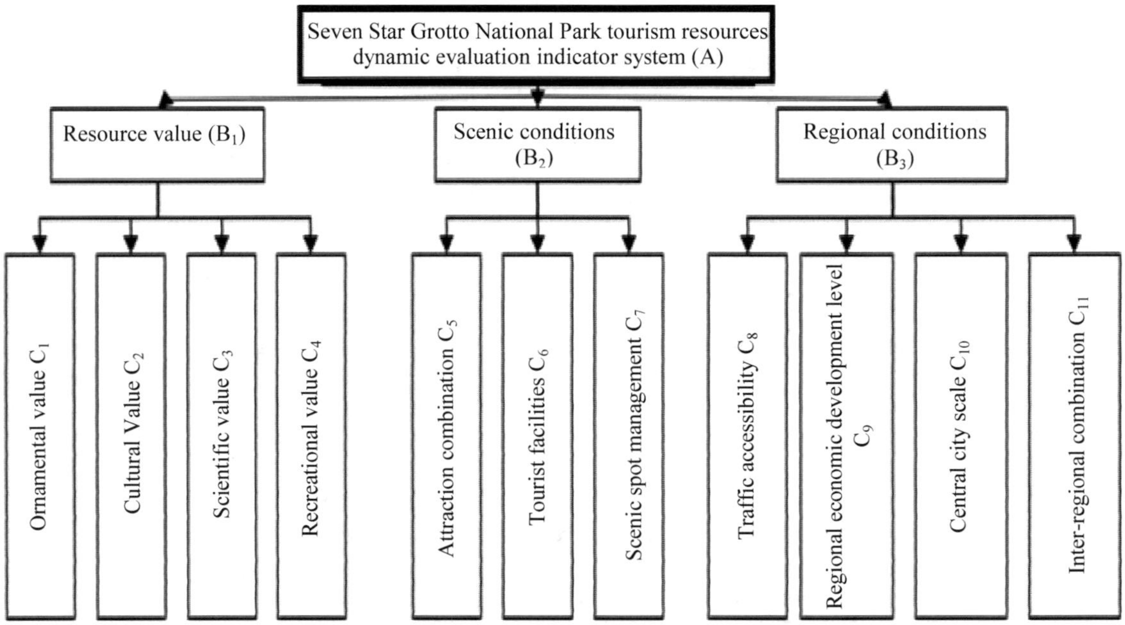

Figure 1. Seven Star Grotto National Park tourism resources dynamic evaluation indicator system

3.2. Data sources

The data and information in this paper are mainly from field trips, looking up relevant yearbooks, works of literature and searching network resources. In addition, due to the particularity of tourism resources, the standards for evaluation of resource value, scenic conditions and regional conditions are different. This paper designs different scoring reference standards according to the collected data, and obtains indicator scores in different ways: resource value scores are mainly taken from the scores given by 10 senior experts, professors and teachers; the scenic condition scores are mainly taken from the scores given by the administrative staffs in different positions of the Seven Star Grotto National Park, Zhaoqing Tourism Bureau, Xinghu Scenic Spots Authority and senior tour guides of travel agencies; the regional condition scores are scored automatically mainly through the concrete data in Zhaoqing Yearbooks.

3.3. Method model

In this paper, through assigning weights and scores to each integrated layer weight, 10 points were assigned to each indicator factor, then the total scores of resource value were 100 points, and the scenic conditions and regional conditions are respectively 50 points, of which the traffic accessibility is given 20 points, because traffic is crucial to the development of the scenic spot, according to which the weights are given as 0.5, 0.25, 0.25, and the comprehensive index of the Seven Star Grotto National Park tourism resources was calculated by time period using the following formula [10-12].

$$B_i = \sum f_i w_i \tag{1}$$

$$A = B1 + B2 + B3 \tag{2}$$

where A is the total value of tourism resources development; i is the ordinal number of indicators; fi is the weight of each comprehensive layer; w is the average score of the assignment of individual indicators; B1, B2, B3 represent the resource value, scenic conditions, and regional conditions respectively [13-14].

Table 1. Annual average scores of the dynamic evaluation indicators of the Seven Star Grotto National Park tourism resources (retain two decimal places)

Target layer (A)	Comprehensive layer (B)	Elemental layer (C)	Indicator layer (D)	Annual average scores In 2001-2005	In 2006-2010	In 2011-2015
Dynamic evaluation of the Seven Star Grotto National Park tourism resources (A)	Resource value (B1)	Ornamental value	Peculiarity	7.38	6.88	6.70
			Enjoyment	7.13	6.75	7.25
		Cultural Value	Historical and cultural value	7.25	7.25	8.38
			Religious sacrificial culture value	7.00	6.38	7.13
			Folk value	6.75	6.25	6.25
		Scientific value	Scientific research value	8.00	8.00	7.13
			Science popularization education value	7.25	7.00	7.88
		Recreational value	Tourism resource portfolio integrity	7.25	8.00	7.63
			Period suitable for tourism	7.75	7.88	7.63
			Environmental capacity	8.50	8.50	7.75
	Scenic conditions (B2)	Attraction combination	Resource combination	6.20	6.60	8.00
		Tourist facilities	Infrastructure	6.20	6.40	8.20
			Project facilities	5.80	6.40	8.00
		Scenic spot management	Institutional setup and staffing	6.00	6.60	7.40
			Environmental	6.40	7.20	9.00

			sanitation			
	Regional conditions (B3)	Traffic accessibility	---	11.00	15.00	17.00
		Regional economic development level	---	5.00	7.00	8.00
		Central city scale	---	5.00	5.50	6.50
		Inter-regional combination	---	6.00	6.50	7.50

3.4. Dynamic evaluation of the development of tourism resources in the seven star grotto national park

Based on the calculated data, this paper will dynamically evaluate it from four respects of the Seven Star Grotto National Park tourism resources: overall development scale, resource value, scenic conditions and regional conditions.

3.4.1. The overall development level of resources has improved, and the conditions of scenic spots are development guarantees.

The assessment shows (see Table 1) that over time, the overall level of tourism resources in the Seven Star Grotto National Park has improved, but at a slower rate. The total value of the three phases of evaluation increased from 51.525 in 2005 to 53.238 in 2010 and 56.75 in 2015. Among them, the increase in resource value is the smallest, even negative growth, and it can be said that it is basically stable at a certain level, which is related to the nature of the Seven Star Grotto National Park itself [15]. Referring to Table 2, the rock itself will not change much, and the time periods researched in this paper are also shorter. The data shows that the increase in regional conditions is the largest overall, which is attributed to Zhaoqing Municipal Government vigorously developing this region and the benefit to the scenic spot. Although the increase of scenic conditions of the Seven Star Grotto National Park is ordinary, the increased margin has an increasing trend, indicating that all parties pay more and more attention to the construction of the scenic spot [16-17].

Table 2. Dynamic evaluation table of the Seven Star Grotto National Park tourism resources

	Scores			Growth rate (%)		
	In 2005	In 2010	In 2015	In 2005-2010	In 2010-2015	In 2005-2015
Resource value (B1)	37.13	36.44	36.85	-1.85	1.13	-0.74
Scenic conditions (B2)	7.65	8.30	10.15	8.50	22.29	32.68
Regional conditions (B3)	6.75	8.50	9.75	25.93	14.71	44.44
Total scores (A)	51.53	53.24	56.75	3.32	6.60	10.14

3.4.2. The value level of tourism resources has declined, and the peculiarity level has decreased year by year.

Referring to Table 3, the assessment shows that the overall development trend of the tourism resources value of the Seven Star Grotto National Park is not clear, and in some time periods the negative growth has occurred. During the five years from 2005 to 2010, the resource conservation and scenic area construction of the Seven Star Grotto National Park is not enough. There is even a suspicion that excessive development has caused its devaluation [18]. Although the situation has been reversed since 2011, due to insufficient strength, the ornamental value, scientific value and recreational value of the tourism resources value of the Seven Star Grotto National Park have shown negative growth during the fifteen years of the research period [19-20].

Table 3. Table of the Seven Star Grotto National Park tourism resources value scores (percentage system)

	Scores			Growth rate (%)		
	In 2005	In 2010	In 2015	In 2005-2010	In 2010-2015	In 2005-2015
Ornamental value	72.50	68.13	69.75	-6.03	2.39	-3.79
Cultural Value	70.00	66.25	72.50	-5.36	9.43	3.57
Scientific value	76.25	75.00	75.00	-1.64	0.00	-1.64
Recreational value	78.33	81.25	76.67	3.72	-5.64	-2.13

3.4.3. The scenic conditions have been greatly improved, and the improvement in scenic spot management has been the largest.

The assessment shows that from 2000 to the present, the scenic conditions of the Seven Star Grotto National Park have always been okay. In the beginning, it is close to the passing score of 60 points. The starting point is relatively high. The space that can be improved is actually slightly less, but in the past ten years, the growth rate of scenic conditions of the Seven Star Grotto National Park is 30%, of which the development of scenic spot management is the largest. Overall, the growth rate in the later period is obviously faster than that in the previous period, indicating that the management of the scenic spot attaches great importance to it and pays more and more attention to it. It can be improved from where it can be improved and from where it is easy to improve.

Table 4. Table of the Seven Star Grotto National Park condition scores (percentage system)

	Scores			Growth rate (%)		
	In 2005	In 2010	In 2016	In 2005-2010	In 2010-2015	In 2005-2015
Attraction combination	62.00	66.00	80.00	6.45	21.21	29.03
Tourist facilities	62.00	64.00	82.00	3.23	28.13	32.26
Scenic spot management	58.00	64.00	80.00	10.34	25.00	37.93

4. Seven star grotto national park is located in the northern suburb of zhaoqing city

Guangdong Province. Zhaoqing City is an inland city in Guangdong Province. Its geographical location is not superior, nor has it been favored by the provincial and municipal governments. There is no policy support, and all aspects of development are lacking, but fortunately located around the developed Pearl River Delta, then the economy and transportation have been developed interactively.

Table 4 shows that Zhaoqing's traffic accessibility and the growth rate of the regional economic level are the largest. In recent years, Zhaoqing has been strengthening traffic construction and improving public transportation and passenger transportation networks. The high-speed railway station of Zhaoqing East Station was also built and put into use a few years ago, which greatly facilitated the surrounding areas and the Guizhou area and the Guangxi area, and strengthened communication, exchanges, and cooperation within and outside the province. And the newly built light rail has greatly facilitated the transportation within the province, making Seven Star Grotto National Park a popular attraction for many tourists on these transportation lines. At the same time, the development of the regional economy has also provided strong support for tourists to go to the Seven Star Grotto National Park. In Table 5, the economic development has further improved the eating, living, transportation, sightseeing, shopping and entertainment related to the tour, which is an extension of the scenic spot, allows visitors not only to be attracted by the Seven Star Grotto National Park but also to be attracted by other things. However, the weak inter-regional combination will drag the hind legs of the Seven Star Grotto National Park. Even if the Seven Star Grotto National Park is extended in other aspects, it will not last long if there is no good and fixed scenic spot to attract tourists as a combination.

Table 5. Table of Seven Star Grotto National Park regional condition scores (percentage system)

	Scores			Growth rate (%)		
	In 2005	In 2010	In 2016	In 2005-2010	In 2010-2015	In 2005-2015

Traffic accessibility	55.00	75.00	85.00	36.36	13.33	54.55
Regional economic development level	50.00	70.00	80.00	40.00	14.29	60.00
Central city scale	50.00	55.00	65.00	10.00	18.18	30.00
Inter-regional combination	60.00	65.00	75.00	8.33	15.38	25.00

5. Conclusion

The research shows that during 2005-2015, the overall development of tourism resources in the Seven Star Grotto National Park was very slow. Seven Star Grotto National Park, as the first batch of key scenic spots in the country, performed poorly in the booming tourism development booms around the country, and the change is little fifteen years ago and fifteen years later, which is worth introspection and discussion. The core product of tourism should be the culture and landscape that people can experience. The value of the tourism resources of Seven Star Grotto National Park is mainly reflected in the value of the natural landscape of the seven rocks and the value of culture, myths, stories, festivals and other cultural landscapes associated with them. The data shows that between 2005 and 2015, the protective development of natural resources landscape of the Seven Star Grotto National Park was relatively slow, and the mountain caves, lake water, animal and plant protection were not in place, and were over-exploited; the human resources were not paid enough attention, and cultural connotation was not excavated and extended with the development of the tourism market, and the cultural value depreciated.

Acknowledgement

The study was supported by "Twelfth Five-year Plan Program of Guangdong Provincial Philosophy and Social Sciences (Grant No.GD15XLS07).

References

[1] Zhang C 2003 *Sci. Geo. Sinica* **3** 372-8
[2] Feng Y 2007 *Research on humanity tourism resources evaluation* Shanxi University
[3] Liao W and He J 2005 *Yangtze River* **4** 68-70
[4] Xu F 2015 *Product identity (PI) research based on the behavior analysis of laptop computers* Southeast University
[5] Zhang R 2005 *Research about Chinese tourism resource combinations* Qingdao University
[6] Lu L 2015 *Knowledge Economy* **3** 63-4
[7] Zhong G Tan Y Chen X and Wu G 2012 *Tropical Geo.* **2** 195-200
[8] Xu Y 2011 *Research on tourism value evaluation and development for the historic district of Qingdao City* Ocean University of China
[9] Li Q 2012 *Evaluation of exploitation potential of tourism resources in the Mount Heng* Hunan Normal University
[10] *Guangdong Local History Editorial Board: Guangdong Yearbook (1980-2013)* The editorial department of Guangdong Yearbook.
[11] Li X 2014 *Social Sci.* **4** 87-91
[12] Zhong G and Chen X 2011 *Chinese Land. Arch.* **2** 44-5
[13] Zhou D 2011 *Econ. Geo.* **31** 514-7
[14] Du Z and Li Z 2011 *J. Northwest Normal U.* **47** 110-5
[15] Li S 2009 *Guangdong Landscape Arch.***5** 61
[16] Chen X Guo Y and Zhong G 2009 *J. Zhaoqing U.* **9** 45-9
[17] Li G 2009 *The develop evaluation research of exploiting the potential for tourism resources - take the Mount Heng for example* Hunan Normal University 22-44
[18] Zeng W and Zeng Z 2006 *History CPC Guangdong* **1** 18
[19] Wang X 2006 *J. Xiangfan U.* **9** 96-9
[20] Lin Z and Lin X 2004 *J. Chinese Urban Forest.* **4** 16-8

Using fuzzy sampling survey to explore the factors influencing regional cooperation: Changsha-Zhuzhou-Xiangtan case

Chaojun Li[1], Kuo Kuang Chang[1] and Shih Ming Ou[2]

[1]Tourism and Historical Culture College, Zhaoqing University, Zhaoqing, China
[2]Department of Logistics Management, Fujian Jiangxia University, Fuzhou, China

E-mail: benao13@gmail.com

Abstract. A comprehensive and profound understanding of the factors influencing regional tourism cooperation is the key to promoting regional tourism cooperation and is the leading condition for deepening the study of regional tourism cooperation economic incentives and cooperation mechanisms. Based on the literature review, a conceptual framework model for analyzing the factors influencing regional tourism cooperation was constructed, and these factors were systematically classified and analyzed. Using fuzzy sampling survey, we distinguished the endogenous and exogenous factors influencing regional tourism cooperation. After the data we collected was processed, we found that economic factors and simple cost-benefit analysis were very important in the start-up phase of regional tourism cooperation; however, as cooperation advanced, the factors that hindered and promoted tourism cooperation should be analyzed in a larger context. At the end of the article, the new organizational economics theory was used to explain this phenomenon, and the basic countermeasures for promoting regional tourism cooperation were put forward.

1. Introduction

Cooperation within and between the public and private sectors of tourism is one of the conditions for tourism destinations to gain a competitive advantage [1]. In recent years, regional tourism cooperation in the West has flourished. For example, the regional tourism committee established by relying on international funds, the regional tourism companies jointly established by the government and private enterprises, and the regional tourism alliances invested by the national tourism administrations have appeared one after another [2]. The economic linkages between regional cities and the microeconomic linkages between tourism products are the economic incentives for regional tourism cooperation and the focuses of early tourism cooperation researches. In addition, the academic community often uses theories of transaction costs, industrial accumulation, resource dependence, relationship exchange, social networks, and public governance to explain tourism cooperation [2-3]. The different international, national and regional levels of tourism cooperation, as well as the dynamic and static analysis views of tourism cooperation, increase the complexity of tourism cooperation [4-5]. The tourism cooperation under the static vision focuses on the challenges and influences of tourism cooperation, and the tourism cooperation under the dynamic vision focuses on the formation and evolution of cooperation. Although previous studies have proposed the decisive factors of tourism cooperation, they all adopt a relatively narrow perspective. They are only indirect research objects of these studies and lack a clear factor type analysis framework (due to many potential factors). This study used the method of integrated research to construct a conceptual framework for the classification of factors influencing tourism cooperation. Case studies confirmed the decisive factors of tourism cooperation and how these factors influenced (promoted and hindered) tourism cooperation.

2. Literature review and theoretical model

2.1. Literature review

Some text. Tourism cooperation is an interactive process where independent stakeholders use shared rules, norms, and structures to jointly address issues related to tourism development [6]. The literature analyses show that there are four different methods used in Western academic circles to analyze the factors influencing of tourism cooperation: the prerequisites influencing cooperation; the key factors for successful cooperative development; the individual ability required for tourism cooperation management; and the factors influencing cooperation effectiveness [2].

The scholars holding precondition views regard the competitive, technical, sociocultural, economic, and political forces that drive interaction and cooperation between tourism development entities as determinants of tourism cooperation. Wang et al. believed that the forces of competition, economic conditions, technology, and risk would lead potential partners to address common problems. Their research also confirmed the organization's support for this particular set of prerequisites, including leadership, sponsorship attitudes, communication, human and financial resources, and competencies [7]. Fyall and Carrod called it the driving force of cooperation, including globalization, international political and trade agreements, equity restrictions, capital accumulation and concentration, public funding, technological advancement, growth in tourism demand, strategic alliances, and increased expectations of tourists [1]. The turbulent environments forced individuals to cooperate to gain a competitive advantage. A large number of works of literature explored the individual motivations of tourism cooperation.

At the same time, the above-mentioned preconditions for tourism cooperation are also seen as the key factors for the success of tourism cooperation development. Through research, Selin and Chavez confirmed the basic characteristics of successful tourism cooperation and classified them as personal and interpersonal factors and organizational and operational factors [8]. The former included strong leadership, common identity, honesty and openness, active listening and adapting to the new situation; the latter included executives, continuity of employment during cooperation, flexible financial accounting systems, and appropriate environments (such as organizational meetings).

Many studies reported on the individual competencies that influenced the success of tourism cooperation. Lemmetyinen and Go proposed four key capabilities for managing regional tourism cooperation: the management skills required to undertake information, interpersonal and decision-making roles; the collaborative network planning abilities that could enhance the commitment of partners; and the abilities that could construct common knowledge to facilitate to enhance learning, reciprocity and dialogue transparency; strong partnership abilities to ensure benefit and efficiency [5].

Some scholars have studied the benefit factors influencing tourism cooperation. Reid et al. believed that effective tourism cooperation should meet the following conditions: making decisions around the overall goal of promoting cooperation; making decisions based on objective research and preset criteria; focusing on the long-term evaluation of cooperation results; efficient management structure; clear resource priority distribution guide [4].

The above-mentioned four research methods of the factors influencing tourism cooperation in the West are intertwined and closely related to each other. It is neither appropriate nor possible to categorize these factors clearly. Although many scholars have carried out researches on tourism cooperation and its influencing factors, the problems of insufficient research on this issue are often mentioned. They call for strengthening the research on the factors influencing tourism cooperation in developing countries and believe that there are nationally or regionally specific factors influencing tourism cooperation [9-10]. Underdeveloped civil society, lacking cooperation experience, lacking learning paradigms, adaptive (non-positive) attitudes of cooperation, and local government financial problems are obstacles to tourism regional cooperation [10].

The domestic focus is mainly on the study of regional tourism cooperation mechanisms. Most researchers have discussed them under the name of "dynamic mechanism" [11]. The theory of integration, the theory of core edge and point axis of economic geography, industrial agglomeration theory and symbiosis theory are the theoretical basis for the formation of regional tourism cooperation driving force mechanism. Renmeng Tu constructed the motive mechanism of regional tourism cooperation with symbiosis, complementation and overall effect [12]. Power, process, operation and spatial role are the four major mechanisms for regional tourism cooperation [11]. Resource sharing, project promotion, problem emergence, and intrinsic value drive are the basic motives of regional tourism cooperation, then resource-driven, project-driven, problem-driven, and value-driven four major cooperation models were formed [2-3].

2.2. Framework model of factors influencing regional tourism cooperation

Based on the literature review, this paper developed an integrated conceptual framework for analyzing the factors influencing regional tourism cooperation, and integrated various factors of cooperation and classified them into three categories: exogenous, endogenous and random factors (Figure 1). The criteria for this classification method are: the source of the influencing factors, the scope of action, and the extent to which the cooperating entity influences it.

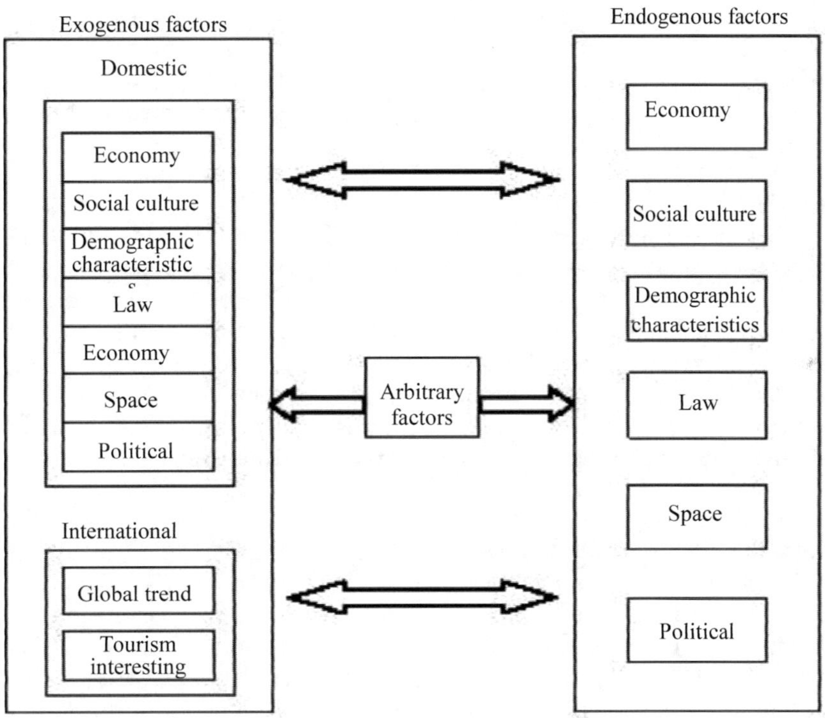

Figure 1. Model of factors influencing tourism cooperation

Exogenous factors relate to a wider range of regions outside this region, such as cross-regional, national and international forces, etc. As these factors are formed outside the region, it is difficult for tourism cooperation participants in the tourism region to influence these factors. Conversely, endogenous factors are produced in the region and are formed by direct and frequent contact between potential or existing collaborators. The randomness factors are the factors that cannot be clearly attributed to exogenous and endogenous factors using the above classification criteria. If they are formed outside the region, they may be considered exogenous factors, such as fluctuations in currency exchange rates. If they are formed within the region, they can be regarded as endogenous factors, such as changes in people's political positions in the region. Random factors are difficult to predict, and systematic analysis is very difficult, so it is not the focus of this study.

Exogenous and endogenous factors can be further subdivided. The former can be subdivided into domestic and international factors influencing tourism cooperation; the latter are all domestic factors and can be further subdivided into economic, political, sociocultural, legal, demographic and geospatial factors. For example, economic factors include macroeconomic and microeconomic factors that are directly related to economic activity and its outcomes, which can be measured and expressed in monetary units, such as economic development levels and financial resources, etc. Spatial factors reflect the geospatial relationship of the collaborators. In essence, only international factors are exogenous, including international economic development trends and tourism trends. International political and economic agreements can promote cooperation. For example, the EU Tourism Collaborative Funding Project has greatly promoted the EU tourism cooperation process. Tourism trends such as the growth of tourism demand and changes in tourism demand have also had a profound impact on tourism cooperation.

It should be pointed out that simplifying the classification of the factors influencing regional tourism cooperation, refining and integrating the conceptual framework model is a need to clearly outline the overall outline of the factors influencing regional tourism cooperation. There may be a strong link between individual factors (including exogenous and endogenous). However, it is difficult to understand how each factor affects tourism cooperation without dividing these factors according to certain classification criteria. Of course, the factors influencing tourism cooperation can also be classified according to the promotion and obstacle factors to form a two-dimensional analysis framework, that is, whether the individual factors are exogenous or endogenous and whether they are promotion or obstacle factors.

3. Research methods

In order to ensure that investigators can obtain the required information about the target market, the development of a well-designed market survey questionnaire will become the most important part of the experimental design. If the traditional sample survey of questionnaires is used, the items are scored according to language variables, which is of a vague nature. It is difficult to reflect the incomplete and uncertain ideas of the interviewee. Therefore, if we can use ambiguity to express the respondent's feelings based on the respondent's own concept, then the sampling result will be closer to the respondent's true thoughts. This is why in this paper, we use the fuzzy sampling survey to explore the main reasons that affect the development of Changsha-Zhuzhou-Xiangtan area.

3.1. Selection of research regions

This study, as a qualitative and exploratory study, aims to disseminate explanatory and evaluative knowledge that is rarely understood by people, and provides a comprehensive picture for understanding the complex issue of regional tourism cooperation. Based on epistemological assumptions, this study faithfully reported the true views of the interviewers. In such studies, the selection of samples should be purposeful, not random [13]. The author used a multi-case research method to specify the use of some cases to explore a problem (the region formed by several cities). The researcher was interested in analytical interpretation rather than statistical induction, the purpose of which was to draw general conclusions by discovering patterns or theories that help explain phenomena [3].

Katarzyna's research established five criteria for selecting research regions: regional cohesion, higher levels of tourism development, and different levels of tourism development in tourism cooperation regions, the experience of collaboration between and within departments, geographical extension and accessibility [3]. Based on the above criteria, we chose the Changsha-Zhuzhou-Xiangtan region. The Changsha-Zhuzhou-Xiangtan urban agglomeration is the core area of Hunan's social and economic development. It relies on the "two-type society" comprehensive supporting reform pilot zone of the Changsha-Zhuzhou-Xiangtan urban agglomeration to promote the integration of the tourism industry and other industries, and to promote the development of agriculture, industry, cultural industry and sports industry with big tourism industry; it takes the opportunity of developing the tourism industry, to improve the level of regional infrastructure, improve the rural natural environment and urban appearance, promote urban-rural harmony, and release the rural resource advantages through tourism industry.

The integration of Changsha-Zhuzhou-Xiangtan has existed for a long time as a vision and issue. In 1997, the Hunan Provincial Party Committee and the Provincial Government formally proposed the integrated development strategy of Changsha, Zhuzhou, and Xiangtan. In 2005, the Provincial Government officially approved the regional planning of the Changsha-Zhuzhou-Xiangtan urban agglomerations. In 2006, the "Changsha, Zhuzhou and Xiangtan Joint Conference Rules" and "Changsha, Zhuzhou and Xiangtan Regional Cooperation Framework Agreement" were adopted. Although the Changsha-Zhuzhou-Xiangtan case is a typical example of regional cooperation, the three municipal administrative divisions led to geographical barriers and interest distinction, and regional tourism development lacks the spontaneity of cooperation and division of work. In recent years, the construction of platforms by the three places has made many attempts to integrate regional cooperation, and there are both successes and failures. Therefore, it is meaningful to explore the success factors and obstacles of regional tourism cooperation with the Changsha-Zhuzhou-Xiangtan case as examples.

3.2. Selection of interviewees in the study
This study used an unconstrained in-depth interview method, and the interviewer asked open questions, and the interviewees were free to express their opinions. In order to expand the knowledge gained from interviews, analyze government documents, and increase the level of research validity, triangulation techniques were used to examine the results. Among them, 10 interviewees came from the governmental tourism work department, and 10 were tourism entrepreneurs (owners and operators), and 5 were researchers of Changsha, Zhuzhou and XiangTan integration. The longest interview time is 75 minutes, and the shortest interview time is 38 minutes, and the average interview time is 45 minutes.

3.3. Data analysis
This research is an exploratory and descriptive study with no empirical features, using naturalistic rather than positivist approaches. The former forms the hypothesis and the latter verifies the hypothesis. This study used Atlas.ti software for coding. Data analysis is a continuous process, is a comprehensive analysis process that integrates data simplification, data display, and data verification. Appropriate analysis units were defined through transcribing the interviews of interviewer and creating the case cards. A piece of information was an information unit, and related information units were organized into related categories to describe the factors influencing tourism cooperation. At this stage, the characteristics of each category were defined to justify the classification of the relevant units into a category.

4. Research findings
Based on the core elements of the initial conceptual framework and the research results, an expanding model for the factors influencing tourism cooperation was formed. The study confirmed that tourism cooperation factors included exogenous factors and endogenous factors, which could be further subdivided into corresponding categories (Figure 2). The study also identified some random and unpredictable factors, such as exchange rate changes, climate changes, and so on. They bring uncertainty to tourism cooperation and make tourism cooperation more difficult.

4.1. Exogenous factors
Consistent with the research framework, exogenous factors include global factors and domestic factors. The former includes the overall socio-economic situation and tourism trends, and the latter includes

economic, social culture, population, the legal system, space, and political factors. Two factors include many individual factors, which are very general in some cases and very specific in some cases.

Studies have shown that due to incomplete economic transformation, insufficient economic development level, and relatively weak market mechanism, they have brought certain obstacles to tourism cooperation in the three cities. Among them, poor governmental financial situation, the unstable financial situation of the travel-related enterprises, and unsatisfactory cooperative income are also its main manifestations. Collaborators often view cooperation as an additional cost and financial burden.

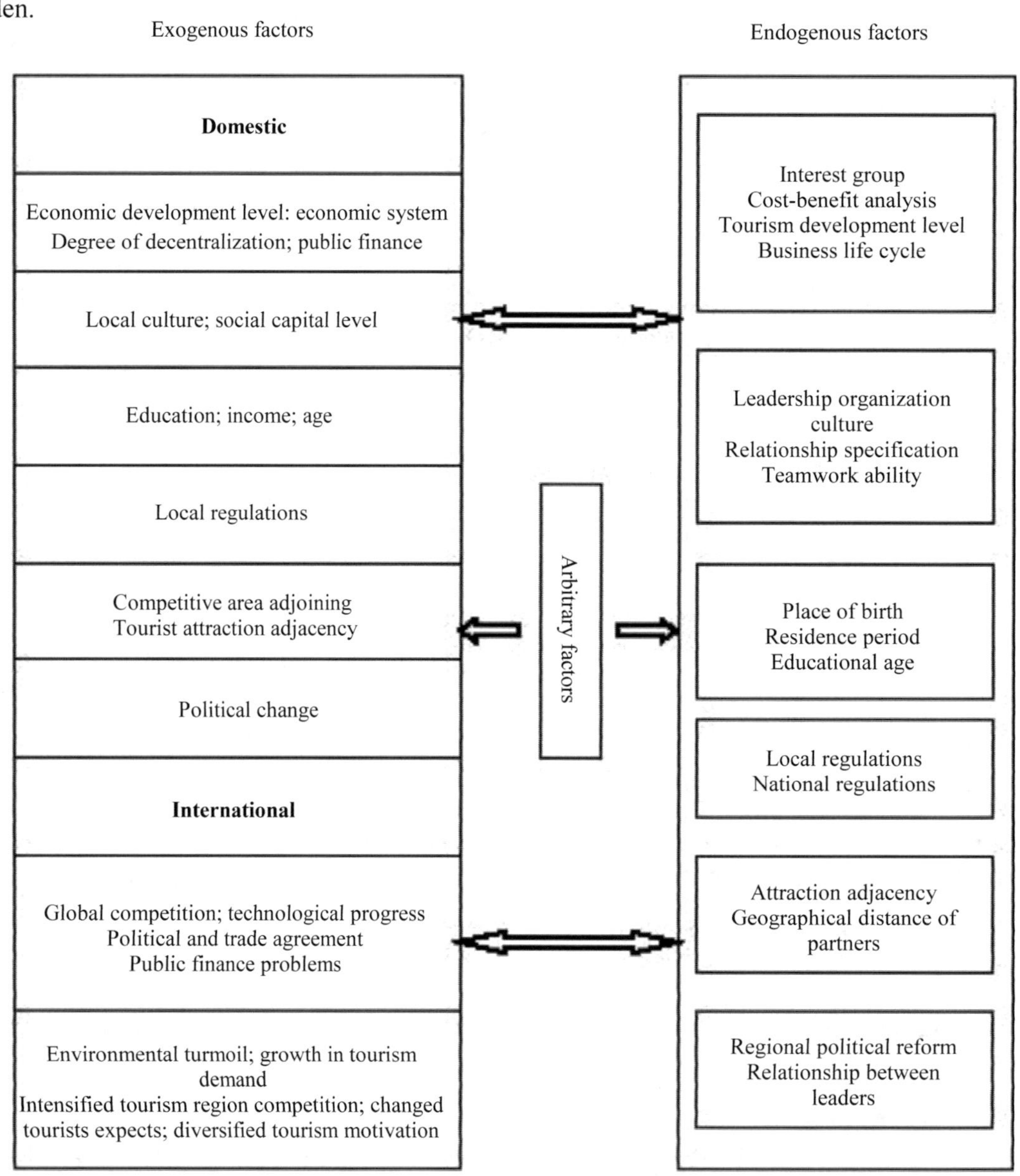

Figure 2. Factors influencing regional tourism cooperation of Changsha-Zhuzhou-Xiangtan

The procrastination of many tourism public and private sectors on cooperation is another major obstacle to regional cooperation. For example, the thinking characteristics of the planned economy are not uncommon in many sectors. Uncertainty leads to an attitude of cooperation that is adaptive rather than proactive, which hinders partners from taking advantage of new opportunities brought about by changes in the external environment. Another important factor is the long-term neglect of the development potential of tourism cooperation. Creative, recreational and leisure travel are not fully utilized before the economic transition is completed. Local governments lack the knowledge and experience to manage tourism development, divide and coordinate roles and structures between governments and partners, which hinders cooperation among tourism development stakeholders. Lack of coordination often leads to repetitive efforts, which wastes the potential benefits of cooperation. The historical development trend and extent of decentralization also hinder cooperation between regions. Studies have shown that the decision-making power of some things (such as tourism investment license) still belongs to the upper-level government, which makes it difficult for entrepreneurs to regard the first-level government-led tourism cooperation investment promotion as an independent and equal business relationship.

Analysis of social and cultural factors shows that national culture, insufficient social capital, and low level of civil society development have all hindered cooperation. It is much easier to start cooperation between simple stakeholders than group stakeholders, especially when group stakeholders are formalized and composed of people who do not know each other. This is the same as the results of Ioanides et al. [14].

The cooperation between the public and private sectors of tourism is also deeply influenced by national cultural factors. The perceived distance between society and government makes the relationship between the two be a relationship between "we" and "they". This kind of relationship understanding and perception is the lingering legacy of the planned economic system. The social environment of mutual distrust also hinders the development of tourism cooperation.

4.2. Endogenous factors

The endogenous factors identified can be divided into six categories: economic, sociocultural, demographic, legal, spatial, and political. Like exogenous factors, each type of factor contains specific factors that may promote regional tourism cooperation or may be an obstacle to regional tourism cooperation.

4.2.1. Economic factors

A large number of stakeholder groups have different goals, roles, and needs, which are important obstacles to regional tourism cooperation. Different cost-benefit analysis methods make cooperation between the tourism public sector and the private sector difficult. Relatively speaking, it is easier for tourism small- and medium-sized enterprises (SMEs) to confirm the time and resources required for cooperation, and it is more difficult for the tourism public sectors to confirm the various costs required for cooperation (because they are brought together with other public resources). In addition, tourism companies expect short-term and quantifiable benefits of cooperation, while the tourism public sectors seek long-term, sustainable benefits of cooperation.

The research shows that the willingness and process of tourism cooperation between Changsha, Zhuzhou, and Xiangtan are affected by the level of tourism development in the three regions. The regions with a high level of tourism development prefer to carry out some activities independently. For example, the level of tourism development in the three regions is different, and it is difficult for the three regions to coordinate in terms of tourism marketing policies and marketing objectives.

The life cycle of tourism destinations and tourism enterprises is a major economic factor affecting regional tourism cooperation. The owners of tourism companies surveyed in the region studied believe that tourism destinations and tourism companies in the initial and recessionary stages are less willing to participate in tourism cooperation projects, and they are more willing to pay attention to their own business than to cooperate with others. The respondents believed that it was too early to consider cooperating with others in the initial stage; cooperating with others during the recession period would add additional costs.

4.2.2. Social and cultural factors

The study also reveals the roles of sociocultural factors such as leadership, organizational culture, and relational rules in regional tourism cooperation. There are many obstacles to this category. The lack of leaders is a major obstacle. Although regional cooperation representatives are considered to be potential leaders, respondents pointed out that they lack leadership qualities such as entrepreneurship, charisma, and neutrality.

Cultural differences between public and private organizations also limit the advancement of regional tourism cooperation. The public sectors of tourism must adhere to the service of society and respect the values of public responsibility, justice, solidarity, and impartiality in tourism development, and try to avoid venture capital or favoring some special interest groups. Non-public organizations, driven by short-term gains, often believe that the public sectors are too conservative and lack initiatives. The public sectors want companies to be proactive and more selfless to participate in cooperation.

Collaboration becomes easier when partners establish relational rules such as flexibility, information sharing, and solidarity. Conversely, failure to comply with these specifications creates transaction costs, which means that transaction costs between entities that do not trust each other increase. Relationship rules are closely related to the ability to cooperate. Of these cooperation abilities, trust is particularly important because people who lack trust do not consider cooperation. Liwicki and Bunker break down trust into three dimensions: calculative, knowledge-based, and emotional [15]. Calculative trust is the lowest level of trust and is a positive outcome of collaborative cost and benefit analysis. Knowledge-based trust is the result of a collaborative experience. When partners like each other, respect each other's similar values and establish long-term relationships, they will form the highest level of trust: emotional trust. In the regions studied, trust is considered by respondents to be a key factor influencing cooperation. Without trust, even the lowest level of calculative trust, uncertainty and transaction costs increase. Lack of trust undermines the willingness and process of cooperation, affecting daily cooperation and formal partnership structures. The lack of knowledge-based trust is the unwillingness of local cooperation in the exchange of promotional materials, the packaging of tourist attractions, and the reduction of prices of tourism products. When there is emotional trust, the partner believes that cooperation is the most successful. But in all cases, the establishment of the highest level of trust is the result of years of cooperation.

4.2.3. Population factors

Among the endogenous demographic factors, the source of residence and the duration of residence in the region determine the network of stakeholders' interpersonal relationships. According to social network theory, private and commercial relationships between entities form a network. Long-term residents of the three regions claimed that they had extensive personal contact networks in the local area and were more likely to participate in cooperation. In these cases, as entities (individuals) understand the characteristics and reputation of other entities (individuals), transaction costs are reduced and collaboration effects may be better. But over-familiarity among stakeholders can lead to suboptimal or inefficient behavior, as their collaboration often focuses on local businesses while ignoring external opportunities.

4.2.4. Legal factors

In the regions studied, local regulations (especially those related to tourism development) have affected public-private partnerships. From a legal point of view, when tourism development is less controlled by government departments, tourism public-private cooperation is easier; otherwise, it is difficult to advance. A favorable investment environment such as tax breaks can stimulate public-private partnerships.

4.2.5. Space and political factors

Studies have shown that space and political factors also affect regional tourism cooperation. The geographical proximity is conducive to the daily cooperation of partners. On the contrary, geographically remoteness will increase costs for partners (especially between the private sectors of tourism) and is an obstacle to cooperation. The level of understanding of local administrative officials is the most important political factor. Government officials often question the possibility of stakeholder cooperation in different regions.

5. Conclusions and implications

5.1. Conclusions

Based on literature review and case analysis, this study constructed a conceptual framework model for analyzing the factors influencing regional tourism cooperation, and systematically classified these factors, and provided a conceptual tool for the analysis of the factors influencing regional tourism cooperation. As an exploratory study, it is also very important to incorporate the factors rarely mentioned in previous studies such as the life cycle, demographic characteristics (mainly source and residence period) of tourist destinations (tourism enterprises), different cost-benefit analyses of public and private sectors, and trust dimensions into the research framework. We also found that economic factors and simple cost-benefit analysis were very important in the start-up phase of regional tourism cooperation; however, as cooperation advanced, the factors that hindered and promoted tourism cooperation should be analyzed in a larger context. At the end of the article, the new organizational economics theory was used to explain this phenomenon, and the basic countermeasures for promoting regional tourism cooperation were put forward.

5.2. Implications

In theory, an integrated theoretical approach is needed to represent and explain all aspects of regional tourism cooperation. New organizational economics is a more appropriate theory. Human rationality is rooted in social, historical and cultural contexts and is governed by formal and informal organizations. For an economy to develop effectively, the two organizations must coordinate and positively influence each other. However, formal organizations are easy to change; informal organizations change slowly subject to people's thinking and behaviors. Therefore, mechanically relocating a well-functioning formal organization in a mature market economy to an immature economy often fails because of a lack of matching informal organizational environments. It is possible to use the new organizational economic theory to explain the factors influencing regional tourism cooperation in non-mature market economies.

In practice, regional tourism cooperation practitioners (especially policy makers) cannot focus solely on economic growth, and improvements in social and human capital are critical to regional tourism cooperation. The simple reference to the cooperation policy of mature market economies cannot guarantee the success of cooperation. The lack of a proactive cooperative attitude and low-level social capital and other planned economic thinking are the main obstacles to regional tourism cooperation. These barriers are mainly exogenous and sociocultural factors that take a long time to change.

Acknowledgement

The study was supported by "Twelfth Five-year Plan Program of Guangdong Provincial Philosophy and Social Sciences (Grant No.GD15XLS07).

References

[1] Fyall A Garrod B 2005 *Tourism marketing: A collaborative approach* Clevedon: Channel View Publications
[2] Feng X and Gao J. 2013 *Tourism Tribune* **28** 57-65
[3] Katarzyna C 2013 *Annals of tourism res.* 40 83-104
[4] Reid L J Smith S L J and McCloskey R 2008 *Tourism Manage.* **29** 581-93
[5] Lemmetyinen A and Go F M. 2009 *Tourism Manage.* **30** 31-40
[6] Thomson A M Perry J L and Miller T K 2009 J. Public Admin. Res. Theory **19** 23 - 56
[7] Wang Y and Fesenmaier D R 2007 *Tourism Manage.* **28** 863-75
[8] Selin S and Chavez D 1995 Annals Tourism Res. **22** 844-56

[9] Beritelli P. Cooperation among prominent actors in a tourist destination [J].Annals of Tourism Research, 2011, 38(2):607-629.
[10] Bornhorst T Ritchie B and Sheehan L 2010 *Tourism Manage.* **31** 572-89
[11] Song Z 2008 *Tourism Tribune* **23** 74-9
[12] Tu R 1994 Social Sci. **5** 83-8
[13] Denzin N K and Lincoln Y1994 *Handbook of qualitative research* Londyn: Sage
[14] Ioanides D Niselsen P A and Billing P 2006 *Tourism Geo.* **8** 122 -42
[15] Lewicki R J and Bunker B B 1996 *Developing and maintaining trust in work relationships. In Kramer R M & Tyler T R (Eds.), Trust organizations: Frontiers of theory and research.* Thousand Oaks: Sage

Hydraulic-mechanical modeling research on coseismic fracture behavior

Chih Cheng Chung, Chi Jen Chen and Yuan Chieh Wu

1000 Wenhua Rd. Jiaan Village, Longtan District, Taoyuan, Taiwan

E-mail: jerrygood100@iner.gov.tw

Abstract. We establish hydraulic-mechanical modeling technique combined with simulation of fracture coseismic behavior based on discrete element method and well accepted equations to investigate the influence of earthquake on pore pressure and transmissivity of fractures. This study concludes 2 models with different hydraulic boundary condition, one designed for stable region with low hydraulic gradient (Mode I) while the other designed for regions with high hydraulic gradient (Model II). Model I shows that pore pressure of fractures interact with aperture and can be summarized into two possible patterns. Besides, the post-seismic transmissivity is 29~138 % of pre-seismic transmissivity depending on location of fractures. Results of Model II indicate that pore pressure is controlled by the inflow of groundwater at the beginning of earthquake and later becomes dominated by opening and closure of aperture. Because of high hydraulic gradient and pore pressure boundary condition set into Model II, post-seismic transmissivity is 102~1989 % of pre-seismic transmissivity. With this technique, well-established source parameter and site-specific hydraulic conditions, we are able to evaluate the possible post-seismic pore pressure and transmissivity of fractures which can be applied to underground facilities designing and evaluation of pollution transportation.

1. Introduction

Earthquake can affect facilities in several ways, including shaking, rock failure and pre-existing fracture. Earthquake can also cause the change of hydraulic conditions, such as discharge, stream flow and water head [1-2]. One of the hypotheses used to explain this phenomenon is the changes in hydro-mechanical properties of fractured rock masses [3-4].

Since the coseismic behavior of rocks and fractures is an important issue for engineering and scientific research, simulation becomes a possible way for us to evaluate the hydraulic and mechanical properties of fractures, especially for fractures underground that is difficult to observe directly. Swedish Nuclear Fuel and Waste Management Company (SKB) and Posiva Oy have published reports about earthquake induced shear displacement of pre-existing fractures based on simulation techniques that include seismic waves and host rock deformation [5-9]. They used 3DEC to construct their model and performed in various source parameters and stress conditions, which allowed them to evaluate the possible impact of seismic activities on fractures. In their studies, normal displacement and hydro-mechanical coupling were not considered, leading to the lack of evaluation of hydraulic properties change. Unlike SKB and Posiva Oy, Lak et al. (2017) [4] used UDEC to investigate the effect of seismic waves on the hydraulic properties of rock masses. Results showed that seismic waves could affect deep earth and increase the permeability and change the flow rate patterns in a fractured rock mass. However, the impact of host rock deformation was not considered in the study.

In this study, we try to improve the simulation technique, combining the hydro-mechanical coupling technique and earthquake simulation that conclude seismic waves and host rock deformation to research the influence of an earthquake on pre-existing fractures. This will help us to calculate the aperture change and the transmissivity change of fractures

2. Model description

We constructed a 191 km × 120 km × 60 km rectangular hexahedron model with a fault in the middle.

Earthquake simulation related parameters required fault geometry parameters, macro-parameters, micro-parameters and crustal structure parameters. The planar fault we created in the model dips 54 degree toward east and was 71 km long, 34 km wide. The seimogenic depth was set to be 25 km with a hypocenter in the center of fault at 12.5 km depth. In this study, we used time-weakening law [10] and generated stress drop by weakening the strength of fault. The magnitude we set was 8.0, which represented an extreme situation that stress accumulated in the fault was released in one earthquake.

In order to make our simulation more reliable and comparable to nature environments, stress regime was input as the equations:

$\sigma xx = 1.55 - 0.0251z$

$\sigma yy = 0.87 - 0.0240z$

$\sigma xy = 1.45 - 0.0024z$

$\sigma zz = -0.0265z$

This stress setting could lead to a strike-slip fault and create compression zones and dilation zones.

In our research, the rocks were homogeneous and isotropic to make seismic wave attenuate linearly. Since the continental crust was mainly composed of granitic rocks, the density and poisson's ratio of host rock was set to be 2700 km/m^3 and 0.25 accordingly. Besides, we are not targeting any specific site, using general parameters will be more appropriate for technique development. The geometry of our model is shown in Figure 1.

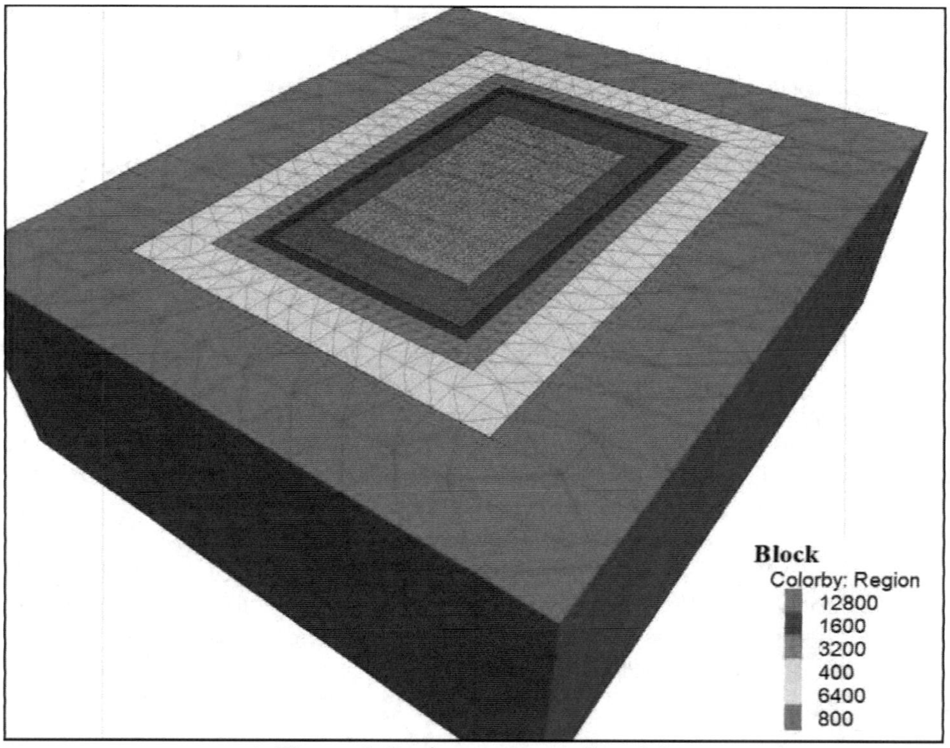

Figure 1. Basic geometry of model

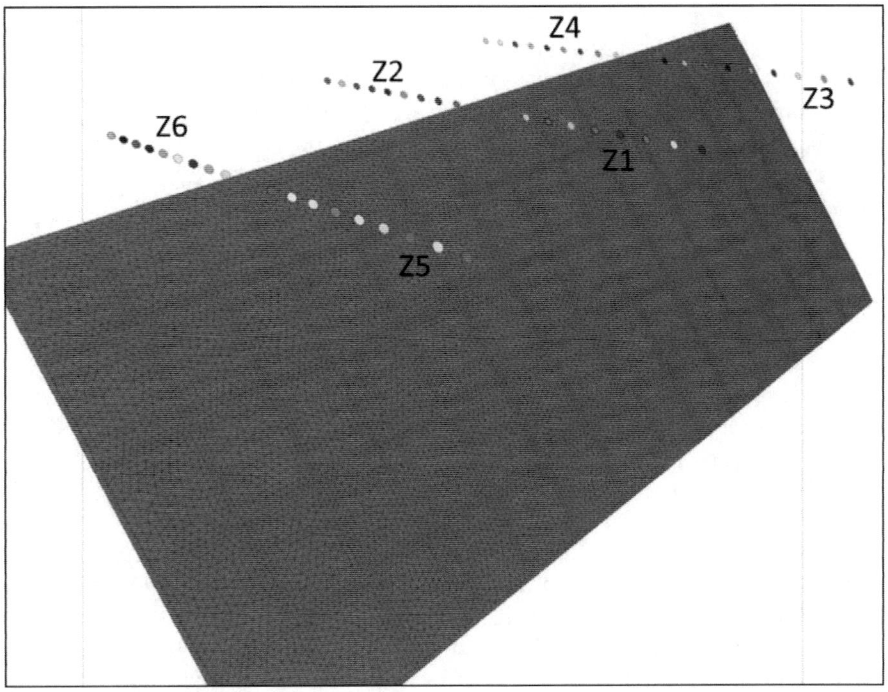

Figure 2. Layout of fractures. Z1 to Z6 are names of the fracture zone. Faults and all fractures locate in the center of the model and meshed with same grid size.

As for the strength of faults, we set 0° to friction angle and dilation angle, 0 MPa to tensile strength. Since that data of normal and shear stiffness was rare, we used 10 GPa/m referenced to SKB and Posiva company [5-9]. The cohesion of fault was set high before seismic to prevent slip and to simulate the lock behavior in inter-seismic period. After the rupture of fault, the cohesion of fault would decrease with time so the rupture could propagate along the fault plane. Based on the depth, we divided our model into three segment 0~20 km, 20~40 km and 40~60km to set up corresponding shear wave velocities, which was 3300, 3980 and 4350 m/s. On the other side, our simulation used a nonspontaneous rupture model and the rupture velocity was 2400 m/s. The setting coincided with the range of rupture velocities of crustal earthquakes collected by Somerville et al. (1999) [11]. The rock mechanic parameters of fractures were also set in the model, including normal stiffness, shear stiffness, friction angle, cohesion, residual friction angle, residual cohesion and dilation angle (Table 1). Besides, strike and dip of fractures were set to be 0° and 54°.

In order to study the importance of the distance from the source earthquake and of the fracture location, we have set 54 target fractures at different area including the central axis (Z1, Z2), dilation zone (Z3, Z6) and compression zone (Z4, Z5) in both hanging wall and footwall (Figure 2). The mechanical properties are listed in Table 1.

Table 1. Fracture parameters

Fracture Parameters	
Normal Stiffness	10 GPa/m
Shear Stiffness	10 GPa/m
Friction Angle	34°
Cohesion	0 MPa
Residual Friction Angle	34°
Residual Cohesion	0 MPa
Dilation angle	0

For the purpose of doing hydro-mechanical modeling to study how water pressure and aperture can change due to earthquake, we designed two models with different boundary conditions of water pressure and fracture flow. Both models share the same geometry and mechanical properties. The differences only occur in the setting of water pressure gradient. In model 1, the water pressure in fractures and adjacent area was all 5 MPa but only the water pressure in fractures can change during earthquake. This model was used to simulate stable areas with low or no hydraulic gradients. We want to know the instant response of fractures during seismic event. In model 2, the water pressure was set to be 50 MPa in south side of fractures, while the north side only had 5 MPa of water pressure. Besides, the water could only flow into the fractures after they failure. This model was designed to simulate areas with high hydraulic gradients and we want to know whether the water pressure will significantly affect the behavior of fractures.

3. Results and discussion

Results showed that the location of fractures was one of the factors that could affect the fracture coseismic behavior, including post-seismic water pressure and aperture. Due to the fluctuation, in this paper we focused on the trend and average aperture.

In model I, the post-seismic fracture transmissivity was calculated from the average aperture of 60-70s after the onset of earthquake. Results are listed in Table 2. In general, water pressure of fractures and aperture influenced each other and can be summarized into two patterns. (1) When the change of aperture was smaller and shear displacement has not occurred, which means that the fracture was not reactivated, water pressure inside the fractures tended to increase due to the closure of aperture. (2) After the reactivation of fractures, both aperture and water pressure changed significantly. In this case, the pattern of water pressure was related to pre-reactivation water pressure gradient and aperture size. If aperture decreased rapidly followed by slow opening, the post-seismic water pressure would be less than 5 MPa. In contrast, if aperture increased rapidly followed by slow closure, the post-seismic water pressure would be higher than 5 MPa. On the other side, fracture transmissivity was controlled by aperture size thus they shared the same pattern and were both location-dependent. Comparing the variation of fracture transmissivity between Z1 to Z6, there were several phenomenon could be observed. Z1 and Z2 located at the central axis of fault trace. The coseismic behavior of fractures was consistently showing a decreasing aperture and transmissivity. Also, the decreasing amount of aperture and transmissivity reduced when fracture – fault distance became larger. The data implied that when considering Mw 8.0 strike-slip earthquake, fractures in Z1 area with fracture – fault distance larger than 16 km might not show significant aperture and transmissivity change. Besides, fracture transmissivity changes were less than 1 % in Z2 area, which meant footwall were less likely to be affected by earthquake events. Z4 and Z5 area were located in the compressional zone. Though the behaviors of proximal and distal fractures were different, most of the transmissivity of the fractures decreased due to the closure of aperture in the compressional zone. Z3 and Z6 area were located at the dilation zone. The behaviors of proximal and distal fractures were different and the change of fracture transmissivity was more obvious in Z3 area. Overall, changes of aperture and fracture transmissivity were more significant in hanging wall of the fault.

Table 2. Aperture and fracture transmissivity of model I

Fracture area	Fracture ID	Initial aperture	Initial fracture transmissivity	Post-seismic aperture	Post-seismic fracture transmissivity	Post/Initial fracture transmissivity
Z1	101	0.00146	0.00252	0.00103	0.00089	35.51%
	102	0.00145	0.00249	0.00101	0.00083	33.43%
	103	0.00145	0.00247	0.00105	0.00093	37.84%
	104	0.00146	0.00252	0.00109	0.00107	42.48%
	105	0.00148	0.00266	0.00122	0.00148	55.46%

	106	0.00146	0.00254	0.00129	0.00175	68.93%
	107	0.00146	0.00255	0.00136	0.00207	81.08%
	108	0.00146	0.00254	0.00145	0.00248	97.68%
	109	0.00146	0.00256	0.00147	0.00262	102.25%
Z2	110	0.00146	0.00256	0.00146	0.00256	99.77%
	111	0.00145	0.00251	0.00145	0.00250	99.68%
	112	0.00146	0.00255	0.00146	0.00254	99.64%
	113	0.00145	0.00250	0.00145	0.00249	99.63%
	114	0.00141	0.00230	0.00141	0.00230	99.64%
	115	0.00145	0.00251	0.00145	0.00251	99.66%
	116	0.00146	0.00252	0.00145	0.00251	99.69%
	117	0.00145	0.00249	0.00145	0.00248	99.72%
	118	0.00148	0.00263	0.00147	0.00262	99.74%
Z3	119	0.001468	0.002582	0.001126	0.001166	45.17%
	120	0.00146	0.002542	0.001022	0.000872	34.33%
	121	0.001468	0.002584	0.000974	0.000755	29.22%
	122	0.001459	0.002536	0.001021	0.000868	34.24%
	123	0.001475	0.002618	0.001097	0.00108	41.23%
	124	0.001465	0.002567	0.001198	0.001405	54.73%
	125	0.001461	0.002545	0.001337	0.00195	76.63%
	126	0.001471	0.002599	0.001472	0.002606	100.26%
	127	0.00147	0.002597	0.001623	0.00349	134.40%
Z4	128	0.001465	0.00257	0.001633	0.003558	138.43%
	129	0.001449	0.002485	0.001184	0.001354	54.50%
	130	0.00147	0.002593	0.001103	0.001096	42.29%
	131	0.001463	0.002556	0.000983	0.000775	30.32%
	132	0.001454	0.002511	0.000979	0.000766	30.50%
	133	0.001467	0.00258	0.000987	0.000785	30.42%
	134	0.001472	0.002603	0.001469	0.002587	99.37%
	135	0.001455	0.002516	0.001452	0.002501	99.42%
	136	0.001476	0.002628	0.001033	0.000901	34.29%
Z5	137	0.001432	0.002397	0.001243	0.00157	65.52%
	138	0.001446	0.002467	0.001273	0.001685	68.32%
	139	0.001448	0.002481	0.001305	0.001813	73.08%
	140	0.001437	0.002422	0.001437	0.002422	99.97%
	141	0.00146	0.002542	0.001485	0.002673	105.16%
	142	0.001436	0.002419	0.001548	0.00303	125.26%
	143	0.001456	0.002521	0.001457	0.002524	100.15%
	144	0.001459	0.002537	0.00146	0.00254	100.14%
	145	0.001473	0.00261	0.001474	0.002613	100.12%

Z6	146	0.001449	0.002487	0.001348	0.002	80.43%
	147	0.001446	0.002469	0.001302	0.001804	73.06%
	148	0.001446	0.002471	0.001532	0.002937	118.85%
	149	0.001443	0.002455	0.001545	0.003015	122.81%
	150	0.001432	0.002396	0.001497	0.002742	114.40%
	151	0.001449	0.002484	0.001476	0.002625	105.66%
	152	0.001441	0.002443	0.001494	0.002723	111.44%
	153	0.001452	0.002499	0.001452	0.002502	100.14%
	154	0.001447	0.002476	0.001448	0.00248	100.17%

In model II, results (Table. 3) indicated that the water pressure in fractures was first controlled by the inflow along the reactivated fractures and then affected by the behavior of aperture. Besides, both fracture transmissivity and aperture became larger after the onset of earthquake. The post-seismic fracture transmissivity is 1.02 to almost 20 times compare to pre-seismic fracture transmissivity. We believed that the high water pressure and pressure gradient was the main responsible for increase of fracture transmissivity. In this model, water was designed to flow into the fractures only when the fractures were reactivated. When cracks occurred along the fractures, high pressure water flow through and applied pressure on the walls of fractures, which enlarged the size of aperture and fracture transmissivity. Based on our simulations, we inferred that earthquakes would increase fracture transmissivity under high water pressure and high pressure gradient environments. The amount of increase would be obvious when close to the faults.

Table 3. Aperture and fracture transmissivity of model II

Fracture area	Fracture ID	Initial aperture	Initial fracture transmissivity	Post-seismic aperture	Post-seismic fracture transmissivity	Post/Initial fracture transmissivity
Z1	101	0.000962	0.000728	0.001853	0.005197	714.04%
	102	0.000956	0.000714	0.001018	0.00086	120.42%
	103	0.000951	0.000703	0.000959	0.000721	102.44%
	104	0.000963	0.00073	0.00097	0.000746	102.32%
	105	0.000989	0.000791	0.001002	0.000822	103.89%
	106	0.000966	0.000736	0.001015	0.000854	115.99%
	107	0.000969	0.000742	0.001028	0.000888	119.73%
	108	0.000965	0.000734	0.001073	0.001008	137.42%
	109	0.000969	0.000744	0.001097	0.001077	144.81%
Z2	110	0.00097	0.000745	0.001204	0.001427	191.44%
	111	0.00096	0.000722	0.001095	0.001071	148.40%
	112	0.000967	0.000739	0.001003	0.000825	111.60%
	113	0.000959	0.000719	0.000967	0.000739	102.79%
	114	0.000919	0.000635	0.000936	0.000669	105.48%
	115	0.000961	0.000725	0.000979	0.000767	105.78%
	116	0.000962	0.000727	0.000981	0.00077	105.90%
	117	0.000956	0.000715	0.000975	0.000757	105.90%
	118	0.000981	0.00077	0.001001	0.000818	106.18%

Z3	119	0.000974	0.000754	0.001623	0.003489	462.56%
	120	0.000966	0.000737	0.001076	0.001017	138.01%
	121	0.000974	0.000755	0.001004	0.000826	109.32%
	122	0.000966	0.000735	0.000986	0.000782	106.39%
	123	0.00098	0.00077	0.001	0.000818	106.25%
	124	0.000971	0.000747	0.000994	0.000803	107.43%
	125	0.000968	0.000741	0.00101	0.000842	113.57%
	126	0.000978	0.000764	0.001108	0.00111	145.30%
	127	0.000976	0.00076	0.001228	0.001512	199.11%
Z4	128	0.000972	0.00075	0.002114	0.00772	1029.54%
	129	0.000956	0.000714	0.001705	0.004048	567.27%
	130	0.000976	0.000759	0.00158	0.003219	424.14%
	131	0.000969	0.000743	0.001115	0.001132	152.46%
	132	0.00096	0.000723	0.000988	0.000788	109.04%
	133	0.000975	0.000757	0.001006	0.000832	109.80%
	134	0.000978	0.000764	0.00101	0.000842	110.21%
	135	0.000962	0.000726	0.000995	0.000805	110.89%
	136	0.000983	0.000775	0.001019	0.000865	111.61%
Z5	137	0.000938	0.000674	0.002038	0.006913	1026.30%
	138	0.000952	0.000705	0.001572	0.003174	450.22%
	139	0.000954	0.000708	0.001405	0.002263	319.53%
	140	0.000944	0.000686	0.001345	0.001989	289.84%
	141	0.000966	0.000737	0.001239	0.001552	210.52%
	142	0.000943	0.000685	0.001125	0.001163	169.81%
	143	0.000962	0.000726	0.001133	0.001187	163.46%
	144	0.000966	0.000736	0.001079	0.001025	139.35%
	145	0.000979	0.000766	0.001046	0.000934	121.90%
Z6	146	0.000956	0.000713	0.00259	0.014184	1989.91%
	147	0.000952	0.000704	0.001782	0.004619	656.30%
	148	0.000953	0.000706	0.001572	0.003173	449.51%
	149	0.00095	0.0007	0.001407	0.002275	325.09%
	150	0.000938	0.000673	0.001296	0.001776	263.88%
	151	0.000955	0.00071	0.001204	0.001426	200.73%
	152	0.000948	0.000696	0.001155	0.001258	180.70%
	153	0.000958	0.000719	0.001144	0.001223	170.12%
	154	0.000953	0.000707	0.001151	0.001244	176.07%

This hydraulic-mechanical modelling can be applied to the fields of engineering design, hydrogeological characteristics evaluation and pollution transportation. For engineering design, we can use this technique to simulate how this site responses to seismic events. Thus we can adjust our design to accommodate the influence of earthquakes. For evaluation of hydrogeological characteristics, our results showed that different hydraulic gradient and water pressure could lead to various respond. With our model, we can assess the hydrogeological characteristics and investigate why water head changes after earthquake. Also, the assessment of post-seismic fracture transmissivity can be helpful for judging the possibility or rate of pollution transportation, which can provide important information for us to decide a reasonable pollutant treatment plan.

4. Conclusions

Our models suggest that different hydraulic conditions can lead to various results of post-seismic fracture behavior. In general, pore pressure can change in two kinds of pattern under stable and low hydraulic gradient conditions. On the other side, aperture and pore pressure generally increase after seismic events under high hydraulic gradient environments. This hydraulic-mechanical modelling technique can help us to evaluate the possible shear displacement, normal displacement, aperture change, fracture transmissivity and water pressure. And this can be applied to engineering design, evaluation of hydrogeological characteristics and transportation of pollutants.

References

[1] Manga M, Brumm M and Rudolph M L 2009 *J. Mar. Petrol. Geo.l* **26** pp 1785-1798
[2] Wang C Y and Manga M 2015 *Nat. Commun.* **6** pp 75-97
[3] Manga M, Beresnev I, Brodsky E E, Elkhoury J E, Elsworth D, Ingebritsen S E, Mays D C and Wang C Y 2012 *Review of Geophysics* **50** RG2004
[4] Lak M, Baghbanan A and Hashemolhosseini H 2017 *Earthq Eng Eng Vib* **16** pp 525-536
[5] Fälth B, Hökmark H and Munier R 2010 *SKB TR-08-11* p 171
[6] Fälth B and Hökmark H 2011 *Posiva Working Report 2011-13* p 74
[7] Fälth B and Hökmark H, 2012 *Posiva Working Report 2012-08* p 41
[8] Fälth B, Lönnqvist M and Hökmark H 2019 *Posiva Working Report 2019—10* p 138
[9] Hökmark H, Fälth B, Lönnqvist M and Munier R 2019 *SKB TR-19-19* p 105
[10] Bizzarri A 2011 *Reviews of Geophysics* **49** RG3002
[11] Somerville P, Irikura K, Graves R, Sawada S, Wald D, Abrahamson N, Iwasaki Y, Kagawa T, Smith N and Kowada A 1999 *Seism. Res. Lett.* **70** pp 59-80

Development and application of an internet of things door lock Network Bridge for classroom access control management

Cheng Hsiu Li, Chi Yu Chen, Xuan Rui Huang, Ting Fu Tsai and Bo Yang Liu

Department of Information Management, National Taitung Junior College, Taitung, Taiwan

E-mail: lcs@ntc.edu.tw

Abstract. School classroom doors are usually managed using conventional machine locks. This can cause inconvenience for the borrowing and returning of keys. In addition, issues on the copying and embezzlement of keys are persistent. To avoid these problems, schools have replaced conventional locks with card readers and magnetic locks to manage classroom access; users need only a RFID card to enter and exit classrooms. However, the hardware and software equipment required for such control schemes is expensive and entails cumbersome installation processes. This study developed an innovative Internet of Things (IoT) door lock network bridge, which can be installed on conventional electronic locks to transform it into IoT locks. The developed bridge can be integrated with a customized web-based classroom access information system and QR code scanning to establish a low-cost classroom access control system. The system can be provided to schools to perform user management, lock management, and lock matching. Each door in the school is assigned with a unique QR code for unlocking. Users only need to use the QR code scanner on their smart phone to send relevant information required for unlocking a door to the web-based classroom access information system, which then performs user and lock matching verification. Verified users can then unlock the door and enter the venue. The proposed classroom access control system does not require card readers or RFID cards. The cable installation of the system is simplified through by using wireless connections, thereby greatly reducing the time, cost, and human resource required for installation. In addition, users and locks can be matched directly using the proposed system. This greatly improves the convenience of classroom management in schools.

1. Introduction

Access to school classrooms is generally controlled with traditional mechanical locks. Apart from the inconvenience of borrowing and returning keys, traditional mechanical locks all provide poor security due to a notable disadvantage: keys are easily copied [1-2]. Electronic locks are common door management devices. Some common electronic locks are electric bolt locks, magnetic locks, electric control locks, silent locks, and electric motor locks. Electric bolt locks are divided into 2-line electric bolt locks, 5-line electric bolt locks, and 8-line electric bolt locks according to the number of wires. The common parts are the red and black wires. The red wire connects to the direct current (+12 V), and the black wire is the ground wire (GND). When the power is cut, the lock retracts, and the door opens. Magnetic locks are also called electromagnetic locks. They are locks that rely on the attraction force generated between electromagnets and iron blocks. They are also electric locks that open when power is cut. Electric control locks have large impact currents; they make an audible "snap" sound when the door opens. The noise is louder, and the installation is less convenient. During installation, installers must be noted that the time delay of opening the door must not be too long, and it is generally set within one second. Due to the loud-noise disadvantage of the electric control lock, the silent lock was designed. Another quiet electric lock is the electric motor lock. It drives a small motor to move the bolt. Electric control locks, silent locks, and electric motor locks can also be opened with a lock knob inside the door or a key. They are mainly used in household doors of communities and combined two-door emplacements in banks. For the type of electric lock required when installing access control, the door type is the crucial point for consideration [3].

Radio Frequency Identification (RFID) is a wireless communication technique that identifies specific targets through radio signals and reads and writes related data. The system does not require mechanical or optical contact between the identification system and the specific targets. The radio signals transmit data from the tags attached to the object by tuning the electromagnetic fields of radio frequencies to automatically identify and track the objects. For a typical identification event, some tags can obtain energy from the electromagnetic field emitted by the identifier and does not require a battery. Some tags have power themselves and can actively emit radio waves (that are tuned to the electromagnetic fields of the radio frequencies). The tags contain digitally stored information that can be identified within several meters. Unlike barcodes, radio frequency tags are not required to be within the scanning vision of the identifier, and they can also be embedded into tracked objects. The RFID technique is used in various industries. In automobile manufacturing, tags are attached to a car that is in production, and the factory can easily track the progress of this car in the production line. In pharmaceutical logistics, locations of medication can be tracked in warehouses. RFID tags can be attached to livestock and pets for identification. RFID cards can allow employees to enter locked doors in buildings with gate control. The radio frequency transponders on cars can also be used to collect the fees for toll roads and parking lots [4-5]. RFID technology offers identification without contact, so card readers and electromagnetic doors are gradually being used in current classrooms to assist management. Each user only requires an RFID card to enter or leave the classroom. However, relative to the costs of other software, hardware, and installation projects, RFID costs are high, and installation is time- and labor-consuming [2].

In 1995, Bill Gates developed his smart home fantasy in his book The Road Ahead, which proposed ubiquitous Internet-connected devices. In 1998, Kevin Ashton, the Auto-ID center director at the Massachusetts Institute of Technology proposed the term Internet of Things (IoT). Globalized Internet infrastructure connects physical objects and virtual data through data extraction and communication for different types of control, detection, identification, and service. Since then, the term has been circulating extensively. Since the introduction of IoT, objects such as cell phones, refrigerators, tables, coffee machines, and weight scales have become smart devices. Consider, for example, a smart refrigerator: when it is out of order, it automatically sends maintenance information. Refrigerators also connect with gas company databases and remind people if the gas has run out. IoT not only makes peoples' lives more convenient but also brings greater safety [6].

Gao He-xuan, the chairman of D-Link Corporation, stated that the most important factor of the rise of IoT is smart phones. Cell phones are the control centers of IoT devices [6]. Barcodes can be seen everywhere in peoples' lives, from the backs of merchandise packages in convenience stores to the tags of registered-mail items. The use of barcodes makes it easier to manage merchandise items. The information can be obtained with barcode scanners. Due to technological advancement, every cell phone has high-pixel camera lenses. Barcodes can be shot and decoded with programs. Thus, cell phones can become barcode scanners that can easily read the information in barcodes.

Consequently, this study aims to develop a brand-new IoT door lock network bridge. After this network bridge has been installed on a traditional electric lock, it can become an IoT door lock that connects the electric lock to the Internet. Thus, the lock can be controlled through cell phones. Then, the device is combined with a customized Web-based classroom access information system and Quick Response Codes (QR codes) to establish a low-cost classroom access system. This system provides the access management functions of user management, door lock management, and door control for schools. This system can replace old classroom access management systems that feature RFID card readers and electromagnetic doors.

2. Literature review

In this study, related RFID literature was examined, and the aspects of hardware design, certification mechanism, virtual key, and establishment cost were discussed as the academic foundation of this study.

Regarding hardware design, Qiu (2018) used Raspberry Pi, RFID, and camera lens modules to develop a card-sensing classroom access system. The system offered the functions of clock punching, photography, and recording, and the records were stored for inquiry [7]. Liao (2010) used the characteristic of super-high-frequency near-field and far-field RFID to read electronic tags; this achieved the goal of access control for people and cars with only one card. When people and cars enter or leave a restricted area controlled by Liao's system, only one access control card is required. The system offers indoor and outdoor access with only one card. It not only avoids extra cost of card manufacture for cars but also integrates access control card management for cars and people [2]. Currently, most access control systems use RFID card access control.

For an access control certification mechanism, Hong, Tsai, and Hong (2005) proposed a RFID smart card. The four techniques of Hash, AES, Random values, and XOR were combined with third-party certification to solve the safety problems of eavesdropping, tracking, and forgery of RFID smart cards. In addition, an active protective system was also added in the structure to further reinforce the safety system. Thus, the risk of attackers using replay attack or DoS attack to overload the safety system was avoided [5]. Chen (2018) proposed a RFID safety certification protocol in the tags with low calculation cost; this protocol was founded on SCR. This smart reader can run the calculations of XOR functions, hash functions, and random number generators while having memory to store the results. The safe qualities of forgery prevention, two-way certification, privacy protection, replay attack protection, and forward safety can be achieved. The proposed certification protocol is more suitable for the application in the IoT operating environment [8]. Liu (2010) used access control keys with RSA and DES encryption algorithms to improve the safety of the data transmission process of access control keys. In addition, the keys are transmitted to cell phones through the PDA wireless Internet function. Then, they are transformed into visualized access control keys of QR codes for the classroom access safety management. The purpose is to allow users to verify their identities and obtain QR code access control keys through mobile devices. For users, the manufacturing cost for lost access control cards is reduced, and the administrative procedure of reissue and cancellation is simplified [9].

For the establishment cost, the cost of the RFID access control used by Yang (2014) is substantially lower than that of manual management. The cost is only 4.78% of the full manual cost, and superior management effects can be achieved. The system records were analyzed, and the average normal door opening probability was higher than 98%. The average incidence rate was lower than 2% [10].

3. System structure and operation model

The classroom access control system (abbreviated as "this system") is composed of three major components. They are the web-based classroom access information system, QR codes, and IoT door locks. The IoT door locks consist of IoT network bridges and electronic locks. The system structure is illustrated in Figure 1. The web-based classroom access information system mainly manages member data, door lock data, access control matches, and QR code decoding. The QR codes transmit the decoding information of lock opening locations. The IoT door locks receive the lock opening information and open the locks.

The system operation model is depicted in Figure 2. Students can open the locks through smart phones and the QR codes on the door locks of every classroom on campus. Assume that a campus has m students ($m > 0$) and n classrooms ($n > 0$). Then, that campus has n QR codes and IoT door locks. Only one IoT server is required to be operational for lock opening verification. If a student wants to open a classroom door lock (door lock #2), the student will scan the QR code on the door lock (QR #2) with a smart phone. The system first guides the student to an account and password login screen for identity verification. After the verification is successful, the location code on the QR code will be checked for location verification. Then, the door lock matching verification of the user, door lock, and usage period is conducted. If the user is in the authorized usage period, the door lock opens and allows the user to enter.

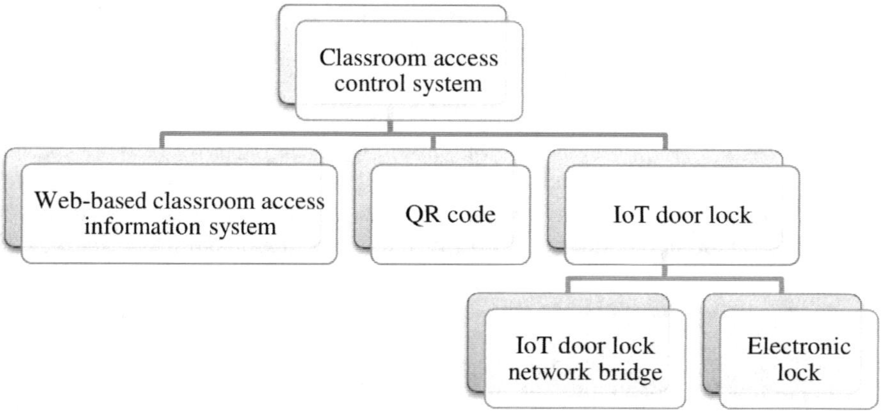

Figure 1. System structure

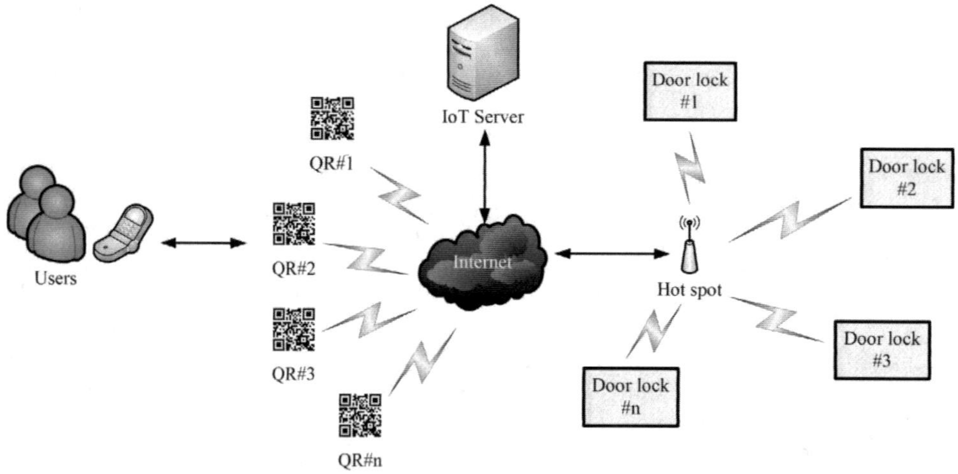

Figure 2. System operation model

4. System design and implementation

4.1. IoT door lock network bridge design and IoT door lock implementation

In this study, IoT technology was used for the design and implementation of the proposed IoT door lock network bridge. This network bridge can connect the electronic lock on the door to the Internet; that connection defines the door lock as an IoT door lock. This door lock can then be controlled through the Internet. Its design concept is schematized in Figure 3. The input end is connected to electrical power, and the output end provides the trigger message to the electronic lock.

Figure 4 presents the hardware structure of the IoT door lock network bridge. It is mainly composed of a Webduino development board and a relay. The Webduino development board consists of an Arduino pro mini development board and a Wi-fi chip. The input end (VCC) is connected to 5V of DC power, and the output end transmits the signals of VCC, GND, and trigger (IN) to the relay. The input end of the relay receives the signals of VCC, GDN, and trigger (IN) from the Webduino, and the normally-open switch of the output end (NO and COM) is connected to the electronic lock. After the input end of the relay receives the trigger signal, the normally-open switch must close, which triggers the electronic lock to connect the circuit and open the door. A photograph of the actual completed system is given in Figure 5.

Figure 3. Design concept of the IoT door lock network bridge

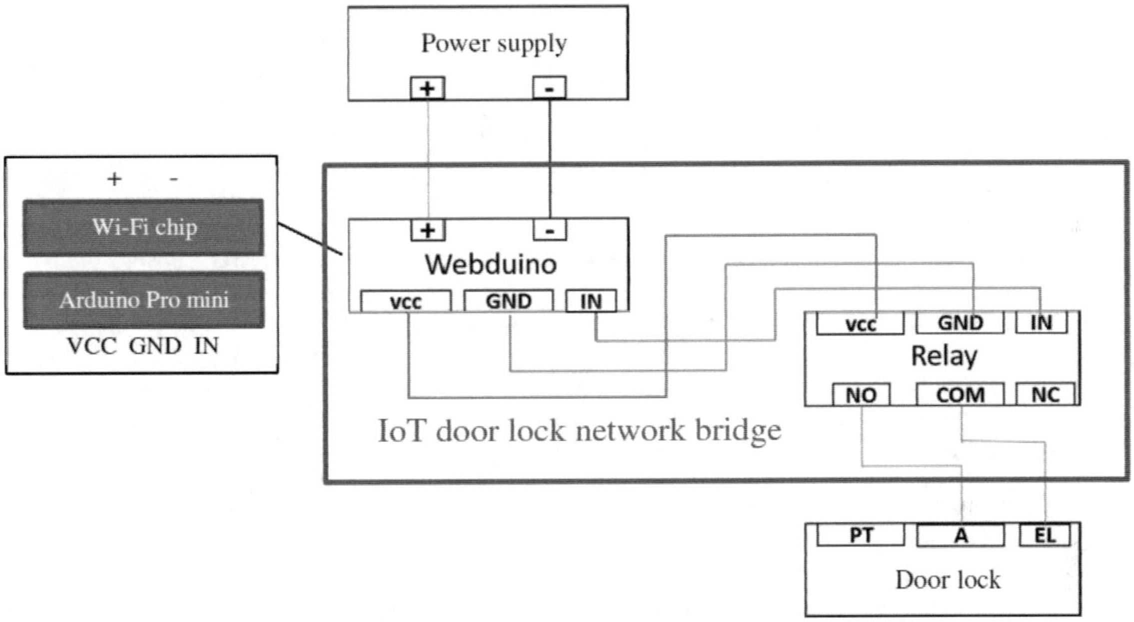

Figure 4. Hardware structure of the IoT door lock network bridge

Figure 5. Implementation of the IoT door lock

IOP Conf. Series: Materials Science and Engineering 1113 (2021) 012022 doi:10.1088/1757-899X/1113/1/012022

4.2. Lock opening mechanism

A triple verification mechanism is adopted. The location match, password match, and door lock match are verified. First, a QR code is scanned by a cell phone to transmit the location decoding information through the Internet for location verification. Then, the web-based classroom access control information system decodes the location and verifies the password. After the location and password verifications are passed, the information system must conduct the door lock match verification. If the user is authorized to use that door lock, and the time is within the authorized usage time, the information system connects over the Internet through the wireless access point to the IoT door lock. The lock opening signal is then transmitted to the IoT door lock to open it.

The web-based classroom access control information system provides the functions of user management, door lock management, and door control matching. Each user must apply for an account and use the related services after logging into the system through a personal account and a password. The services include checking the authorized classroom door locks, door lock usage records, and checking or changing personal information. Managers can manage user information, door locks (creating doors), door control match (designating doors), and viewing all lock opening records. When a door is created in door lock management, the IoT door lock network bridge number (Webduino device ID) and the name of the classroom are set (through the interface presented in Figure 6). After users and doors have been recorded in the database, the door control matching of users and doors can be conducted. The authorized usage periods of door locks must be set (as presented in Figure 7); these periods specify which users are allowed to enter which classrooms during which periods.

The verification of location mainly verifies if the user is at the door to avoid opening locks by mistake. The system has multiple methods to verify the location. Passive infrared sensors or image identifiers can be used to sense whether the user is at the door. However, such sensors increase the costs of hardware, wiring, and development. In this study, the location verification was conducted with software. The Internet was the medium between QR codes and IoT servers. To avoid counterfeiting in the transmission process and verify that the user is indeed at the location of the lock, a random code was added referencing the randomized access control method in this study. The website address of the server (such as http://IoTwebsize) is embedded into the lock opening key (key=randomcode) into a QR code. The QR code transmits a coded website address (such as http://IoTwebsize/index.php?key=20200410081112) to the information system for decoding. After the lock opening key has been verified to be the correct random code, the lock opening screen is opened so that the user can open the lock. Because the lock opening key in the QR code of every classroom is randomly generated, users cannot know it in advance or enter the information system through a remote connection. They must scan the QR code at the classroom door to transmit the lock opening key. Thus, the location is verified. In addition, using software for verification not only substantially reduces the cost of hardware purchase and installation but also simplifies the construction. The process of QR door lock coding and decoding is presented in Figure 8.

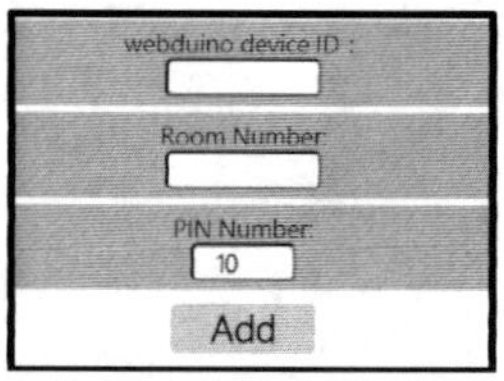

Figure 6. Door lock management

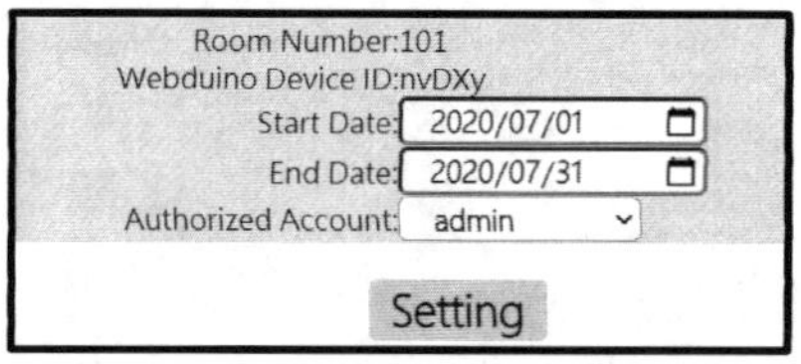

Figure 7. Door control match

Figure 8. Coding and decoding process of a QR door lock

5. System test and efficacy evaluation

5.1. System test

1. Test environment and subjects

Almost every student on campus has a cell phone, so the campus has a sufficient number of mobile devices. Classrooms are one of the best locations for the establishment of this access control system.

In this study, the management department classroom in National Taitung Junior College was selected as the test environment, and 10 students of the management department were randomly selected as the test subjects.

2. System test

In this study, 10 test subjects were randomly selected for the system test. The test is presented in Figure 9.

According to the test results listed in Table 1, the average lock opening time was 9.2 seconds, and the lock was successfully opened at a 100% rate.

(1) Scanning the QR code

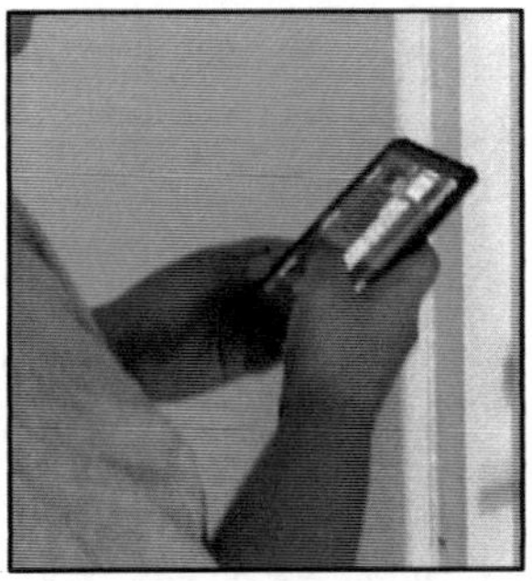

(2) Account and password verification

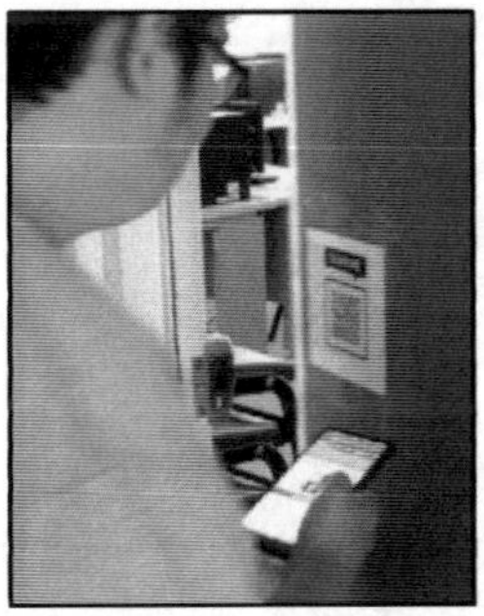

(4) Door lock opening complete

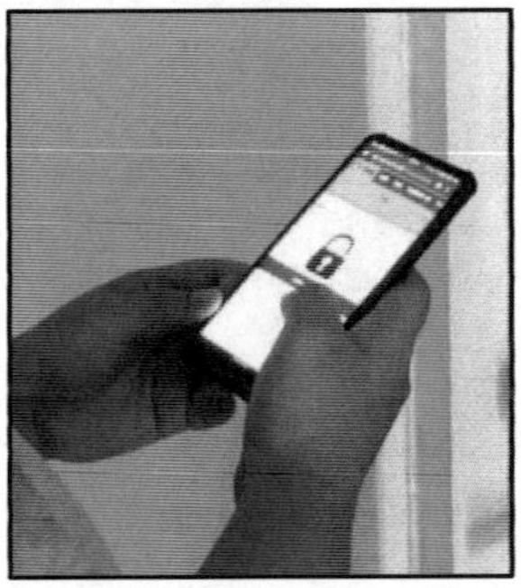

(3) Opening the door lock

Figure 9. IoT door lock test

Table 1. Door lock opening time test

Test subject number	Opening time (second)	Successful opening or not
Student #1	10	Yes
Student #2	9	Yes
Student #3	11	Yes

Student #4	10	Yes
Student #5	11	Yes
Student #6	10	Yes
Student #7	12	Yes
Student #8	9	Yes
Student #9	10	Yes
Student #10	9	Yes

5.2. Establishment cost analysis

An investigation indicated that the price of general smart electronic locks (including cloud functions) (reference https://my-best.tw/21229) on the market was between 9,690 NTD and 20,760 NTD (at the time of writing). The price of the IoT door lock developed in this study was 2,000 NTD; the IoT door lock cost analysis is listed in Table 2. If the original environment already has an electronic lock, only a 1,000 NTD IoT door lock network bridge is required. Consequently, the equipment cost can be substantially reduced by a factor ranging from 4.8 to 10.3 (with the price of 2,000 NTD for comparison). The installation cost can be substantially reduced.

Table 2. IoT door lock cost analysis

Number	Item name	Cost	Note
1	Webduino development board	840	
2	Relay	35	
3	Electronic lock	1,000	(optional)
4	Miscellaneous	125	Wire and 3D printing materials
	Total	2,000	

6. Conclusions

In this study, a new IoT door lock network bridge was successfully developed. Combining the network bridge with an electronic lock can produce an IoT door lock. A low-cost classroom access control system can be established by combining an IoT door lock with a customized web-based classroom access control information system and QR codes. According to the system test results, the average lock opening time was 9.2 seconds, and 100% of users successfully opened the lock.

This classroom access control system does not require card readers and RFID cards. The wireless connection also simplified the installation. According to the analysis, typical equipment costs 4.8 times to 10.3 times as much as the proposed hardware. In addition, if the original campus already has electronic locks, only the IoT door lock network bridges are required. The proposed system can substantially reduce the cost, time, and manpower required for lock deployment and management. The classroom management can be done with door control that matches directly through the system. The convenience of classroom management can be substantially improved.

References

[1] What is the Difference between the Door Lock and the Electronic Lock? Safety Comparison of 4 Door Lock Types 2016 https://decomyplace.com/newspost.php?id=4580&ref=sg
[2] Liao D Y 2010 Application of UHF RFID Technology in Security Access Control https://www.digitimes.com.tw/iot/article.asp?cat=130&id=0000179326_ed94ito01nf3be7h3snxc
[3] Graphic Classification and Installation of Access Control Electric Lock 2018 https://kknews.cc/zh-tw/news/q5pjary.html
[4] Weis S A 2007 *RFID (Radio Frequency Identification): Principles and Applications* (MIT CSAIL)
[5] Hung Y C, Tsai C W and Hong C H 2005 Discussion on Security Framework of RFID Access Control Identity Authentication System RFID *In Proceedings of 2005 Taiwan Academic*

IOP Conf. Series: Materials Science and Engineering 1113 (2021) 012022 doi:10.1088/1757-899X/1113/1/012022

Network Conference(TANET 2005)
[6] Wong S T 2015 30 Keywords! Let You Understand the Internet of Things (IoT) https://www.bnext.com.tw/article/34549/bn-article-34549
[7] Chiu S W 2018 *College of Science, National Chiao Tung University* (Unpublished Master Thesis)
[8] Chen Y C 2018 *Department of Computer Science and Information Engineering, Chung Hua University* (Unpublished Master Thesis)
[9] Liu C Y 2010 *Institute of Information Technology and Application, College of Informatics, Kao Yuan University* (Unpublished Master Thesis)
[10] Yang M H 2014 *Department of Industrial Engineering and Management, National Chiao Tung University* (Unpublished Master Thesis)

IOP Conf. Series: Materials Science and Engineering 1113 (2021) 012023 doi:10.1088/1757-899X/1113/1/012023

Integration of intelligent diagnosis system and augmented reality for electric motors

Yu Ching Lin[1], Heng Chuan Kan[1], Jian Ming Lu[1], Chih Min Yao[1], Yu Chia Liao[2], Chun Hui Chung[3], Kai Chung Shih[3] and Mi Ching Tsai[3]

[1]National Center for High-Performance Computing, Hsinche, Taiwan
[2]National Tsing Hua University, Hsinche, Taiwan
[3]National Cheng Kung University, Tainan, Taiwan

E-mail: 1203043@narlabs.org.tw

Abstract. With the rapid development of industrial automation technology, electric motor control has also become an important technology. However, motor failure has many possible causes, and it is not easy to detect in advance. Therefore the fault diagnosis is an important issue for motors. The augmented reality (AR) technology is also developing towards various industrial and educational application in recent years. This research is focused on developing the intelligent diagnosis system and AR application for electric motors. The intelligent diagnosis system applies cloud data management and machine learning methods to predict the health status of electric motors. And the time-recurrent Long Short-Term Memory (LSTM) neural network algorithm is applied to establish a motor health diagnostic model. The experimental results show that the motor diagnosis method can predict the health status of motor effectively. The proposed system will also provide more industrial application services in the future.

1. Introduction

Many devices in our life are related to the electric motors, such as the engine of a car, or electric fan and so on. It is also widely used in the machinery industry, electrical machinery industry, medical industry, etc., and also used in the field of automation control and electromechanical integration. The motor can convert electrical energy into mechanical energy to drive for vibration, rotation, and linear motion. There are many different types of motors, which are mainly composed of stators and rotors. The stator is stationary in the space, while the rotor can rotate around the shaft and is supported by bearings. One of the stator and the rotor can also be a permanent magnet.

When the motor is running, it will be accompanied by vibration. When the motor runs with abnormal vibration, it indicates that the faults may occur. For the motor, when it faults, it can be diagnosed through various vibration analysis methods. There are some researches discussed about motor fault diagnosis methods [1-2], and the authors proved that the sequential vibration signals are useful for the motor faults detection. The permanent magnet motor is another type of motor which commonly used in our daily life. Demagnetization fault is one of the faults existing in permanent magnet motors. There are also some demagnetization diagnosis methods have been proposed [3-4], such as finite element analysis (FEA) and time-frequency analysis (TFA), but these methods are difficult to use for real-time analysis.

In the meantime, the rapid development of smart phones and various applications that can be used for life, work, and entertainment are constantly being developed and improved. With the massive developments of software and hardware, there are more Augmented Reality (AR) and Virtual Reality (VR) devices and applications, which have generated the motivation for users to enjoy new interactive experiences. At the same time, driven by this wave of AR, it is changing the landscape of operating modes in the fields of education, industry, medical, manufacturing and others [5]. In addition, due to the rising trend of the Industry 4.x, artificial intelligence (AI) technology will rapidly boost applications through AR, which will promptly increase the various types of services. The key technologies of AR are developing towards various applications as well. How to integrate AR services in cloud computing is also a hot research topic.

In addition, machine learning is the key to the next industrial revolution, and it can be used for motor fault diagnosis effectively. Machine learning can replace routine operations and the use of machine learning technology can instantly monitor equipment and predict maintenance needs [6-7]. Massive data systems can collect and analyze various possible variables and build a model to predict the behavior of the entire system. Using a large amount of sensor data in this way has a good help on the device's operating status and maintenance costs. The machine learning technologies can help to analyze and predict the possible faults of electric motors accurately.

This research is mainly based on the development of a machine learning based diagnosis system and AR technology for electric motors. The proposed method is applied to prevent the fault occur for higher power or permanent magnet motors, and combines the AR technology to provide the innovative application for users. The intelligent diagnosis system uses the LSTM neural network model [8] to establish an electric motor health diagnostic model. Based on the differences between the parameters such as the vibration, voltage, temperature, and flux linkage collected during the electric motor operating and the predicted response of the diagnostic model to determine whether the current system is healthy or not. The intelligent diagnosis system also uses cloud data server to monitor and predict the health and potential faults of motors, which provides a more convenient and efficient management and maintenance mechanism, and presents the combination of reality and virtualization operating experiences.

The rest of the paper is organized as follows: the related technologies of augmented reality and machine learning are described in section 2, and the electric motor diagnosis system design is presented in section 3, and the development of augmented reality application is presented in section 4. Finally, the conclusion and future works are presented in section 5.

2. Augmented reality and machine learning

2.1. Augmented reality

Industry and academia in various countries around the world have invested in AR research and development for many years [9]. How industry experts turning requirements into specifications, designs, and manufacturing industry-compliant applications remains an innovative challenge.

Augmented Reality is a technology that augments virtual information into real space. It does not replace real space, but adds a virtual object to the real space to augment the virtual information into reality. In space, emphasis is placed on the interpretation and interaction of AR space with virtual information. Through the combination of camera recognition technology and computer programs, when a set graphic appears in the lens, a virtual object will appears on the screen. Recently, the booming development of mobile technology has also promoted the increased availability of AR applications.

AR consists of technologies that combine the real world and virtual goods, and allows users to use actual applications and system interaction in the real environment. This will increase user focus and motivation. AR is mainly used to detect the user's 3D spatial position in terms of visual presentation, and then to make a corresponding visual representation of virtual and real objects [10]. One of the most important technologies is spatial 3D positioning. Some researches proposed the 3D graphic display technologies, and to combine the virtual objects with 2D or 3D images. Precise spatial 3D positioning displays the pre-completed AR information on the device, and generates a virtual and real picture through visual overlay effects. The AR technology provides a 3D virtual object with oblique viewing angles and the real-world scene, which can experience the combination of reality and virtualization during the interactive operation.

2.2. Machine learning and LSTM neural network

Machine learning is a multi-disciplinary domain that has emerged in recent years. It can automatically analyze from a large amount of data to find the regularity and use the regularity of the data to predict the future. Deep learning is a part or subset of machine learning, and it is an algorithm that attempts to do high-level abstraction of data by using multiple processing layers. Deep neural networks are currently the most common deep learning models, such as the Convolutional Neural Networks (CNN) is suitable for image recognition; and RNN is suitable for sequence to sequence translation, including converting speech to text or translating different languages. RNN is a very powerful dynamic system, but due to the problem of gradient disappearance/explosion, it is restricted to solve the problem of long-term dependence effectively. To solve this problem, the LSTM model with special implicit units is proposed.

LSTM is a kind of time-recurrent neural network, and its unique design structure is suitable for processing and predicting longer and more delayed events in time series. The LSTM contains a special unit called a memory cell that is used to memorize values of varying lengths of time, similar to accumulators and control neurons. LSTM has been proven to perform better than traditional RNN in speech recognition and machine translation applications [11].

LSTM network has three control gates: forget gate, input gate and output gate respectively. These three control gates confirm the feasibility of using gradient-based optimization methods and avoid the diffusion and explosion of gradient. The structure of LSTM neural network is shown as Figure 1. X_t represents the input at the current moment, C_{t-1} represents the long-term memory vector before this moment, h_{t-1} represents the output of the previous unit, C_t represents the long-term memory vector at the current moment, and h_t represents the output at the current moment. The operation processes of LSTM are as follows: (1) Combine the output vector at the previous moment and the input vector at the current moment into a vector σ; (2) Use the activation function to matrix multiply the vector σ with the parameters of different states; (3) f_t and C_{t-1} at the previous moment are combined to realize the memory or forgetting effect of the state at the previous moment; (4) Use *tanh* activation function to operate C_t, and combine with the O_t vector to control the output vector and get h_t; (5) Input Ct and h_t as the current state vector and output vector into the next LSTM unit model.

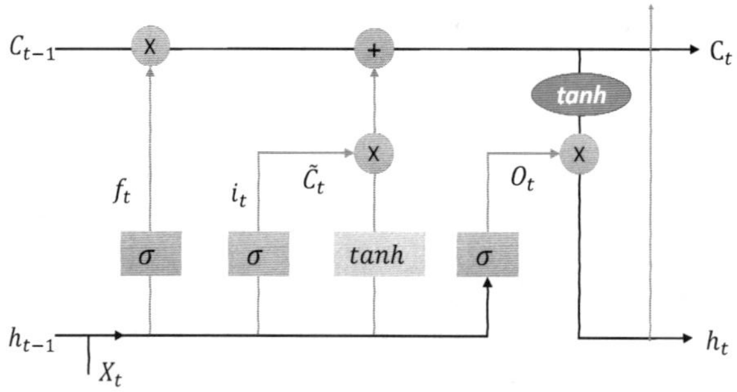

Figure 1. The structure of LSTM neural network

3. Design of electric motor diagnosis system

The intelligent diagnosis system uses the LSTM model to establish an electric motor health diagnostic model. Based on the differences of motor parameters between the current motor operating status and the predicted response of the diagnostic model to determine whether the motor is healthy or not. The accelerometer sensors are developed on each component of motors, and the operational status data are corrected by diagnosis platform through CAN (Controller Area Network) bus or Bluetooth interfaces. These data such as vibration, voltage, and flux linkage are sent to the motor diagnosis model to analyze and predict the healthy status of the motor. Two hidden layers with LSTM are chosen in our diagnosis model.

The algorithm of the proposed diagnostic model has three main steps: data augmentation, model training, and evaluation. In the first step, the purpose of data augmentation is to reduce computation time and improve the convergence speed. The second step is training the diagnosis model with training dataset. The third step is validating the model with testing dataset, and the accuracy of prediction is the important evaluation index.

When the motor diagnostic model is chosen, it is important to decide the parameters of input layers and output layers for LSTM model. The variable speed transmission system can be referred to the motor diagnostic model. The transmission medium of variable speed transmission system is power, and the system has an obvious vibration reaction. In order to discuss the vibration behavior effectively, the limited main degrees of freedom is supposed to descript the dynamic behavior of structural system. The motion equation can be defined as eq. (1).

$$\mathrm{M}\ddot{x} + C\dot{x} + Kx = f \tag{1}$$

the parameter M, C and K represents as mass matrix, damping matrix and stiffness matrix respectively; x is displacement vector, and f is external force vector.

Eq. (1) is a description model of mechanical vibration system, and the model can be descripted according to the parameters such as: displacement, velocity and acceleration. Unfortunately, the displacement, velocity and acceleration parameters are unable captured at the same measurement node at the same time. To solve the problem, the centered difference scheme is applied as eq. (2).

$$\begin{cases} X_{t+\Delta t} = X_t + \dot{X}_t\Delta t + \frac{1}{2!}\ddot{X}_t\Delta t^2 \\ X_{t-\Delta t} = X_t - \dot{X}_t\Delta t + \frac{1}{2!}\ddot{X}_t\Delta t^2 \end{cases} \Rightarrow \begin{cases} \dot{X}_t = \frac{X_{t+\Delta t} - X_{t-\Delta t}}{2\Delta t} \\ \ddot{X}_t = \frac{X_{t+\Delta t} - 2X_t + X_{t-\Delta t}}{\Delta t^2} \end{cases} \tag{2}$$

and then we can get eq. (3) according to the eqs. (1) and (2):

$$\left(\frac{M}{\Delta t^2} + \frac{\mathrm{C}}{2\Delta t}\right) X_{t+\Delta \mathrm{t}} + \left(\frac{2M}{\Delta t^2} + \mathrm{K}\right) X_t + \left(\frac{M}{\Delta t^2} - \frac{\mathrm{C}}{2\Delta t}\right) X_{t-\Delta \mathrm{t}} = f_t \tag{3}$$

To measure the displacement response, it requires a motionless reference point, and one observation parameter is needed to describe the function. The acceleration parameter is applied to describe this system. Therefore, find the second derivative of time for centered difference eq. (2), and we can get the eq. (4).

$$\left(\frac{M}{\Delta t^2} + \frac{\mathrm{C}}{2\Delta t}\right) \ddot{X}_{t+\Delta \mathrm{t}} + \left(\frac{2M}{\Delta t^2} + \mathrm{K}\right) \ddot{X}_t + \left(\frac{M}{\Delta t^2} - \frac{\mathrm{C}}{2\Delta t}\right) \ddot{X}_{t-\Delta \mathrm{t}} = \frac{f_{t+\Delta t} - f_{t-\Delta t}}{2\Delta \mathrm{t}} \tag{4}$$

by simplifying the eq. (4), it will get the time series model as eq. (5). The values of Φ_1, Φ_2, Θ_0 and Θ_2 are calculated as eq. (6).

$$\ddot{X}_t = \Phi_1\ddot{X}_{t-\Delta t} + \Phi_2\ddot{X}_{t-2\Delta t} + \Theta_0 f_t + \Theta_2 f_{t-2\Delta t} \tag{5}$$

$$\begin{cases} \Phi_1 = -\left(\frac{M}{\Delta t^2} + \frac{\mathrm{C}}{2\Delta t}\right)^{-1}\left(\frac{2M}{\Delta t^2} + \mathrm{K}\right) \\ \Phi_2 = -\left(\frac{M}{\Delta t^2} + \frac{\mathrm{C}}{2\Delta t}\right)^{-1}\left(\frac{M}{\Delta t^2} - \frac{\mathrm{C}}{2\Delta t}\right) \\ \Theta_0 = \frac{1}{2\Delta t}\left(\frac{M}{\Delta t^2} - \frac{\mathrm{C}}{2\Delta t}\right)^{-1} \\ \Theta_2 = \frac{-}{2\Delta t}\left(\frac{M}{\Delta t^2} + \frac{\mathrm{C}}{2\Delta t}\right)^{-1} \end{cases} \tag{6}$$

The eq. (5) is the autoregressive exogenous (ARX) time series model. In this model, the current time responses are outputs, and the external interference at current time and previous time are inputs. The inputs and outputs of the time series model can be applied as the input layer and output layer of LSTM neural network. The prediction model for acceleration response can be built up according to the neural network architecture [12]. By comparing the actual measured response with the predicted response, the degree of difference between the current state of the system and the initial health state can be obtained. If the degree of difference exceeds the alert value, determine that the system is in an unhealthy state. We can use roots mean square error (RMSE) to quantify the degree of difference between the current state and the initial state of the system, the estimation function is represented as eq. (7).

$$\sum_{\bar{n}=1}^{n}\sum_{\bar{\zeta}=1}^{\zeta}\left[a_{\bar{\zeta}}(t+\bar{n}\Delta t) - \bar{a}_{\bar{\zeta}}(t+\bar{n}\Delta t)\right]^2 \tag{7}$$

$a_{\bar{\zeta}}(t+\bar{n}\Delta t)$ indicates the actual measured response, and $\bar{a}_{\bar{\zeta}}(t+\bar{n}\Delta t)$ indicates the predicted response of the system.

The variables between the stator circuit and the rotor circuit of the permanent magnet AC motor are coupled with each other and have a time-varying nonlinear equation relationship. If the mathematical model is represented by the three-phase axis, the analysis of dynamic characteristics is complicated and difficult to control. Therefore, the two-axis coordinate conversion method is used to convert the stator and rotor rated voltage, current and flux linkage from the three-phase reference coordinate axis with a phase angle of 120 degrees to the two-phase reference coordinate axis with a 90-degree phase difference. The motor mathematical model of the three-phase coordinate system can be transformed into the two-axis static coordinate system of the stator through Clarke Transform, and then transformed into the synchronous rotating coordinate system of the rotor through Park Transform [13]. The motor electrical formula can be represented as eqs. (8) and (9):

$$V_d = R_s I_d + \frac{d}{dt}\lambda_d + E_d \tag{8}$$

$$V_q = R_s I_q + \frac{d}{dt}\lambda_q + E_q \tag{9}$$

where, V_d and V_q are the stator voltage of *d-q* axis; I_d and I_q are the stator current of *d-q* axis; R_s is the stator phase resistance. Ed and Eq are the counter electromotive force generated by the stator coil of *d-q* axis, which can be represented as eqs. (10) and (11):

$$E_d = -\omega_e \lambda_q \tag{10}$$

$$E_q = \omega_e \lambda_d \tag{11}$$

where ω_e is the electrical speed. λ_d and λ_q are the flux linkage of *d-q* axis, which can be represented as equation (12) and (13):

$$\lambda_d = L_d I_d + \lambda_m \tag{12}$$

$$\lambda_q = L_q I_q \tag{13}$$

where λ_m is the permanent-magnet flux linkage. The electrical formula of the permanent magnet motor can be simplified as eq. (14):

$$\begin{bmatrix} V_d \\ V_q \end{bmatrix} = \begin{bmatrix} L_d s + R_s & -\omega_e L_q \\ \omega_e L_d & L_q s + R_s \end{bmatrix} \begin{bmatrix} I_d \\ I_q \end{bmatrix} + \omega_e \begin{bmatrix} 0 \\ \lambda_m \end{bmatrix} \tag{14}$$

and then the torque equation of permanent magnet motor can be represented as eq. (15):

$$T_e = \frac{3}{2}\frac{N_P}{2}\left(\lambda_m I_q + \left(L_d - L_q\right) I_d I_q\right) \tag{15}$$

Since the current variation speed cannot be measured in the motor electrical formula, the central difference method is introduced as eq. (16):

$$\begin{cases} I_{t+\Delta t} = I_t + \dot{I}_t \Delta t + \frac{1}{2!}\ddot{I}_t \Delta t^2 \\ I_{t-\Delta t} = I_t - I_t \Delta t + \frac{1}{2!}\ddot{I}_t \Delta t^2 \end{cases} \Rightarrow \begin{cases} \dot{I}_t = \frac{I_{t+\Delta t} - I_{t-\Delta t}}{2\Delta t} \\ \ddot{I}_t = \frac{I_{t+\Delta t} - 2I_t + I_{t-\Delta t}}{\Delta t^2} \end{cases} \tag{16}$$

and then the motor electrical formula can be discretized and represented as eq. (17):

$$\begin{bmatrix} V_{d,t} \\ V_{q,t} \end{bmatrix} = \begin{bmatrix} R_s & -\omega_e L_q \\ \omega_e L_d & R_s \end{bmatrix} \begin{bmatrix} I_{d,t} \\ I_{q,t} \end{bmatrix} + \frac{1}{2\Delta t} \begin{bmatrix} L_d & 0 \\ 0 & L_q \end{bmatrix} \begin{bmatrix} I_{d,t+\Delta t} - I_{d,t-\Delta t} \\ I_{q,t+\Delta t} - I_{q,t-\Delta t} \end{bmatrix} + \omega_e \begin{bmatrix} 0 \\ \lambda_m \end{bmatrix} \tag{17}$$

after calculating, we can get the time varying autoregressive with exogenous input (TV-ARX) model as eqs. (18) and (19):

$$\begin{bmatrix} I_{d,t} \\ I_{q,t} \end{bmatrix} = \sum_{i=1}^{2} \Phi_i \begin{bmatrix} I_{d,t-i\Delta t} \\ I_{q,t-i\Delta t} \end{bmatrix} + \Psi_1 \begin{bmatrix} V_{d,t-\Delta t} \\ V_{q,t-\Delta t} \end{bmatrix} + \Theta_1 \begin{bmatrix} 0 \\ \omega_{e,t-\Delta t} \end{bmatrix} \tag{18}$$

$$\begin{cases} \Phi_1 = \left[\frac{1}{2\Delta t}\begin{bmatrix} L_d & 0 \\ 0 & L_q \end{bmatrix}\right]^{-1} \begin{bmatrix} R_s & -\omega_e L_q \\ \omega_e L_d & R_s \end{bmatrix} \\ \Phi_2 = \begin{bmatrix} 1 & 0 \\ 0 & 1 \end{bmatrix} \\ \Psi_1 = \left[\frac{1}{2\Delta t}\begin{bmatrix} L_d & 0 \\ 0 & L_q \end{bmatrix}\right]^{-1} \\ \Theta_1 = \left[\frac{1}{2\Delta t}\begin{bmatrix} L_d & 0 \\ 0 & L_q \end{bmatrix}\right]^{-1} \begin{bmatrix} 0 & 0 \\ 0 & \lambda_m \end{bmatrix} \end{cases} \tag{19}$$

According to this TV-ARX formula, a model for predicting motor current response based on the data of current, voltage, speed, etc. can be established. The LSTM neural network of electrical response can be constructed by substituting the input and output values determined by this model. And then compare the current response obtained from this model with the actual current data, it can determine whether the magnetic characteristics of the motor have made a difference or not. Figure 2 is the LSTM neural network for permanent magnet motor diagnosis system which is determined by the electrical equation of the permanent magnet motor.

Since the demagnetization of the motor is an irreversible chronic deterioration phenomenon, it is not easy to verify the actual data collected by the actual motor. If the data is generated by finite element analysis software, the results are not easy to accept. Therefore, through the hardware-in-the-loop system (HIL), the working condition of the real motor is simulated online, and the magnetic health state of the motor is adjusted in real time by changing the parameters of the operating interface. And then upload the data to cloud server for modeling and analysis.

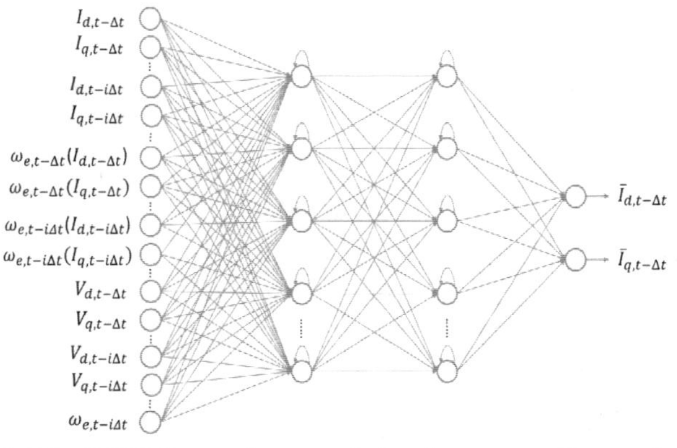

Figure 2. The LSTM neural network for the motor diagnosis system

There are some hyper-parameters that have to be setup and adjusted while building the LSTM model. We can find the best value for hyper-parameters by evaluating the LSTM model with the test data to. The batch size and learning rate are taken as hyper-parameters of our LSTM model. In the process of back propagation, the value of weight and bias are adjusted continuously until the number of iterations reaches the specific value. To accelerate the training process and prevent the local optima, the mini-batch gradient descent algorithm is applied, in which the batch size of training samples is selected for iteration. The parameters of last iteration of training process was updated and used for testing datasets. The learning rate defines the updating speed of some parameters such as weight and bias. The results of accuracy rate in the testing dataset for different hyper-parameters are shown in Table 1. The best configuration with the batch size of 80% and the learning rate of 0.003 presents a correct prediction.

Table 1. Accuracy rate in testing dataset for different hyper-parameters

		Batch Size				
		20%	40%	60%	80%	100%
Learning Rate	0.001	97.3	97.4	97.6	98.2	98.1
	0.002	97.2	97.3	97.8	98.3	98.2
	0.003	97.6	97.8	98.3	98.6	98.4
	0.004	97.5	97.7	98.1	98.5	98.2
	0.005	97.2	97.6	98.1	98.3	98.1

All samples for the experiment will be represented the health conditions with length 256 and each condition has 1680 samples, in which 1344 samples are randomly selected for the training dataset, and the rest 336 samples are for the testing dataset. There are 256 units for the input layer of LSTM model, and 64 units for the first hidden layer of LSTM, and 32 units for the second hidden layer of LSTM. A max iteration value is set as 80, but if the loss of training dataset is still minor changing after a few iterations, the training phase will be ended. To evaluate the effectiveness of prediction model, the root mean square error (RMSE) is employed for evaluation. The average RMSE of the diagnosis model is 1.6253, and the average R-squared value of the testing model is 0.86. The accuracy and loss rate in the training dataset of the model are plotted in Figure 3. The accuracy rate represents the effectiveness of the diagnosis model, and the accuracy rate is the number of correct prediction divided by total number of dataset. The training phase is convergence after about 48 iterations, which proves it can reduce the training time and get a good result.

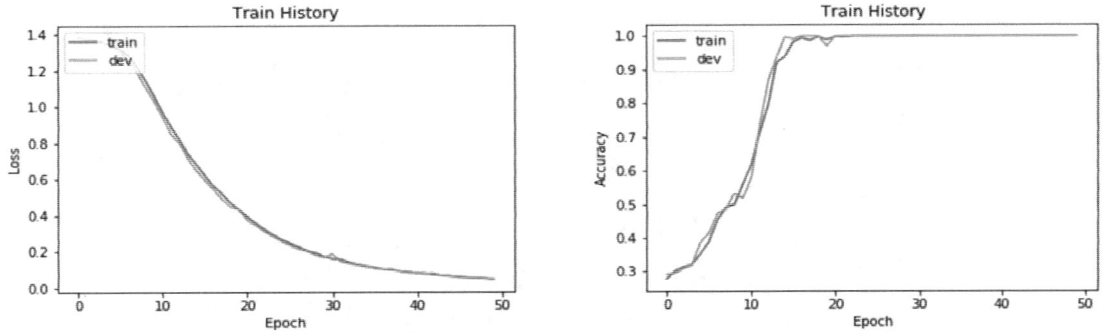

Figure 3. The accuracy and loss rate of the diagnosis model

Compared with traditional techniques of motors fault diagnosis, the advantages of using the deep learning networks are comparably fewer iterations and the computing time is acceptable. The proposed health diagnosis method provides the prediction of abnormal vibration and demagnetization for electric motors. The diagnosis model can be applied to the higher power and permanent magnet motors. According to the system response diagnosis model, the accuracy of fault prediction is up to 98%. The permanent magnet motors not only shows the difference caused by motor demagnetization, but it also demonstrates that the degree of demagnetization is positively correlated with the amount of difference. Therefore, this motor demagnetization diagnosis method can indeed present the health status of the permanent magnet motors.

4. Development of augmented reality application

The motor 3D AR display application is developed with Unity [14] software on mobile device platform. Unity is a cross-platform 2D/3D game engine developed by Unity Technologies. The platform supported by Unity also extends to the HTML5 web platform based on WebGL technology. The graphical expression of the augmented objects became more precise by using Unity's built in shades.

Vuforia [15] AR development tools have been added as a unified workflow across AR devices to provide customized resources that turn immersive vision into reality. Vuforia is a set of software development tool from Qualcomm for mobile device augmented reality applications, which provide good augmented reality development tools for mobile devices. It allows developers to place virtual objects through the camera viewfinder and adjust the position of objects on the solid background in front of the camera.

The augmented reality object display and user interactive operation combines with the motor design platform at the back-end cloud server. The back-end server controls the animation display of the rotation speed of the 3D electric motor model at the mobile device, allowing users to experience the 3D models and animations of various electric motor designs during the interactive operation.

The mobile device application is set to scan specific graphics to load the electric motor 3D module. This 3D electric motor model consists of several components, including a rotating shaft, a rotor, a stator, etc. These 3D motor components can be displayed or hidden by selecting from the program menu. The various components of the electric motor can be exploded or closed to see details of the electric motor model, which is shown in Figure 4. The user not only can control the operation modes of the electric motor model to zoom in, zoom out, rotate up, down, left and right, etc. from the application interface, but also read the speed, voltage, temperature, and magnetic data from the cloud computing server to control the rotation speed of the electric motor model or display possible abnormal conditions. The operating interface provides a convenient management function for users.

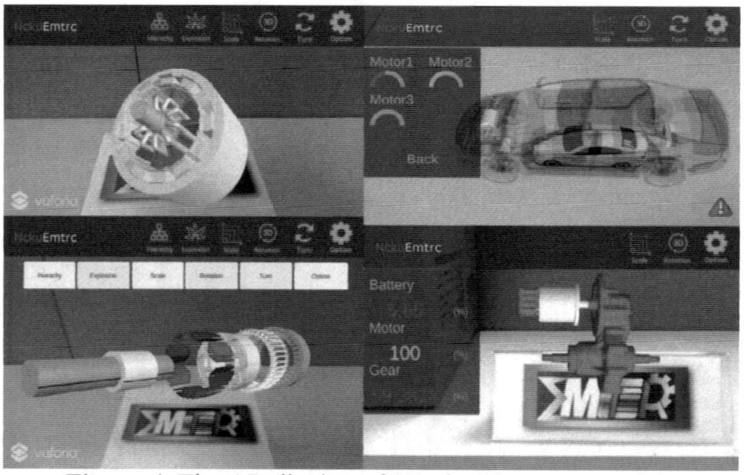

Figure 4. The AR display of 3D electric motor models

If an abnormality is detected or a possible abnormal condition is predicted by the diagnosis system, the components of motor unit will appear different color in the application. The diagnosis system can also be applied for the automotive system. The virtual components of automotive model can be displayed by scanning a car model. The motors of the car can be monitored and diagnosed by the intelligent diagnosis system, and displayed the fault components of the car on the screen of mobile device.

The schematics of electric motor AR diagnosis system is illustrated as Figure 5. In addition to maintaining the necessary parameters generated by the electric motor design, the cloud AI computing server writes relevant data generated by the operation of the electric motor into the database, including the speed, voltage, vibration, temperature, and flux linkage, etc. of the electric motor. Then, AI computing module executes the fault prediction for electric motors, and the results are transmitted to the application program on the mobile device. The AR program on the mobile device displays the operating status of the electric motor. After obtaining the electric motor-related parameters on the server, the operating status of the AR electric motor model is displayed on the screen. And the information such as the operation, execution of the motor module, or setting of the motor parameters is synchronously returned to the server. By updating the database, and sending information back to the electric motor controller to set the operating parameters of the motor.

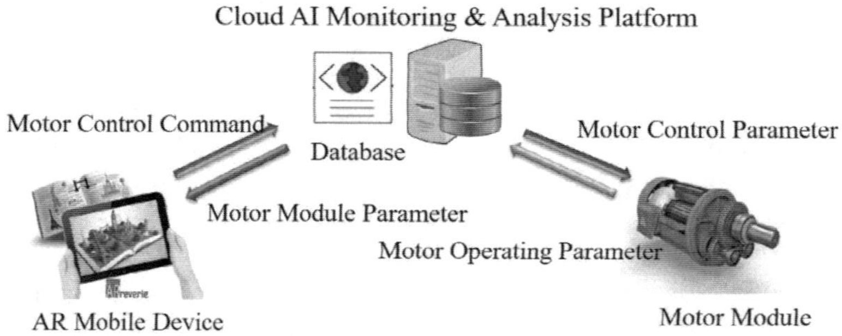

Figure 5. Architecture of AR diagnosis system

The application on the mobile device has the ability to identify different modules of the electric motor and switch to the electric motor model to be simulated or monitored. The switching method can be used to load specific motor modules without image recognition, or use the screen menu to manually switch between different electric motor modules. The operating parameters of the electric motor model can also be set and controlled by the application, and then sent back to the database of the cloud server to update the electric motor status synchronously. It also allows users to experience 3D digital electric motor models and animations of various electric motor designs during the interactive process.

In terms of the electric motor design and industrial applications, using AR software to control the parameter design and operating status of motor equipment can improve the work flow and efficiency of staff inspection equipment. Based on accumulated data on the cloud server and machine learning technology, the system can determine the status of this equipment is abnormal or not, and display early warning to reduce the risk of equipment failure.

5. Conclusions and future works

This research provides an intelligent diagnosis system which integrates the augmented reality technology for the electric motors. The intelligent diagnosis mechanism of electric motors applies LSTM machine learning algorithms to establish the motor health diagnostic model. By combing with cloud data management and machine learning technologies, an AI cloud platform is developed to predict the health status and detect possible abnormalities of the electric motor. The AR technology can also displays the alarm signals of the digital electric motor model on the mobile devices immediately to provide a combination of reality and virtualization experience during the interactive operation. In the future, we will enhance the diagnosis system for electric motors, and compare with various machine learning algorithms to get the better diagnosis model and improve the accuracy of prediction results. Besides, the AR display system will scan the real solid motors and vehicles to display virtual information on a real object, and provides real intuitive interactions for users.

Acknowledgement

This research was funded by National Center for High-Performance Computing, National Applied Research Laboratories and the Ministry of Science and Technology, R.O.C., grant number MOST 109-2622-8-006-005.

References

[1] Lu S, Zhou P, Wang X, Liu Y, Liu F and Zhao J 2018 Condition monitoring and fault diagnosis of motor bearings using undersampled vibration signals from a wireless sensor network *Journal of Sound and Vibration* **414** pp 81-96

[2] Prudhom A, Antonino-Daviu J, Razik H and Climente-Alarcon V 2017 Time-frequency vibration analysis for the detection of motor damages caused by bearing currents *Mechanical Systems and Signal Processing* **84** pp 747-762

[3] Kao I, Wang W, Lai Y and Perng J 2019 Analysis of permanent magnet synchronous motor fault diagnosis based on learning *IEEE Trans. Instrum. Meas.* **68**(2) pp 310-324

[4] Zhu M, Hu W and Kar N 2017 Torque-ripple-based interior permanent-magnet synchronous machine rotor demagnetization fault detection and current regulation *IEEE Transactions on Industry Applications* pp 2795-2804

[5] Masood T and Egger J 2019 Augmented reality in support of Industry 4.0—Implementation challenges and success factors *Robot. Comput.-Integr. Manuf.* **58** pp 181-195

[6] Huang Y C and Zheng J H 2018 Ball nut preload diagnosis of the hollow ball screw through support vector machine *Adv. technol. innov.* **3**(2) pp 275-277

[7] Lee D, Lee S and Seo Y 2020 Application of recent developments in deep learning to ANN-based automatic berthing systems *Int. j. eng. technol. innov.* **10**(1) pp 75-90

[8] Pan H, He X, Tang S and Meng F 2018 An improved bearing fault diagnosis method using one-dimensional CNN and LSTM *J. Mech. Eng.* **64** pp 443-452

[9] Daling L, Abdelrazeq A, Haberstroh M and Hees F 2019 Usability evaluation of augmented reality as instructional tool in collaborative assembly cells *International Conference on Advances in Computer-Human Interactions* pp 199-205

[10] Manuri F, Pizzigalli A and Sanna A 2019 A state validation system for augmented reality based maintenance procedures *Appl. Sci.* **9**(10)

[11] Khan T, Alekhya P and Seshadrinath J 2018 Incipient inter-turn fault diagnosis in induction motors using CNN and LSTM based Methods *IEEE Industry Applications Society* pp 23-27
[12] Zhu Z, Peng G, Chen Y and Gao H 2019 A convolutional neural network based on a capsule network with strong generalization for bearing fault diagnosis *Neurocomputing* **323** pp 62-75
[13] O'Rourke C J, Qasim M M, Overlin M R and Kirtley J L 2019 A geometric interpretation of reference frames and transformations: Dq0, clarke, and park *IEEE Trans. Energy Convers.* **34**(4) pp 2070-2083
[14] https://docs.unity3d.com/Manual/index.html, Unity 3D
[15] https://developer.vuforia.com, Vuforia developer

IOP Conf. Series: Materials Science and Engineering 1113 (2021) 012024 doi:10.1088/1757-899X/1113/1/012024

Development of a model for the mass loss rate of chemical erosion in fracture intersection with the deposition hole

Jia Sheng Chiou[12], Chung Yue Wang[1], Yuan Chieh Wu[2]

[1]Department of Civil Engineering, National Central University, Taoyuan, Taiwan
[2]Chemical Engineering Division, Institute of Nuclear Energy Research, Taoyuan, Taiwan

E-mail: Jschiou@iner.gov.tw

Abstract. The erosion of bentonite buffer in a fracture intersecting a deposition hole will affect the safety function. The purpose of this study was to build up a model which was to predict the rate of erosion of buffer/backfill, and this research focused on the KTH model which was developed by SKB. The KTH model consists of fluid pressure, sodium concentration, and smectite volume fraction, which is governed by Darcy's Law and advection-diffusion equations respectively, and the mutual dependencies of these variables were considered. This model was implemented using the COMSOL Multiphysics, and the results show the different behavior of sodium concentration and smectite in various seeping water velocity, the erosion rate increases with the seeping water velocity, but the distance to the smectite rim border decreases with the water velocity. The KTH considers the relationship between sodium and smectite, and with scalability. it lays a foundation for the evaluation buffer/backfill mass loss by erosion.

1. Introduction

The buffer/backfill erosion problem was first considered by The SR-Can report [1], and seeping water did not affect erosion rate with advection, which was dominated by diffusion. it means that bentonite intrudes into the fracture by diffusion effect, and the critical coagulation concentration (CCC) was used to evaluate stability and formation of the smectite. Colloids were assumed not to form when the ion strength in seeping water was greater than CCC. The CCC was assumed to be 1 mM. SKB assumed the repository will overlain by ice sheet in the future, and the ion strength of seeping water below the CCC, which could flow into the repository there is erosion at gel/sol/water interface and cause mass loss of bentonite. Liu and Neretnieks used the conceptual model discussed above to estimate bentonite erosion rates. and give a method to estimate the rate of loss of bentonite due to erosion, which depends on the rate of loss of calcium and the concentration of calcium in the bentonite [2]. These findings provide methods for estimating the rate of erosion of bentonite, but there is a lack of understanding of the mechanism and process of colloidal release from deposition hole and erosion by seeping water. Therefore, it's necessary to develop a model that describes the bentonite expands and intrudes into the fracture and the colloidal release process in the correct way [3]. Therefore, many experiments, numerical modeling, and in-situ tests have been conducted in Sweden and internationally to address the chemical erosion of buffer/backfill.

To describe the phenomenon of the bentonite expansion and extrusion into the fracture, Liu et al. developed a dynamic force balance model based on the force balance of Brownian force, drag force, van der Waals force, diffuse double layer force, buoyant force, and gravitational force. In addition, the numerical model was verified by the free expansion experiment of the bentonite [4-6]. However, in addition to extrusion and expansion, erosion of the bentonite also involves the formation of colloids and the erosion process carried away by the seeping water. Therefore, Moreno describe the flow characteristics of colloids, taking into account the viscosity coefficient equation related to the smectite volume fraction and the sodium concentration in the seeping flow which were based on the dynamic force balance model [7].This model contains three physical quantities, which are the pressure, the volume fraction of smectite, and the concentration of sodium. with Darcy's law for the pressure, and advection-diffusion equation for smectite volume fraction and sodium concentration respectively which is called the KTH model [8-9]. Based on the results of the KTH model, the erosion rate of the bentonite is directly proportional to the velocity of the seeping water but inversely proportional to the depth of penetration of the smectite.In addition to this, Moreno also conducted extensive parameter research and proposed an equation to evaluate the erosion rate of the bentonite, where the erosion rate is proportional to the seeping water velocity to the power of 0.41 and the fracture aperture, and this equation was also used to estimate the time to reach advective conditions in the buffer [10]. However, on the subject of buffer/backfill erosion, SKB concluded in the SR-Site report that demonstrates the phenomenon can not be ruled out in the assessment of long-term safety. Furthermore, there is still uncertainty about the modeling of colloid formation and erosion of bentonite, and using larger bentonite mass may offer a slightly longer time before advective conitions are created, but it's not a solution to the erosion issue, hence continued study on buffer erosion mechanisms is needed [11].

2. The KTH model

The KTH model was developed by Moreno for mass loss rate evaluation of buffer and backfill under fracture intersecting the deposition hole which describes the bentonite expansion and the diffusion of sodium under the seeping water. The fluid flow described by the Darcy equation with viscosity function which depending on sodium concentration and smectite volume fraction, the sodium ions transport represented by advection-diffusion equation, and the bentonite expansion and intrusion into fracture, described by the dynamic force balance model.

2.1. Fluid flow

The seeping water flow in the fracture is modeling by the Darcy equation, the KTH model modified the hydraulic conductivity to account for the viscosity of colloid which strongly dependent on the co-volume fraction of bentonite [9].

$$\rho S\frac{\partial p}{\partial t}+\nabla\cdot\left(\rho u\right)=Q_m \tag{1}$$

In this equation, ρ is the fluid density (kg/m^3), S is storage coefficient (1/Pa), p is the fluid pressure (Pa), Q_m is a mass source term (kg/m^3-s) and u is the Darcy velocity (m/s), which can be expressed as

$$u=-\frac{K}{\rho g}\left(\nabla p+\rho g\nabla D\right) \tag{2}$$

Where K is the hydraulic conductivity (m/s), g is the gravitational acceleration (m/s^2) and ∇D is a unit vector in the gravity direction, which could set to zero in 2D simulation. The hydraulic conductivity is assumed to depend on the viscosity of the fluid flow

$$K=K_w\frac{\eta_w}{\eta} \tag{3}$$

Therefore, the seeping water velocity can be calculated from the volumetric flow rate, Q (m^3/s) and its cross area, A (m^2)

$$\bar{u} = \frac{Q}{A} = \frac{Kb \cdot \bar{i}}{\delta} \tag{4}$$

Here $\bar{i}$ is hydraulic gradient (m/m), δ and b represent the fracture aperture (m) and aquifer thickness (m) respectively, in this study, the fracture aperture is equal to the aquifer thickness.

In Eq. (3), η is the viscosity of fluid (Pa-s) and the subscript w indicates the fluid is water. In addition, the relative viscosity of the fluid is defined as

$$\frac{\eta}{\eta_w} = 1 + 1.022\phi_{cov} + 1.358(\phi_{cov})^3 \tag{5}$$

Where ϕ_{cov} is the co-volume fraction of the fluid. The co-volume fraction is defined as follow when the smectite particle was assumed a coin-like particle

$$\phi_{cov} = \frac{2}{3}\frac{(D_p + 2m\kappa^{-1})^3}{D_p^2 \delta_p}\phi \tag{6}$$

Here δ_p is the particle thickness of smectite (m), D_p is the diameter of smectite particles (m), m is a fitting parameter which set to be 1, besides, the Debye length (m) represented by sodium concentration c (mol/m^3) could be written as

$$\kappa^{-1} = \sqrt{\frac{\varepsilon_0 \varepsilon_r RT}{2cz^2F^2}} \tag{7}$$

Where ε_0 and ε_0 is the vacuum permittivity and the relative permittivity respectively, R is the gas constant (J/K/mol) and T is the absolute temperature (°C), F is the Faraday constant (C/mol) and z is the valence of the sodium ions which is 1. The relative viscosity as a function of the smectite volume fraction in different ion concentrations is plotted in Figure 1. As shown in the figure, as smectite volume fraction or ion concentration increases, the fluid becomes less mobile, therefore, the fluid velocity is affected by smectite volume fraction and ion concentration in the rim border of expansion bentonite. Which becomes slower when the smectite volume fraction or ion concentration decreases.

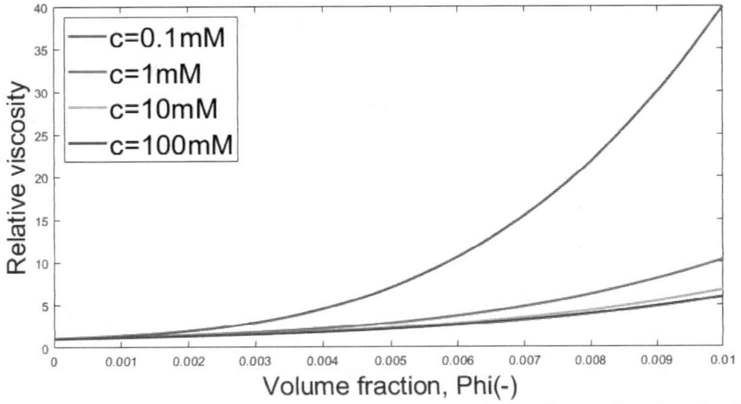

Figure 1. The relative viscosity as a function of the smectite volume fraction in different sodium concentrations

2.2. *Smectite expansion*

The advection-diffusion equation is used to simulation the smectite expansion in a fracture aperture, in which the diffusivity is modeled by the dynamic force balance model [4]

$$\frac{\partial \phi}{\partial t} = -u\nabla\phi + \nabla\cdot\left(\frac{\chi}{f}\nabla\phi\right) \tag{8}$$

Where ϕ is the volume fraction of smectite and the ratio of χ to f can be considered as diffusivity, in which χ is the sum of the energy of the particles (J), where $f=f_{fr}/(1-\phi)$ is the effective friction coefficient which consider the smectite moving into the fracture and water is moving in the opposite direction and the friction coefficient f_{fr}(Ns/m) is defined as

$$f_{fr} = 6\pi\eta_w\left[\frac{\frac{\delta_p}{2}\sqrt{\frac{4S_p}{\pi\delta_p^2}-1}}{\tan^{-1}\left(\sqrt{\frac{4S_p}{\pi\delta_p^2}-1}\right)}\right] + V_p k_0 \tau^2 a_p^2 \eta_w \frac{\phi}{(1-\phi)^2} \tag{9}$$

Here η_w is the viscosity of water (pa-s), S_p is the particle surface area of smectite (m^2), V_p is the volume of coin-like particles (m^3), ap is the specific surface area per unit volume of particles (m^2/m^3) and $k_0\tau^2$ is Kozeny's constant. In addition, the function χ is defined as

$$\chi = k_B T + (h+\delta_p)^2\left(\frac{\partial F_A}{\partial h} - \frac{\partial F_R}{\partial h}\right) \tag{10}$$

Where k_B and T is the Boltzmann constant (J/K) and absolute temperature (K), respectively, h is the separation between particles is given by

$$h(\phi) = \delta_p\left(\frac{\phi_{max}}{\phi} - 1\right) \tag{11}$$

Explicitly, the derivative terms in Eq. (10) are given by

$$\frac{\partial F_A}{\partial h} = -\frac{A_H S_p}{2\pi}\left[\frac{1}{h^4} - \frac{2}{(h+\delta_p)^2} + \frac{1}{(h+2\delta_p)^4}\right] \tag{12}$$

$$\frac{\partial F_R}{\partial h} = -4\kappa c RTS_p \tanh y^m \left[\cosh y_\infty^m \sinh\left(\frac{y_\infty^m}{2}\right) + \frac{1}{\kappa h}\sinh y_\infty^m + \frac{2}{(\kappa h)^2}\sinh\left(\frac{y_\infty^m}{2}\right)\right] \tag{13}$$

Where A_H is the Hamaker constant (J), and

$$y^m = \sinh^{-1}\left[2\sinh y_\infty^m + \frac{4}{\kappa h}\sinh\left(\frac{y_\infty^h}{2}\right)\right] \tag{14}$$

$$y_\infty^m = 4\tanh^{-1}\left[\tanh\left(\frac{y_\infty^0}{2}\right)\exp\left(-\frac{\kappa h}{2}\right)\right] \tag{15}$$

$$y_\infty^h = 4\tanh^{-1}\left[\tanh\left(\frac{y_\infty^0}{2}\right)\exp(-\kappa h)\right] \tag{16}$$

and

$$y_\infty^0 = 2\sinh^{-1}\left(\frac{zF\sigma^0}{2\varepsilon_0\varepsilon_r\kappa RT}\right) \tag{17}$$

Where σ_0 is the specific charge on the particle surface (C/m^2).The effective diffusivity D_F could be calculated depends on χ and f, and the diffusivity as a function of smectite volume fraction at different sodium concentrations is plotted in Figure 2. As shown in the figure, the diffusivity is constant at different sodium concentrations when the volume fraction less than 10^{-4} and above 10^{-1} but quite non-linearly between the range.

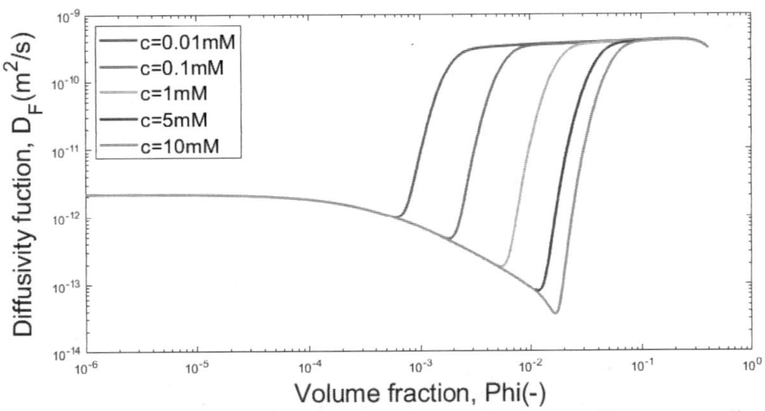

Figure 2. Diffusivity as a function of smectite volume fraction at different sodium concentrations

2.3. Sodium concentration

The sodium ions transport in the fluid is described by the advection-diffusion equation which is given by [9]

$$\frac{\partial c}{\partial t} = -u\nabla c + \nabla\cdot(D\nabla c) \tag{18}$$

In this equation, u is the Darcy velocity (m/s), t is time (s), c is the sodium concentration (mol/m^3) and D is the diffusion coefficient for sodium in seeping water (m^2/s), which varies as a function of smectite volume fraction with D_0 equals to 2×10^{-9} (m2/s)

$$D = D_0(1-\phi)^{1.6} \tag{19}$$

3. Case study-based model validation

The solver validation was performed by using a case study of bentonite expansion and sodium ions migration with fracture intersecting deposition hole. The purpose of this study is to establish the KTH model and evaluate the mass loss rate of bentonite under the seeping water effect,in addition, investigate the distribution of sodium ion concentration and smectite volume fraction under different seeping water conditions.

3.1. Computational details

A fracture intersects the deposition hole with 10m×5m geometry was modeled and implemented using the COMSOL Multiphysics. The cylindrical deposition hole has a radius of 0.875m. The fracture aperture and the hydraulic gradient is modeled with constant values which are 10^{-3} m and 0.1 respectively. The hydraulic conductivities are 1×10^{-6},3×10^{-6}, and 1×10^{-5} (m/s), therefore the seeping water velocities are 3.15, 9.45, and 31.5 (m/year) respectively according to Eq. (4). The hydraulic head is assigned to be 1 m at the left-hand side boundary and inflow seeping water with sodium concentration 0.1 mM but no flux of smectite. The outflow boundary condition and constant hydraulic head 0 m at the right-hand side are imposed. In addition, assigned no flux boundary conditions at the fracture mouth but the volume fraction of smectite is 0.4, and sodium concentration is 10 mM. The initial condition of sodium concentration is set to be 0.1 mM, and the smectite volume fraction at fracture mouth is set to zero.

The fracture water flow is calculated using Darcy's law in COMSOL, which describes the different seeping water velocities by varying the hydraulic conductivities. Also, the dynamic force balance model used to describe the expansion of smectite was developed using the transport of diluted species interface, in which the diffusivity represent by the ratio of χ to f. Besides, the same interface solved the sodium concentration, but the diffusivity is set as a function of smectite volume fraction directly.

3.2. Simulation results

The results of the seeping water velocity of 3.15, 9.45, and 31.5 (m/year) are shown in Figures 3 to 5, respectively. The contours represent the sodium concentration (mM), the color bar represents the distribution of smectite volume fraction (-), and the Darcy flow field is represented by the vector arrow (m/s).In addition to the release of sodium ions from the edge of the deposition hole into the fracture, the upstream also inflow the seeping water with 0.1 mM sodium ions concentration.

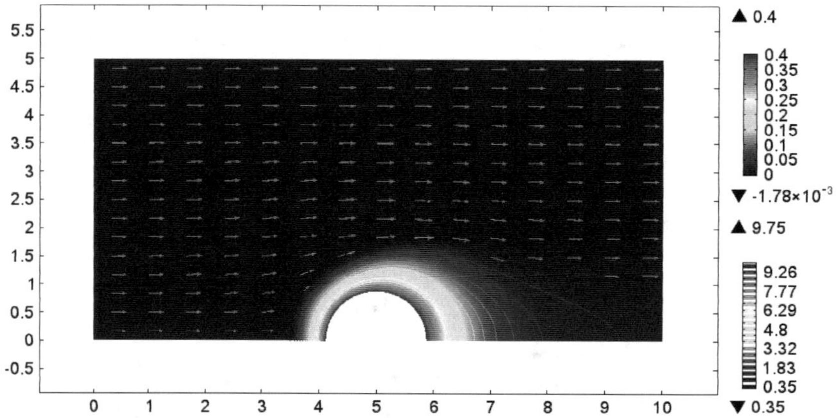

Figure 3. Results of KTH model for the case of 3.15 m/yr of seeping water (c: contour plot, U: arrows, phi: color bar)

The results of the ions concentration analysis are affected by both the seeping water and the deposition hole geometry. Under the effect of diffusion and advection, the ions concentration spreads outward uniformly from the edge of the deposition hole, and the distribution of contours has a higher density at the front of the disposal hole and a smaller outward spread. Also, the ions concentration near the low-pressure zone at the right-hand side of the deposition hole expands significantly outward and forms a wake flow of smectite that extends to the outlet of the computational domain. Besides, the diffusion effect causes the smectite to expand outward along the edge of the deposition hole. The expansion of smectite at the left-hand side of the deposition hole is suppressed under the influence of geometric and advection effects, while the smectite wake flow with a certain thickness is formed at the right-hand side of the deposition hole, and the smectite volume fraction decreases as the distance from the edge of the deposition hole increases.

The smectite spreads outward mainly by diffusion, there is lower mobility around the edge of the deposition hole. As the volume fraction decreases, the effect of advection becomes more obvious, and its mobility is increased gradually. Therefore, the mobilities of the smectite volume fraction mainly control the shape variation, but without any change, its mobility properties still follow the description of Eq. (5).

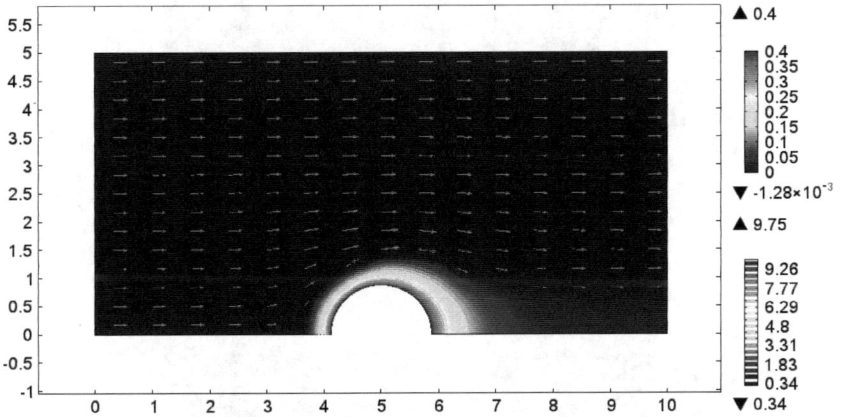

Figure 4. Results of KTH model for the case of 9.45 m/yr of seeping water (c: contour plot, U: arrows, phi: color bar)

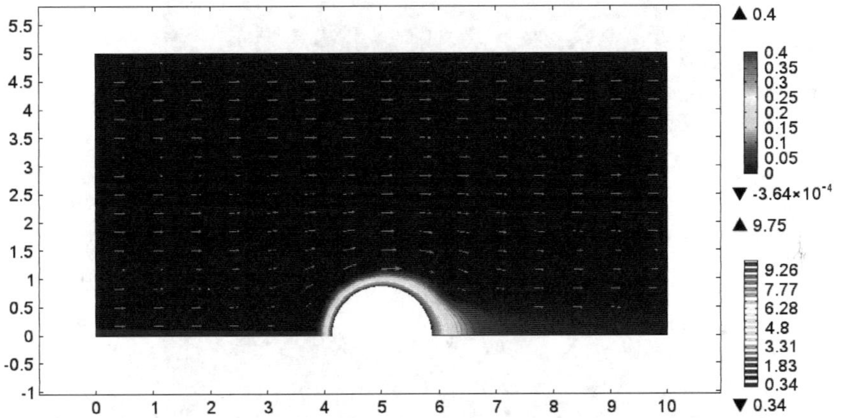

Figure 5. Results of KTH model for the case of 31.5 m/yr of seeping water (c: contour plot, U: arrows, phi: color bar)

To quantify the results of the analysis of the chemical erosion rate of the buffer material, the mass flow rate of the buffer material through the right-hand side boundary of the fracture can be calculated as the product of the Darcy velocity and the smectite volume fraction.

$$N = \rho_s \delta_{fr} \int_0^{\infty} u(y)\phi(y)dy \tag{20}$$

The product of the Darcy velocity and the smectite volume fraction in the above equation can be represented as the mobility of smectite. The product of darcy velocity and smectite volume fraction with different seeping water velocities u=3.15, 9.45, and 31.5 (m/year) are shown in Figures 6 to 8, respectively. The figure shows that "flowing" smectite exists only in the edge region at the interface between the seeping water and the bentonite. There is no smectite volume fraction outside this area, while the Darcy velocity inside is close to zero, which means that the erosion rate of the buffer material is mainly contributed by the "flowing" smectite in the marginal area.

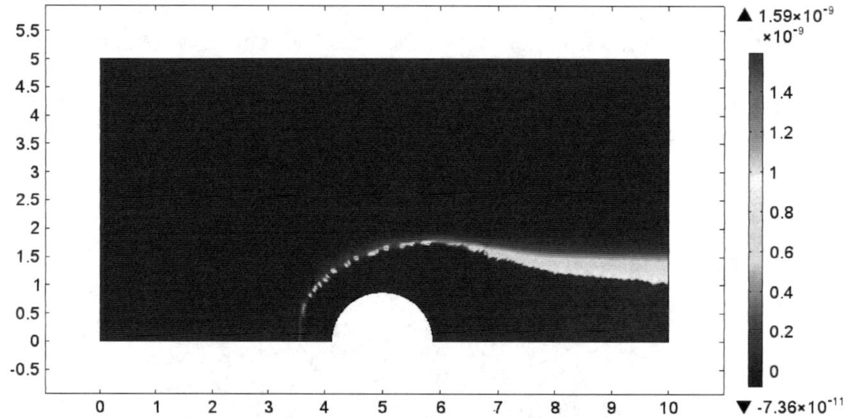

Figure 6. The product of Darcy velocity and smectite volume fraction with seeping water of 3.15 m/yr

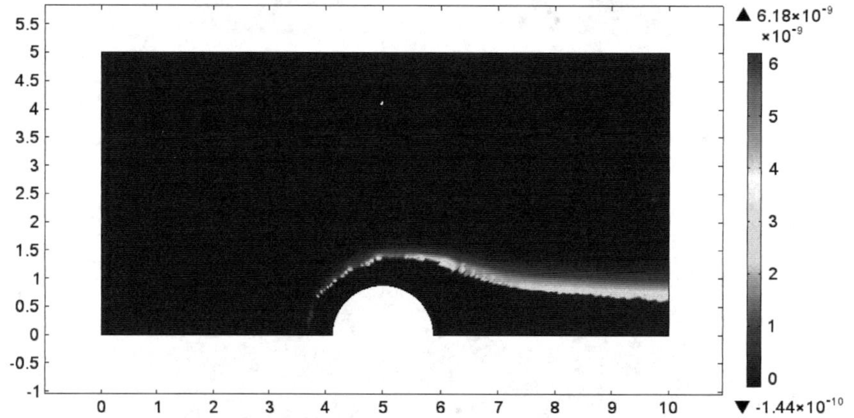

Figure 7. The product of Darcy velocity and smectite volume fraction with seeping water of 9.45 m/yr

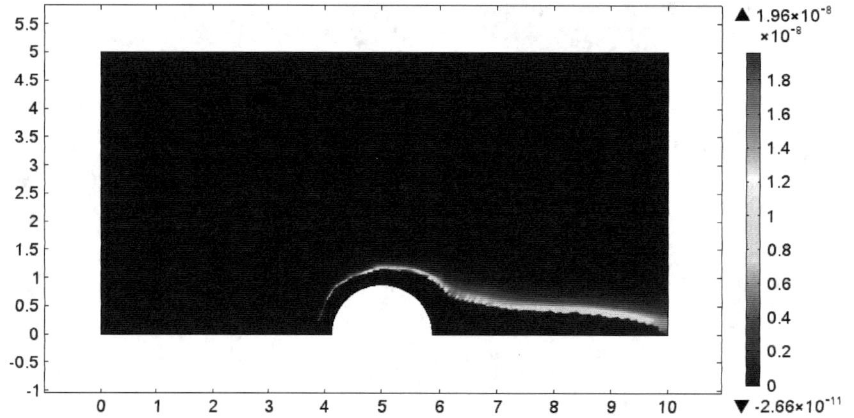

Figure 8. The product of Darcy velocity and smectite volume fraction with seeping water of 31.5 m/yr

3.3. Mass loss rate evaluation

To evaluate the results of the mass loss rate, the mass flow rate of the smectite through the right-hand side boundary of the fracture can be calculated by the Eq. (20), in which, the fracture aperture δ_{fr} is 10^{-3} m, and the specific density of smectite is 2750 kg/m^3.The corresponding mass loss rate is obtained by multiplying the integral of Darcy velocity and smectite volume fraction with the specific density of smectite and the fracture aperture in each case. The results are shown in Table 1 for the mass loss rate calculations at seeping water velocities u=3.15, 9.45, and 31.5 m/yr and compared with the reference [8]. As shown in the results, the model of this study is better validated in the case of u=3.15 and 9.45 m/yr, with relative errors of -6.42% and 1.54%, respectively, but 46.58% in the case of u=31.5 m/yr. This is because the edge region of the buffer will change with time, a high-resolution mesh should be used to solve the problem to provide a good description of the temporal and spatial behavior of the rim border of the smectite.

Finally, in Figures 3 to 5, the mass loss rate of smectite increases with the increase of seeping water velocity, but the thickness of the wake flow of smectite decreases with the increase of the seeping water velocity can be observed. The phenomenon of reduced thickness and increased mass loss means that the erosion rate of the bentonite is dominated by the mobile smectite, which, despite the reduced thickness of the wake flow of smectite, can transport more mass per unit of time at higher seeping water velocity.

Table 1. Results of the mass loss rate evaluation for various seeping water velocity

Seeping water velocity (m/yr)	Mass loss rate (kg/yr)		Relative Error (%)
	SKB	Present	
0.95	0.026	—	—
3.15	0.043	0.040	-6.42
9.45	0.071	0.072	1.54
31.5	0.117	0.172	46.58
94.5	0.18	—	—
315	0.292	—	—

4. Conclusions

The KTH model is a numerical model based on different physical parameters, and then the dependencies between the parameters are used to achieve the coupling calculation. And the fluid flow is described by Darcy's equation, and the hydraulic conductivity is modified by using the relative viscosity to describe the flow characteristics of both the fluid and the smectite. The sodium concentration and smectite volume fraction are governed by advection-diffusion equations, in which, the diffusion effect is defined by the corresponding diffusion coefficients, while the advection effect is solved by the coupled Darcy velocity. The KTH model has good extensibility. The gravitational effect can be considered to analyze the effect of a non-horizontal fracture or the condition of different shapes or multiple fractures. Furthermore, it is possible to replace the Darcy equation with the Navier-Stokes equation to account for higher Reynolds number flow conditions or turbulence, but due to the non-linear relationship between the physical quantities, it can be expected that will take more time to compute than the original model. It can also be observed from the KTH model that the flux of smectite through the right-hand side boundary of the fracture increases with the increase of the seeping water velocity, but the thickness of the smectite wake flow decreases with the increase of the seeping water velocity. The phenomenon of reduced thickness and increased mass loss means that the erosion rate of the bentonite is dominated by the mobile smectite, which, despite the reduced thickness of the wake flow of smectite, can transport more mass per unit of time at higher seeping water velocity. Finally, to avoid the lack of mesh resolution, which makes it impossible to properly describe the process of smectite volume fraction change over time and space, future research work should include the study of mesh convergence analysis.

References

[1] SKB, 2004 *Interim process report for the safety assessment SR-Can* SKB Report R-04-33.
[2] Liu J, Neretnieks I, 2006 *Physical and chemical stability of the bentonite buffer* SKB Report R-

06-103.
[3] SKB, 2006 *Long-term safety for KBS-3 repositories at Forsmark and Laxemar – a first evaluation* SKB Report TR-06-09.
[4] Liu L, Moreno L and Neretnieks I, 2009 A Dynamic Force Balance Model for Colloidal Expansion and Its DLVO-Based Application *Langmuir* **25** pp 679-687.
[5] Neretnieks I, Liu L and Moreno L, 2009 *Mechanisms and models for bentonite erosion* SKB Report TR-09-35.
[6] Dvinskikh SV, Szutkowski K and Furo I, 2009 MRI profiles over very wide concentration ranges: Application to swelling of a bentonite clay *Journal of Magnetic Resonance* **198** pp 146-150.
[7] Liu L, 2011 A model for the viscosity of dilute smectite gels *Physics and Chemistry of the Earth* **36** pp 1792-1798.
[8] Moreno L, Neretnieks I and Liu L, 2010 *Modelling of erosion of bentonite gel by gel/sol flow* SKB Report TR-10-64.
[9] Moreno L, Liu L and Neretnieks I, 2011 Erosion of sodium bentonite by flow and colloid diffusion *Physics and Chemistry of the Earth* **36** pp 1600-1606.
[10] SKB, 2010 *Corrosion calculations report for the safety assessment SR-site* SKB Report TR-10-66.
[11] SKB, 2011 *Long-term safety for the final repository for spent nuclear fuel at Forsmark main report of the SR-site project* SKB Report TR-11-01.

IOP Conf. Series: Materials Science and Engineering 1113 (2021) 012025 doi:10.1088/1757-899X/1113/1/012025

Using improved neural network for the risk assessment of information security

Zne Jung Lee[1], Zhao Yun Yang[2], Chou Yuan Lee[1], Zhi Hao Chen[1] and Wen Bing Wu[1]

[1]School of Technology, Fuzhou University of International Studies and Trade, China
[2]School of Economics and Management, Fuzhou University of International Studies and Trade, China

Email: lrz@fzfu.edu.cn

Abstract. Because of the information age, protecting information is very important to satisfy the three main aspects of information security, namely confidentiality, integrity and availability. In this case, information security has become one of the most important problems in information technology. Information security is a very important activity and risk assessment is the kernel of information security. However, most of the current risk assessment activities are comparatively subjective and the performances are not good enough. To understand this problem, we propose the improved neural network for the risk assessment of information security. Basically, it is processed under back-propagation neural network (BPN). Moreover, particle swarm optimization (PSO) is used for fine parameter optimization of BPN. The experimental results show that the proposed algorithm has the best performance among these compared approaches.

1. Introduction

The progress and rapid development of information technology have changed the way of life of human beings. With the wide application of information technology in daily life, information security has become an important issue under the conditions of information service. Information security is a very critical activity, and the kernel core of information security is risk assessment. However, most of the current risk assessment activities are conducted manually. The evaluation process might be subjective, such as participant analysis, retrospective analysis, historical comparison method, Delphi method and analytic hierarchy process (AHP), because they depend on judgment and experience [1]. A number of researchers used fuzzy methods, data mining and regression models to analyze the risk assessment of information security [2-3]. These above methods have ameliorated the subjective nature of risk assessment, but these results are not good enough. Recently, neural networks have successful applied to solve problems in many fields. In this paper, a new type of neural network is used to evaluate the risk assessment of information security. In the proposed improved neural network, these parameters are optimized to improve performance by PSO.

This paper is summarized as follows. Section 2 introduces the concept of neural network. The third section introduces the proposed algorithm. In section 4, the simulation results of the proposed algorithm are introduced. From the simulation results, it is found that the proposed algorithm is superior to other approaches. The concluding remarks are presented in section 5.

2. The concept of neural networks

This paper focuses on the use of improved neural network for the risk assessment of information security. The objective of neural network is to simulate human brain. It lets the computer through the learning process find the features, not from the human brain to determine the characteristics. For neural network, it mostly uses multilevel structures and widely used in many fields such as function approximation, classification, and data processing. At the same time, the rapid progress of the hardware can shorten the execution time required by the computer learning algorithm and become an important reason for the attention of neural network in recent years [4-5]. The neural network can be divided into different models according to variant characteristics such as convolution neural network (CNN) and BPN. The applications of these kind of neural networks are very extensive, among which supervise learning network is the most widely used [6]. BPN is the most widely used neural network model, whose network behavior is based on input-output learning pairs(x_j, d_j) where $j = 1, \cdots, m$. BPN trains the neural network by identifying a set of weights to minimize errors [7]. There are two main stages in BPN, forward propagation and backward propagation. A three-layer of BPN is shown in Figure 1. The function of forward propagation is to transmit the input samples to produce the current output. For a neuron q in the hidden layer, the input sample of net is shown as Eq. (1),

$$net_q = \sum_{j=1}^{m} v_{qj} x_j \tag{1}$$

and its output is shown as Eq. (2):

$$z_q = a(\sum_{j=1}^{m} v_{qj} x_j) \tag{2}$$

where $a(\sum_{j=1}^{m} v_{qj} x_j)$ in Eq. (2) is called the activation function, and v_{qj} is the weight between input x_j and neuron q. For the output neuron i, it is shown as Eq. (3),

$$net_i = \sum_{q=1}^{l} w_{qj} z_q = \sum_{q=1}^{l} w_{iq} a(\sum_{j=1}^{m} v_{qj} x_j), \tag{3}$$

and its output is shown as Eq. (4):

$$y_i = a(net_i) = a(\sum_{q=1}^{l} w_{iq} a(\sum_{j=1}^{m} v_{qj} x_j)) \tag{4}$$

The function of backward propagation is to backward propagate the error to update the weights by gradient descent method. The weights between the hidden layer and output are updated as shown in Eq. (5).

$$\Delta w_{iq} = \eta (d_i - y_i) a'(net_i) z_q \tag{5}$$

where η is learning constant. The weight between the input layer and the hidden layer will be updated as Eq. (6):

$$\Delta v_{qj} = \eta \sum_{i=1}^{n} ((d_i - y_i) a'(net_i) w_{iq}) a'(net_q) x_j \tag{6}$$

3. The proposed algorithm

The standard of ISO/IEC 27005 is mainly to explore the steps and methods for the risk management of information security. It proposes to use information assets, threats, and vulnerability for risk assessment. In this paper, we also use these 3 aspects to evaluate risk for information security. The use of Internet services, information systems, and data transmission can generally be regarded as valuable that is called information assets. For these information assets, like other important valuable assets, should be properly protected. The most important aspects of information assets are confidentiality, integrity, and availability. In addition, there could also involve some qualities such as authentication, accounting, non-repudiation, and reliability [1]. The assessment of confidentiality is shown in Table 1. The assessment of integrity is shown in Table 2. The assessment of availability is shown in Table 3. After the completion of above 3 assessments, the maximum value of these assessments shall be taken as the value of the information asset.

In this paper, we propose the improved neural network for the risk assessment of information security. The flowchart of the proposed algorithm is shown in Figure 2. In the flowchart, we use BPN with one hidden layer as the main architecture for the risk assessment of information security. For BPN, there are many parameters that need to be well adjusted to achieve good risk assessment performance, such as the number of neurons of hidden layer, the weights between input and the hidden layer, and the weights between hidden layer and output neuron. Particle swarm optimization (PSO) is easy to implement with few parameters, and it has been successfully applied in many fields. In the proposed algorithm, we use PSO to adjust these parameters. For PSO, each particle has a position and velocity vector. In n-dimension, the position P_i^n and velocity V_i^n of the ith particle can be explicitly described as $\vec{P_i} = (P_i^1, P_i^2, \cdots, P_i^n)$ and $\vec{V_i} = (V_i^1, V_i^2, \cdots, V_i^n)$ respectively. The PSO algorithm is represented as follows [8].

$$V_i^n = \Omega * V_i^n + \theta_1 * \pi * (pbest_i^n - X_i^n) + \theta_2 * \pi * (gbest^n - X_i^n) \quad (7)$$

$$P_i^n = P_i^n + V_i^n \quad (8)$$

where Ω is called inertia weight, θ_1 and θ_2 are real numbers, π is a random number in [0,1], $pbest_i = (pbest_i^1, pbest_i^2 \cdots, pbest_i^n)$ is the best position for the ith particle, and $gbest = (gbest^1, gbest^2 \cdots, gbest^n)$ is the global best position.

4. Simulation results

For the risk assessment of information security, it is related to the value of information asset, threat, and vulnerability. Threat refers to the damage caused by the information security of the harmful event. It is shown in Table 4. Vulnerability is the weakness of using information assets. It is shown in Table 5. To calculate the final risk value, it is done by the product of the value of information asset, likelihood of occurrence, and level of impact [1].

This paper uses the proposed algorithm to calculate the accuracy for the risk assessment of information security. There are 22 attributes and 84 records for the risk assessment of information security. We use 54 records for train data and 30 records for test data. Three learning algorithms of decision tree (DT), support vector machine (SVM), and BPN are implemented for comparing results in this paper. The decision tree, known as the classification tree, is an inductive learning method that can analyse and classify data from entropy. It takes the concept of a tree into account and produces interpretable rules [9]. Based on the principle of structural risk minimization, SVM is mainly for classification issues. It finds a hyperplane in high dimensional states to divide data into different classification to get the best accuracy [10-11]. The result of classification accuracy is shown in Table 6. In Table 6, it finds that the improved neural network has the best classification accuracy.

5. Conclusions

This paper proposed the improved neural network for the risk assessment of information security. In the proposed method, we use BPN as the main architecture to classify data and use PSO to adjust the best values of parameters of BPN. A comparison of classification accuracy with DT, SVM, and BPN. It demonstrates that the proposed algorithm can conduct risk assessment and effectively control the information security risk.

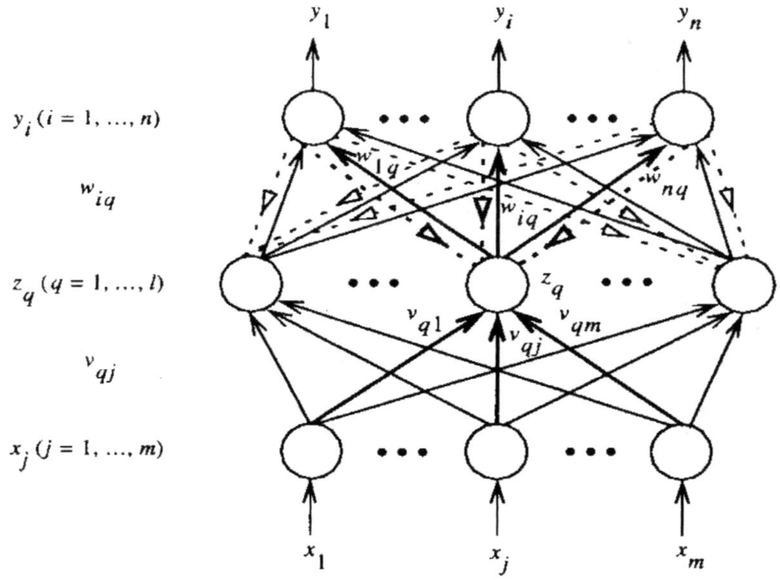

Figure 1. A three-layer BPN neural network

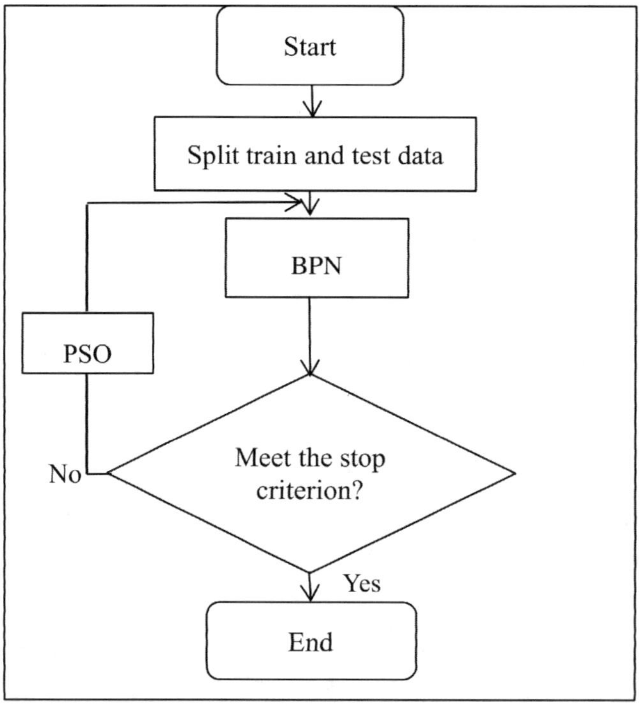

Figure 2. The proposed algorithm

Table 1. The assessment of confidentiality

Number	Item	Value
1	Confidential information assets are not required.	1
2	Information assets containing sensitive information, but no confidentiality	2

	requirements, limited use of internal staff.	
3	Limited use of information assets within relevant department staff.	3
4	Information assets in information, including confidential information regulated by relevant laws or regulations.	4

Table 2. The assessment of integrity

Number	Item	Value
1	The integrity requirements of information assets are very low.	1
2	Information assets have integrity requirements, but will not be damaged by integrity.	2
3	Information assets have integrity requirements that can cause harm if integrity is compromised.	3
4	Information assets have integrity requirements that can disrupt business due to a breach of integrity.	4

Table 3. The assessment of availability

Number	Item	Value
1	Information assets allow for failures of more than 3 working days and do not need to be repaired or alternatives are needed in real time.	1
2	Information assets allow for failures of more than 8 working hours and 3 working days or less, without real-time repair or search for alternatives.	2
3	Information assets allow for failures of more than 4 working hours and less than 8 working hours, without real-time repair or search for alternatives.	3
4	Information assets allow for failures of more than 4 working hours, without immediately repair or to find alternatives.	4

Table 4. The assessment of threats

Number	Item	Value
1	Possibility of it happening once a season.	1
2	Possibility of it happening once a month.	2
3	Possibility of it happening once a weak.	3

Table 5. The assessment of vulnerability

Number	Item	Value
1	This weakness is not easily exploited by threats.	1
2	This weakness is vulnerable to threats.	2
3	This weakness is very vulnerable to threats.	3

Table 6. The classification accuracy of test data

Number	Method	Classification accuracy
#1	DT	83.3%
#2	SVM	89.1%
#3	BPN	90.%
#4	The proposed algorithm	93.33%

Acknowledgment
This research was supported by Fuzhou City research Grant No. 2019-SG-6. It was partially supported by 2020 and 2019 Fujian Province research Grant No. FBJG20200371 and FBJG20190284. It was also supported by Fuzhou University of International Studies and Trade research Grant No. 2018KYTD-02 and FWB19003.

References

[1] Chang L Y and Lee Z J 2013 Applying fuzzy expert system to information security risk

Assessment-A case study on an attendance system *2013 IEEE International Conference on Fuzzy Theory and Its Applications* p 346
[2] Bojanc R and Jerman-Blažič B 2013 A quantitative model for information-security risk management *Engineering Management Journal* **25** 25
[3] Lee Z J and Chang L Y 2014 Apply fuzzy decision tree to information security risk assessment *International Journal of Fuzzy Systems* **16** 265
[4] El-Khatib M J *et al* 2019 Glass Classification Using Artificial Neural Network
[5] Roblek D *et al* 2019 *U.S. Patent No. 10,467,493*. Washington, DC: U.S. Patent and Trademark Office
[6] Asuntha A *et al* 2019 A hybrid feature extraction approach for the detection of melanoma using neural network *International Journal of Research in Pharmaceutical Sciences* **10** 1836
[7] Maru A *et al* 2020 Effective Software Fault Localization Using a Back Propagation Neural Network *Computational Intelligence in Data Mining* p 513
[8] Dabhi D and Pandya K 2020 Enhanced velocity differential evolutionary particle swarm optimization for optimal scheduling of a distributed energy resources with uncertain scenarios *IEEE Access* **8** 27001
[9] Lee Z J and Lee C Y 2020 A parallel intelligent algorithm applied to predict students dropping out of university *The Journal of Supercomputing* **76** 1049
[10] Huang Y C and Zheng J H 2018 Ball Nut Preload Diagnosis of the Hollow Ball Screw through Support Vector Machine *Adv. technol. innov.* **3** 94
[11] Lee Z J *et al* 2020 A hybrid system for imbalanced data mining *Microsystem Technologies* **26** 3043

AUTHOR INDEX

Institute of Physics
Dirac House, Temple Back
Bristol BS1 6BE UK

ISSN: 1757-8981
ISBN 978-1-7138-3402-1

9 781713 834021

CURRAN ASSOCIATES INC.
proceedings.com